The Institution of
StructuralEngineers

Conceptual design of buildings

Authors

J Norman	PhD MEng CEng MICE FHEA (University of Bristol) *Lead*
O Broadbent	MEng MChem (Constructivist Ltd)
J F Carr	BEng MPhil CEng FIStructE FHEA (University of Sheffield and Jon Carr Structural Design)
R De'Ath	MEng CEng MIStructE MICE (University of Bristol and Arup)
R Harpin	BEng CEng MIStructE (University of Sheffield)
G Knowles	BEng CEng MIStructE (University of Bath)
I Lloyd	PhD MSc(Eng) BSc(Geol) CGeol FGS FHEA (University of Bristol)

Reviewers

G Evans	BSc(Hons) PhD CEng FICE FIStructE MBCS (Constructex) *Technical Products Panel*
T Ibell	FREng PhD BSc(Eng) CEng FIStructE FICE FHEA (University of Bath)
J Lord	MEng CEng MICE (Whitby Wood)
N Russell	BSc CEng FIStructE FICE FASCE MCMI (Perega)

Publishing

L Baldwin	BA(Hons) DipPub (The Institution of Structural Engineers)
R Thomas	BA(Hons) MCLIP (The Institution of Structural Engineers)

Published by The Institution of Structural Engineers
International HQ, 47–58 Bastwick Street, London EC1V 3PS, United Kingdom
T: +44(0)20 7235 4535
E: mail@istructe.org
W: www.istructe.org

First published (version 1.0) April 2020
This version 1.1 (published April 2021) includes minor amendments/additions to the following pages: 45, 75, 78, 89, 95, 170, 172, 174, 180, 191, 217, 222 and 250

978-1-906335-42-7 (print)
978-1-906335-43-4 (pdf)

Contents

Notation (for Chapter 10)

Term	Definition
A_c	Cross-sectional area of concrete
A_{chord}	Cross-sectional area of truss chord
A_f	Area of flange
A_s	Cross-sectional area of reinforcement
A_v	Shear area
b	Width of section
b_f	Width of flange
c_f	Outstand length of flange
d	Depth of truss
d_{eff}	Effective depth
d_{sect}	Depth of section
δ	Deflection
e	Eccentricity
E	Modulus of elasticity
$E_{0.05}$	Fifth percentile value of modulus of elasticity
E_{min}	Minimum value of modulus of elasticity
f_{all}	Allowable stress in steel section
f_{ck}	Characteristic compressive cylinder strength of concrete
$f_{c,0,d}$	Design compressive strength parallel to grain
$f_{c,0,k}$	Characteristic compressive strength parallel to grain
$f_{m,d}$	Allowable bending strength parallel to grain
$f_{m,k}$	Characteristic bending strength parallel to grain
f_y	Yield strength of steel
f_{yk}	Characteristic tensile strength of reinforcement
$F_{b,0,d}$	Design buckling resistance parallel to grain
$F_{c,0,d}$	Design compressive resistance parallel to grain
g_d	Design uniformly distributed load due to permanent loads
γ_m	Partial material factor
h	Height of arch
h_w	Height of web
i_y	Radius of gyration, y-y axis
I	Second moment of area
I_{truss}	Second moment of area of truss
k	Modification factor for timber section
k_{yy}, k_{zy}	Interaction factors for steel columns
L	Span length
L_{cr}	Buckling length
λ	Slenderness
λ_{rel}	Relative slenderness
$M_{b,Rd}$	Design lateral torsional buckling resistance
M_{Ed}	Design bending moment

$M_{c,z,Rd}$	Design moment resistance, z-z axis
$N_{b,Rd}$	Design buckling resistance
$N_{c,Rd}$	Design resistance to axial compression
N_{cr}	Euler buckling resistance
N_{Ed}	Design axial force
$N_{c,Ed}$	Design axial compression force
$N_{t,Ed}$	Design axial tension force
$N_{pl,Rd}$	Design plastic resistance to axial forces
q_d	Design uniformly distributed load due to variable loads
r	Radius of circular arch
R_H	Horizontal reaction
R_V	Vertical reaction
S	Swept length of arch
t_f	Thickness of flange
t_w	Thickness of web
V_{Ed}	Design shear force
v_{Ed}	Design shear stress
$V_{pi,d}$	Design plastic shear resistance
$V_{Rd,c}$	Design shear resistance
w_d	Uniformly distributed load (UDL)
W_d	Design point load
W_{el}	Elastic modulus
y	Distance from neutral axis to centroid of member
z	Lever arm of internal forces

Foreword

I wish I had this book when I was a student! It would have put into perspective so beautifully at the time what it really is to be a structural engineer. To dream a little and have ideas. To rely on deep technical skills to prioritise some of these ideas, and to work them up into reality such that the outcome enhances somebody's life. Wonderful. This is structural engineering, and a reflection of its power. This book oozes with reasons why our profession is so special.

It lays out the story for students and graduates about the realities of the day job right through to our contributions to humanity, and the excitement which these responsibilities provide. It is written by highly experienced authors whose communication skills ensure total accessibility to students and graduates in explaining clearly the entire process of the creative structural design of a building.

The aspect of the book which I like most is the desire by the authors for the book to become outdated quickly. Our climate emergency has placed our profession in the spotlight, given the embodied-carbon issues inherent in construction. This book reflects current best practice, but it also asks the big 'What if?' questions. What if we had no cement? What if we had to design according to an inventory-constrained palette of re-used components? What if we could rely on technology to mitigate risk of overload in buildings? If you ever wanted students and graduates to make the link between our commitments to the climate emergency 'declare' initiatives and the day job, the 'What if?' questions highlighted in this book provide just this inspiration. It takes our profession out of the spotlight and into the limelight.

This book reflects the extraordinary skills which structural engineers possess, and how they might think about deploying them. Additionally, it challenges us to be even better in the future. The guidance is priceless for those entering our fabulous profession.

Prof. Tim Ibell
Department of Architecture and Civil Engineering, University of Bath, UK

The authors

James Norman – University of Bristol
James has 12 years design experience working for Ramboll and Integral Engineering Design. He has nine years' academic experience, including a PhD at the University of Bristol. He has designed buildings out of mud, timber, steel and lots and lots of concrete, and worked for a year on the facade of the extension to the Tate Modern. James authored *Structural timber elements: a pre-scheme design guide* and is Associate Professor of Sustainable Design.

Oliver Broadbent – Constructivist Ltd
Oliver is Founder of Constructivist Ltd, and specialises in helping engineers develop their creativity. He is a Royal Academy of Engineering Visiting Professor at Imperial College and hosts Eiffel Over, a podcast about engineering, creativity and practical philosophy.

Jon Carr – University of Sheffield and Jon Carr Structural Design
Jon is a Senior University Teacher in Structural Design at the University of Sheffield, as well as running Jon Carr Structural Design, as a sole practitioner. Jon previously worked for Anthony Hunt Associates from 1988 to 2010, specialising in education and sports and leisure sector projects. His notable projects include the KCOM Stadium in Hull and, at the other end of the scale, the 'Hen House' in Sheffield.

Rachael De'Ath – University of Bristol and Arup
Rachael has more than 16 years' design experience working for Arup, and has recently joined the University of Bristol to teach design, alongside her work in industry. She prefers working on re-use projects, where the existing structure is creatively re-imagined into something new. She was named as one of the Women's Engineering Society 'Top 50 female engineers' in 2018.

Richard Harpin – University of Sheffield
Richard is a University Teacher in Structural Design at the University of Sheffield. He was previously a Lecturer in Structural Engineering and Architecture at Nottingham Trent University and, before this, spent 16 years working for Arup. Significant projects include Citibank European Headquarters at Canary Wharf, Pallant House Gallery in Chichester and the School of Theatre, Film and Television at the University of York.

Gavin Knowles – University of Bath

Gavin studied Civil Engineering at Oxford Brookes University and graduated in 2001. Since working in practice he gained his professional chartership with the Institution of Structural Engineers. He was an Associate with Bath-based engineering firm Integral Engineering Design, and is now a full-time lecturer at the University of Bath. Gavin's previous projects include many education and office buildings, along with conservation and refurbishment projects, interweaved with diverse structures, such as rammed chalk-walled houses, recycled material stages at WOMAD Festival and the odd sculpture.

Isobel Lloyd – University of Bristol

Isobel has 20 years' design experience with BuroHappold, Atkins and Mott MacDonald, mostly in the UK but also in Europe, the Middle East and Hong Kong. She has six years' experience of working for various contractors including ground investigation companies, and has spent eight years in academia. Significant projects include the Globe Theatre, Royal Armouries Museum in Leeds, Valentine Bridge in Bristol, Extension to British Library and many school buildings.

Acknowledgements

Permission to reproduce the following has been obtained, courtesy of these individuals/organisations:

Cover © Simon Smith (Smith & Wallwork)
Figures 3.4, 3.9 and 5.1 © Integral Engineering Design
Figures 3.8 and 3.24 © Hatcher Prichard Architects
Figure 3.10 © Redenbrow.com
Figures 3.20 and 5.3 © David Grandorge
Figures 3.25–3.28 © E3 Consulting
Figures 4.10, 8.1, 10.5 and 10.10a–b © Arup
Figures 7.1 and 7.2 Contain British Geological Survey materials © UKRI [2020]. Base mapping is provided by ESRI
Figures 7.3–7.5 Contain British Geological Survey materials © UKRI [2020] accompanying the record
Figure 7.8 *Foundation design and construction*, M.J. Tomlinson and R. Boorman, 7th ed, 2001.
Reprinted by permission of Pearson Business
Figures 8.2, 8.27–8.28 and 8.35–8.37 © Bond Bryan
Figures 8.3 and 8.4 © steelconstruction.info
Figure 8.5 © Waugh Thistleton Architects
Figure 8.6 © Curtins
Figure 8.7 © Dema Formwork
Figure 8.8 © Daniel Shearing (Photographer)
Figure 8.9 © Hadley Steel Framing
Figure 8.10 © K K Law (Photographer)
Figure 8.11 © Acton Ostry Architects
Figure 8.18 © Robert Bird Group
Figure 8.20 © Jon Shanks
Figures 8.21 and 8.22 © F P McCann
Figure 8.23 Courtesy of Bentley SIP Systems
Figure 8.24 © Portakabin
Figure 8.25 © Kier
Figures 8.26 and 8.33 © Tony Hunt
Figures 8.29–8.30, 8.32 and 8.34 © SKM (now Jacobs)
Figures 9.6 and 12.7a–b © Ramboll
Figure 10.2a © Dominic Beer
Figures 10.2b and 10.6a–b © Stephen Fernandez (Arup)
Figure 10.2c © Smith & Wallwork
Figure 10.2d © MCW and CH2M (now Jacobs)
Figure 10.3 © Tom Page [CC BY-SA 2.0]
Figure 10.4 Courtesy of Cullinan Studio
Figure 10.6c © Nottingham Trent University
Figure 10.8 © K C Kong [CC BY-SA 3.0]
Figure 10.9a © Tim Green [CC BY 2.0]
Figure 10.14a © Vlatka Rajcic
Figure 10.14b © New Steel Construction
Figure 10.21 © Focchi
Figure 10.32 © British Land
Figure 10.39a–b © Paul Denning (Allerton Steel)
Figure 12.6 © Feilden Fowles
Table 7.8 © Wiley
Tables 7.12 and 7.14 *Foundation design and construction*, M.J. Tomlinson and R. Boorman, 7th ed, 2001.
Reprinted by permission of Pearson Business

Permission to reproduce extracts from British Standards is granted by BSI Standards Limited (BSI). No other use of this material is permitted. British Standards can be obtained in PDF or hard copy formats from the BSI online shop:
www.bsigroup.com/Shop

James Norman
University of Bristol

1 Introduction

Every project requires conceptual design. Taking the idea or problem from the imagination of the client, and creating a solution that is feasible. Asking lots of questions along the way.

Architects discuss the conceptual design phase like an artist. It is the creative spark. The moment of magic. When you read books about buildings, they never talk about the window schedules, the waterproofing details, the toilet setting-out drawings, and yet these represent a large proportion of their time and effort. They discuss the moment of inspiration, how the design has used the historical context for the building, or the abstract inspiration which has led to its curves or jagged design.

Engineers are the opposite. We talk about the detail. The trials of designing a 1m deep post-tensioned transfer beam for disproportionate collapse. We fail to talk so much about our involvement in the creative part of the project, the art galleries we visited, which inspired us to suggest a folded plate solution, or the book we read that created a reference to the rhythm of the structure, or the abstract maths equation we have been chewing over that created the sculptural doubly-curved roof of the building. However, engineering is required at the conceptual design phase on every project whether we talk about it or not.

On some projects the engineering conceptual design is done by the architect. The design is delivered fully-realised to the engineer, with the simple requirement to prove it stands up. We have all worked on these projects. At best, they are dull to work on as all we have to do is crunch the numbers. At worst, the design is flawed, and the engineer becomes the enemy of creativity as they start to add columns to support 20m cantilevers, in the process ruining the 'Parti' (the central idea or concept)[1] of the building.

It is always more fun to work on projects where we get to play with the architect, or even on our own, allowing our technical ability to be used in creative and unusual ways. Solving complex problems through divergent thinking. These are the projects we remember working on for all the right reasons and the projects we feel most proud of.

So, if you are a graduate engineer who has carried out multiple load-takedowns, and designed numerous beams and columns, then this book is for you. It is designed to help you engage in the next step of your design career, the conceptual design. It will build on your innate understanding of how buildings work and your library of solutions to design problems, but hopefully it will also challenge you to both draw inspiration from further afield and learn to communicate your ideas. This book is designed to be used by anyone, anywhere. We are aware that our collective personal experience is based mostly from designing buildings in the UK, but hopefully the book is universally applicable. However, some examples and finer details (such as codes referenced) are UK-specific.

If you are a student, we hope that this book will also be a great aid for you. While your ability to carry out conceptual design grows as you experience more design, it is never too early to start practicing, and we hope this book will offer helpful advice.

1.1 From blank page to complete building

Every project starts off as a blank page. It may first exist in the client's head. A desire or business need that has to be fulfilled. At some point the client decides to take the plunge and employs some professionals. An architect. An engineer. A quantity surveyor. A building services engineer. Maybe an acoustician. A landscape architect. A project manager. A CDM coordinator. The list goes on. Generally, these people are referred to collectively as the 'design team' (Figure 1.1), as designing large complex projects is always a team activity. The design at this point may be a set of business needs, or it might be a sketch on a napkin. It may be fully realised in the client's head, or there may be a problem that needs solving without a preconceived idea of what the solution should be.

Figure 1.1: A typical design team

The conceptual design stage is the most creative, most fast-paced and most ill-defined of any stage of a project. As an engineer, we may be involved from the very beginning, or we may be involved long after the initial conceptual design has been completed. Whatever stage the project has reached, we will need to start to create a structure that will support the building and prove it will stand up.

This book is designed to help you successfully navigate, and make the most of, the conceptual design stage. The conceptual design stage itself (Figure 1.2) is very short – and the moment when key decisions get made often lasts just a few hours compared to a project that could last for years. When it happens, you will not have time to learn to draw, read widely, create narrative or carry out detailed design of sections. You will either be ready, or you won't! The aim of this book is to make sure you are ready. To prepare you for the moment an architect pops by to discuss a project they've been thinking about. To prepare you for when a client mentions in passing about a new project they've been dwelling on. To prepare you for when you sit your IStructE Chartered Membership exam. The aim of this book is to prepare you, as best we can, for these moments. These are milestone events in your career, for as you move on from just being able to carry out detailed design and become a competent conceptual designer, your skill set becomes bigger, the opportunities ahead of you become greater and you are ready to be a chartered structural engineer.

A small note of advice right at the very start. When you start a conceptual design, you are frequently confronted with a blank page. It can feel daunting to put a mark on it, to come up with a design. But as with any creative endeavour the best way to overcome the blank page is to start. Start drawing. Start making models. Start asking questions (and never stop). Start researching. Start spending time in the space where the building will be. Start talking to people. The (possibly metaphorical) blank page should not stay blank for long.

Figure 1.2: Concept design process

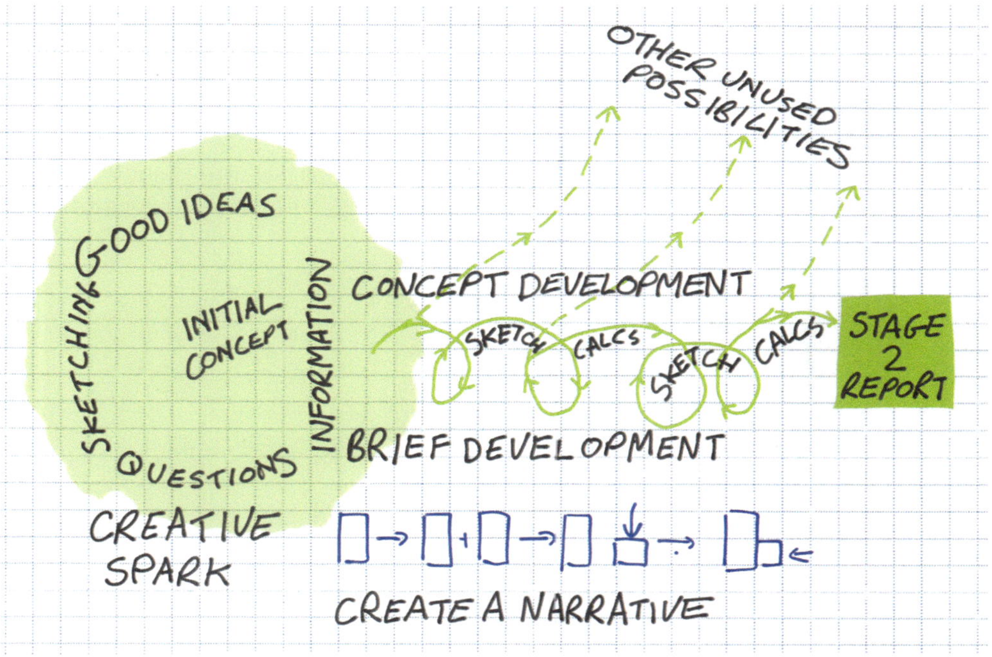

1.2 The importance of 'good' concept design

There is a phrase bandied around in some design offices:

'Right first time'

The idea is that you do everything once and only once. If you have worked on a design project you will have designed the same beam a number of times as the loading changes, or the column position moves, or the cladding changes. This is all a natural part of the design process, and on large and complicated projects it is inevitable that the design iterates and converges on the right solution. It is not possible to get the full design right first time. It is too complicated a system.

However, I am also sure we have all worked on projects where it feels like decisions made earlier could have made the project so much simpler. From an ill-conceived stability system to an irregular grid which feels like it could/should be rationalised — but the design has got so far down the line, that to change now would jeopardise the project. I am sure we have all scratched our heads and wondered how have we got here? If you have never had this experience, someone in your office is getting the conceptual design right.

A good conceptual design should offer flexibility without needing to start all over again. It should offer a rational solution that works not just for the engineer or for the architect, but for everyone. Sometimes this design is not easy to come by. The requirements of the client mean that finding a rational grid requires much thought and problem-solving. But if you get it right, the benefits further along the design process are huge (Figure 1.3).

Occasionally we will be carrying out the design process on our own, as the only professional involved in the project, but more often than not we will be working as part of a team. As part of a team, it is not only important to get the conceptual design right, we need to be able to communicate it. This communication will be verbal, written and through drawings. Drawings often form the heart of the conceptual design, reflecting the requirements of the client, and articulating a solution that fulfils a client's brief. As engineers, we need to be able to draw in a number of ways, from rapid sketching to transferring ideas during a meeting, to more considered sketches being bounced backwards and forwards by email, as we try and converge on a solution, to neat and detailed sketches, or even CAD drawings, issued at the end of the conceptual design process. Often it is the information on these early drawings that defines the success of the project and removes the complicated negotiations later on in the design. Making drawings

Figure 1.3: Degree of influence on design versus how long the project has been running

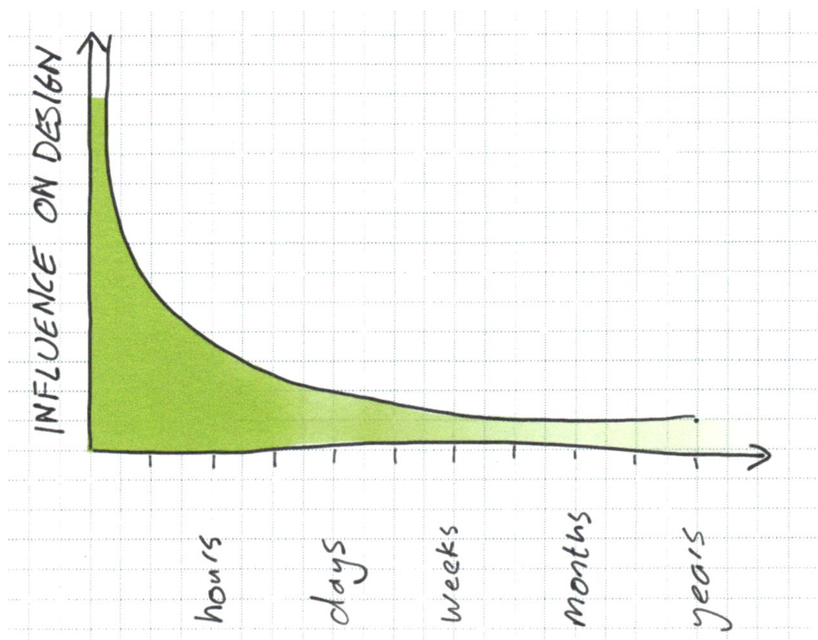

rich with information, may feel time consuming when all you have been asked for is a beam size, but articulating key information, like expected deflection and connection sizing and tolerance can help solve many problems later on.

You may be wondering where BIM fits in with this process. And the simple answer is, it doesn't! Up to the end of the conceptual design process, drawings are sketches, the information is in constant flux and a fat felt tip pen is your friend as its lack of accuracy leads to a sensible level of approximation. As you converge on a solution and prepare to issue information at the end of conceptual design, it is quite possible that you will issue 3D drawings, often based on overlays from other practices. At this stage, it is unlikely that you are formally achieving BIM standards, but at the same time it is quite probable that you will be considering this for the next step as you enter detailed design. As we said, ideally it is best to get it right first time and, as such, you may well work closely to BIM protocols. This avoids having to do everything again at the next stage. As a result, while BIM is only mentioned in passing in these pages, it is worth considering it, just as you would the detailed design of your stability.

In every aspect, a successful conceptual design will make the later stages of the design process easier, but it cannot completely remove the risk of the brief being changed later, and will hopefully lead not just to a great design but also a happy client and architect, increasing your chances of working with them again in the future.

The term 'great design' is used here rather flippantly. The successful outcome of a design is complex and multi-faceted. Defining what we are trying to achieve is often key to this and we look at 'developing the brief' in Chapter 5.

1.3 Building design process and the RIBA stages

The Royal Institute of British Architects (RIBA) Plan of Work 2020[2] reflects the different stages of a design project. These stages can be loosely described in three sections (Figure 1.4):

- Conceptual design stage (Stage 0–2)
- Detailed design stage (Stage 3–4)
- Construction and use (Stage 5–7)

If you are not already familiar with the Plan of Work, it is free to download from the RIBA website.

Figure 1.4: The RIBA design stages and the degree of influence an engineer can have at each stage

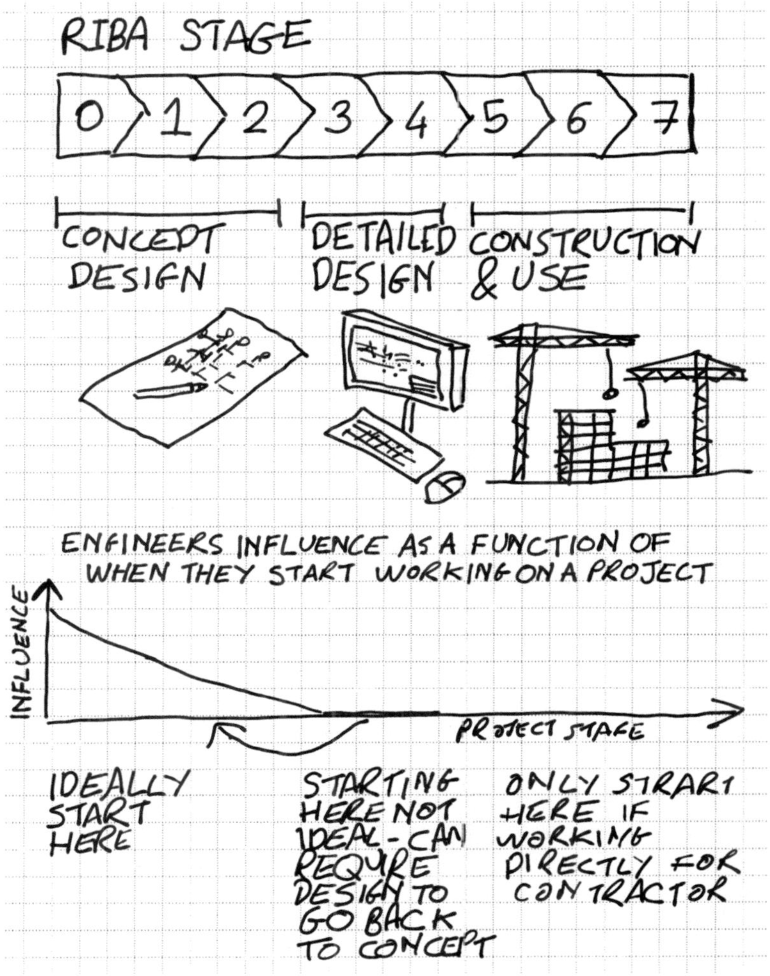

The conceptual design occurs between Stages 0 and 2. Stages 0 and 1 are the 'Strategic definition' and 'Preparation and briefing' and are covered in detail in Chapter 5. Stage 2 is the 'Concept design' and is covered in more detail in Chapters 7–11. At the end of this phase, the design team is normally required to produce a report and a set of drawings.

It is possible that the project has gone beyond Stage 2 before you are approached to work on the job. This does not mean that you are no longer required to carry out the concept design — you still need to go through the process, it just means that the design is far more constrained and that your options may well be limited. The frame may already be a steel frame with precast planks. However, there is a risk that the design is so constrained that a solution is not possible, forcing the design back into Stage 2 before it can develop.

It is our belief that the earlier the engineer is involved in the project, the better both for the client, who will get a better building with less problems later on, and for the engineer who will enjoy working on the job far more. However, there are some architects who are very good at considering the engineers needs without involving an engineer. At times in these situations, it may feel like the engineer is adding little value to the process.

1.4 How to use this book

This book is here to act as a guide. To challenge you to ask questions you may not have thought to ask, but to also have the confidence to make decisions you may not have felt confident making. However, it is very important to state that this book does not replace both your own experience and the experience of your colleagues. Every project is different and while the authors have, in our combined 100 years of design experience, seen many different situations, we haven't seen them all.

We would recommend you view this book as being similar to the advice you might get from a senior engineer. You'll still want to weigh it against your own experience, and you should still have your information reviewed by someone more senior, such as a director or chartered engineer, prior to issue. Ultimately, talking your design problems through with someone, especially someone more knowledgeable than yourself, will always be more helpful than reading a book.

It is also important to be aware of your employer's attitude to risk. Different organisations have different approaches to risk and the attitude of yours may be affected by previous issues they have encountered. This may also depend on where the liability lies regarding the design. If your director's house is at risk if there was a problem, they may have a different attitude to risk than that of an employee of a limited company. These different views of risk do not mean that the building itself will be more (or less) safe, rather that different solutions may be viewed with more, or less, scepticism. With all conceptual design, there is a risk that things might need to change later on.

The book should be viewed in three ways:

Linear
It is written to be read linearly and is divided into three sections.

The first part looks at idea generation (Chapter 2), communicating and developing ideas through both drawing (Chapter 3) and writing (Chapter 4), developing a brief (Chapter 5) and the questions we must ask (Chapter 6).

The second part looks at the concept design process, taking you through the site constraints and ground conditions (Chapter 7), developing a structural scheme (Chapter 8), robustness and stability (Chapter 9), the quick sizing of elements (Chapter 10), and finishing with some worked examples (Chapter 11). It should be noted that the conceptual design process (Chapters 7–9) should not be seen as a linear process. Instead, the consideration of site constraints, geotechnical issues, structural solutions, materials, construction considerations, stability, prefabrication and all the other rich questions you might want to ask about the design should be seen as acting simultaneously. In many ways, we will gather as much information as possible and then throw it all up into the air and grab hold of the solution in one rapid movement, much like a skilled circus trick (Figure 1.5).

Figure 1.5: The early concept design process involves throwing everything in the air and catching it all at once

The third part (Chapter 12) looks at how we communicate the brief, design process and proposed solution/s and the other deliverables we may want to produce at the end of the conceptual design process.

For reference
The book is also designed to be a reference. If you want to choose a location for concrete shear walls or want to get a quick idea of the advantages of different materials, you can dive straight in. Every section is written to be self-contained, but it may occasionally refer to ideas contained elsewhere.

As part of a building designer's library

Finally, this book is not designed to provide all the information you require to carry out concept design. Instead, it is intended to supplement and support a number of excellent texts which already exist. We would suggest that every engineer should have access to the items in Table 1.1.

Table 1.1: Suggested books to have in your library

Title	Comments
Structural engineer's pocket book: Eurocodes (3rd edition)[3]	This book is a toolbox containing a vast array of useful information in one place
Structural timber elements: a pre-scheme design guide (2nd edition)[4]	These three books all deal with rapid sizing of elements in their selected material. They provide a large array of data covering beams, slabs, columns and walls in timber, steel and concrete respectively
Steel building design: design data (SCI Publication P363) [also known as the 'Blue book' and freely available as a fully interactive version][5]	There is also a plethora of load span tables available online. We have avoided specifying brand-specific tables but if you require load span tables for precast concrete planks, a simple web search will give you all the information you require
Economic concrete frame elements to Eurocode 2: a pre-scheme handbook for the rapid sizing and selection of reinforced concrete frame elements in multi-storey buildings designed to Eurocode 2[6]	
Conceptual structural design: bridging the gap between architects and engineers (2nd edition)[7]	This book is aimed primarily at architects and what architects need to know about the engineering side of conceptual design. However, it is also a useful reference for engineers
101 things I learned in architecture school[1]	This book is short and sweet, but will introduce you to some architectural ideas such as the 'Parti' and shape-space theory, which will help you in better understanding what the architect is thinking and will enable you to communicate more freely with architects. It should be seen as a very small introduction and we would recommend delving deeper into architectural literature as part of your own development in conceptual design

1.5 One hundred years of design experience

This book is written by a group of engineers who have gathered over 100 years of design experience between them. Each chapter includes personal experience and anecdotes. With concept design there is no right answer, just experience. Don't be surprised if we provide occasionally conflicting opinions. While this at first may be frustrating, in time you will discover a richness to this discourse. I can recall a number of times when my colleagues at work disagreed on the approach to a problem and, as the engineer, I had to work out the best way forward. I also wouldn't be surprised if you and your colleagues don't always agree with us either. That's fine too. Ultimately, as a practicing engineer, it is your responsibility to make the best decision you can, based on the available information.

1.6 'Good enough'

The information provided in this book is designed to be used for quick conceptual design calculations only. As a result, much of the information is 'good enough' for this purpose but should *never* be used for detailed design of buildings. For all design post-Stage 2 of the RIBA Plan of Works 2020, the latest codes of practice and most up to date information available at the time should be used.

Oliver Broadbent
Constructivist Ltd

2 How to have ideas

2.1 Why do engineers need good ideas?

Engineers need good ideas because:

- the world is constantly changing, so we need new ways to meet society's needs, particularly in the context of climate breakdown
- otherwise, the creative thinking would be left to other people who don't necessarily have our technical understanding
- having good ideas gives us a greater opportunity to shape the outcomes of the projects we work on
- in a future where computers do most of the analysis work, it will still be the role of the engineer to dream up what to analyse

I believe that the ability to have good ideas is a critical skill for engineers; yet, idea generation barely features in traditional civil engineering courses in the UK.

I started my work in the field of engineering and creativity working with Ed McCann and Chris Wise at Think Up, from whom I have learnt a great deal, and I now continue this work at Constructivist Ltd. Over the last few years I have worked with hundreds of engineers to help them develop their idea generation skills. What I have learnt is that everyone's personal practise, whether they are aware of it or not, is different — my aim is always to help people understand what they do, so they can do it better. My approach is to offer theoretical models, to which they can map personal experiences where they align, and to encourage individuals to apply techniques which emerge from these models reflectively, to see what works for them.

The suggestions that I have put forward in this chapter are aimed towards structural engineers, but you will see that almost all of them can be applied to any situation that requires creative thinking.

2.1.1 Four principles for idea generation

Four considerations form the foundation upon which the idea generation strategies in this chapter are built:

What is an idea?

In the context of design, I like the answer provided given by James Webb Young in his book *A technique for producing ideas*[8]:

"an idea is nothing more or less than a new combination of old elements"

From this simple grounding we can build up to what it means to have an idea.

The Kalideascope

Along with his simple definition, Young also provides us with a useful, visual model for the process of having an idea. The model is that having an idea is like using a kaleidoscope — the multi-coloured, different-shaped bits of glass at the end represent the existing elements in the mind. Turning the Kaleidoscope changes the relationship between the elements to create new patterns — each new pattern is analogous to a new idea formed from old elements.

My contribution to the history of thought is to update this model by inserting a pun (which I have always found supports retention and recall) — I call it the 'Kal**idea**scope'. Importantly, this model reveals two factors in the idea generation process that we can influence:

- What pieces of glass we put into the Kalideascope i.e. what information we have in our minds
- How we turn the Kalideascope i.e. what actions we take to stimulate new connections

Subconscious thought

The subconscious plays an important role in idea generation. It busily creates new connections between existing elements in our mind, most of which we are unaware of, but some of which emerge as a bolt out of the blue. Unlike our conscious mind, we can't directly control how our subconscious behaves, but we *can* prime it for action e.g. by choosing how we organise our work and rest time, and by controlling what we let distract us.

The creative system

We now extend our consideration from the creative individual to the social and cultural context, within which ideas emerge and are propagated using a simplified version of Mihalyi Csikszentmihalyi's 'Systems model for understanding creativity' (Figure 2.1)[9]. The system consists of three elements:

- The creative individual
- The culture of existing ideas upon which the individual draws in the creative process
- The society within which the individual operates

Figure 2.1: An illustration of Csikszentmihalyi's 'Systems model for understanding creativity'

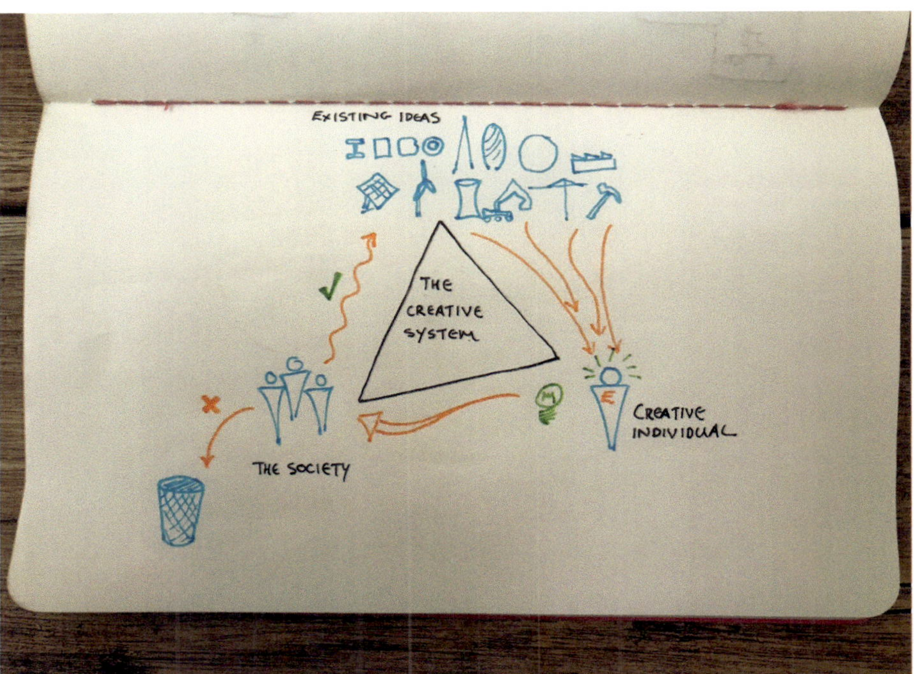

In this model, new ideas are built upon previous cultural ideas that have been accepted by society. A new idea must be acceptable to society for it to be admitted to the culture. If an idea is not accepted, it is not propagated in the culture, and doesn't influence further creative thought. This model works on many levels.

There may be pinned to the wall of a design office different design options for a bridge scheme. An engineer may see these designs and include elements of these designs in their ideas for a new project. If their ideas are deemed acceptable in an internal design team meeting, their sketches may get pinned up on the wall, at which point they become part of local culture, upon which other engineers might develop their own ideas.

At a different scale, a building may be awarded a prestigious architectural prize and get lots of attention in the press. In this case, the jury has acted on behalf of society to say this idea is worthy of high acclaim, and as a result, the project takes a highly visible place in the culture, and is likely to influence many other designs.

We can summarise these considerations as four principles of idea generation:

1. An idea is a new combination of old elements.
2. We can exercise some control over what elements i.e. information, we put into our minds to begin with, and how we stimulate our minds to form new combinations between all these elements.
3. We can prime our subconscious to make the most of its creative power.
4. We can manage the creative system in which we operate, to ensure that we draw upon a wide range of existing elements and that our ideas propagate effectively between the people with whom we are working.

2.2 Tools for idea generation

What information do we need to have ideas?

Principle 2 tells us that we need to gather information in order to have ideas. In this and the following sections, I lead you through sources of information for idea generation, and techniques for mixing this information to generate new ideas. In reality, these processes happen together:

- The act of researching a problem generates ideas
- The act of playing around with a problem reveals new areas of research that are necessary to develop the idea

Therefore, it is not only the information that you gather that warrants attention, but the process you go through in order to gather it. The question of how you gather, collate, display and share information impacts the creative system, as we shall see later in this chapter.

The information that we need in order to have ideas can be split into two categories (Figure 2.2):

1. Information in the moment — which we actively seek, when we know what problem we want to solve
2. Information over time — which builds up over time and we draw upon when we have ideas

Figure 2.2: Possible sources of information to support idea development

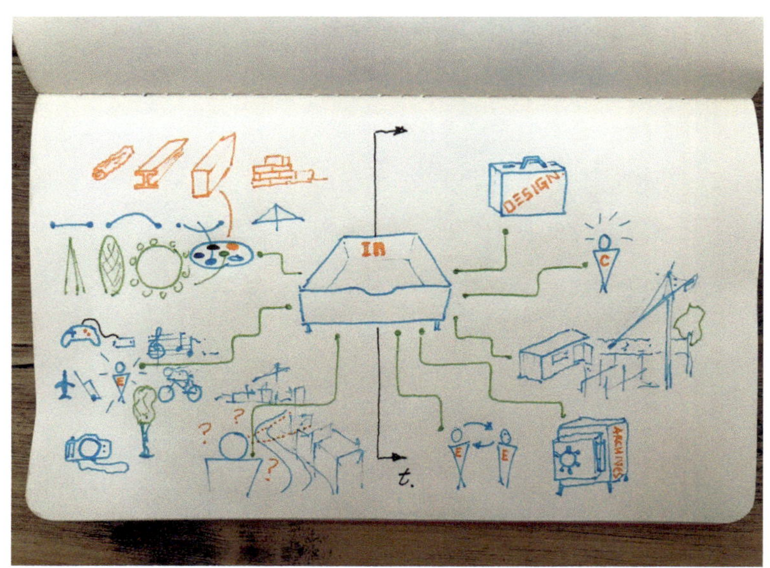

2.2.1 Information in the moment

When a new design brief calls for some idea generation, there are several sources of information specifically relevant to that problem that can be immediately tapped.

The design brief

For any design brief, we can pose a long list of questions that will yield a wealth of information we can use in our ideas. Questions lead to more questions, so here are a few places to start.

- Who are the users? Are they all the same? What are their differing needs?
- Who are the key stakeholders? What are their preferences?
- Who are the other designers on the project? How does their design process work?
- What is the budget? What is the programme?
- What are the materials? What is their relevance? Why were they chosen?
- What are the necessary compatibilities with other systems?
- What is the history of the project?

Write down the questions, pin up the answers, share what you've brought in and ask what next to find out.

The client

The client probably knows more about the context in which you are designing than anyone else, but you won't necessarily get that information from the design brief, for two reasons:

- The client may have no experience of commissioning the sort of project you are working on and so may not know what information to provide
- According to McCann's 'Designer's paradox', the client doesn't know what they want until they know what they can have[10]

To unlock that wealth of information, talk to the client. The aim is to gather information, and so the conversation must be client-focused, rather than to talk about ideas, which would be designer-focused. One technique is to use Blake and Mouton's 'catalytic questioning'[11], which aims to help the client discover for themselves things about the project that they hadn't realised. This sort of approach helps build trust with the client, and opens up better dialogue between client and designer.

Catalytic questioning takes practise. As my colleague Nick Zienau recommends, to get started:

- Ask the client open questions about the project
- Follow the client where they want to go, rather than directing them towards what you want to cover
- Offer quick summaries every so often to show that you are listening and to give the client the opportunity to correct any misunderstandings
- Don't take notes, as this shows you aren't fully listening — write down the salient points afterwards, and share them with the design team

Go to site

The brief may provide you with some details of the project, but to really get inside the project and get information that sticks, you need to visit the site for yourself. According to the science of embedded cognition, humans perceive and understand the world by moving through it. As Matthew Crawford describes in his book *The world beyond your head*[12], our faces are a sensory array attached to a multidirectional moving armature, that can climb up, get inside and walk around *"in ways that are characteristic of the particular sort of bipedal animals that we are"*.

Go to site, walk around, get down and smell the earth, lay down and look at the sky, get up high and gaze down from above, close your eyes and look with your ears. Embody the information that the site has to offer and let it seep deep inside. Capture what you learn with photos, videos and sketches (because to draw you really need to see, not just look), so that you can share this information with the rest of the team.

Previous projects
According to the creative system model, the individual draws on the body of existing ideas to create something new. In the context of a design office, a significant channel for existing ideas is the organisation's existing portfolio of projects. The advantage of drawing on existing work is that these are artefacts that your organisation already knows a great deal about.

Trawl the back catalogue, sketch the key diagrams and pin them on the wall.

Go and ask someone
The information you need may be in someone else's head, so you need to go and ask for it. I have been surprised by how rich the responses have been when I have posted a note in the office kitchen or a messaging channel to say, "I am working on so-and-so, has anyone got any useful information on it?"

In my experience, people feel good about sharing information. They will put extra effort into sharing, saving you time, and you will feel grateful and are more likely to help them — a positively reinforcing process.

2.2.2 Information over time
The other source of information for our Kalideascopes is the knowledge that we have been building up over time, including through our professional experience, our ongoing professional development and our education. It also includes knowledge and experience gained through our outside interests, and from the simple fact of being a human being growing up and living in an engineered world.

Of course, we can only reap what we sow, therefore as well as suggestions on how to harvest this information in the midst of a project, this section also includes ideas on how to plant and cultivate a diverse crop of inputs.

Professional knowledge
Practising engineers build up highly organised arrays of knowledge, grouped around themes like construction materials, foundation designs, structural typologies, construction methods, analytical methods, example projects etc. These pockets of knowledge can be delved into for information to feed into idea generation.

For a new project, I recommend choosing a related knowledge theme and quickly writing down any information that comes to mind in the context of this project. For example, in generating ideas for a new structure that needs to span some distance, draw the boundary conditions on a sheet of paper, and quickly sketch out a compression structure, a suspension structure, a bending structure, and any other structural form you think relevant. By listing out the full range of options, you bring a wider set of inputs into the Kalideascope than simply the most straightforward answer. Apply this same technique to draw out information from other existing bodies of knowledge, to bring them into explicit consideration.

This technique relies on having built up a diverse range of inputs over time, and recall is enhanced when we spend time actively structuring the information e.g. by keeping a scrapbook of example projects or a box of material samples. At the same time, ask yourself what you don't know about — as these may be blind spots in your thinking — and seek out something new. Lay down rich deposits first so you can mine them later.

Outside interests
While professional knowledge provides an important set of inputs to our Kalideascopes, these inputs are likely to be fairly similar for a group of engineers practising in a similar field. In order to increase the diversity of inputs, and therefore the range of outputs, we need to bring in some new colours. A great place to start is our outside interests — the things other than our jobs, about which we are passionate.

From our personal interests, we might be able to draw on:

- Different patterns that we are used to dealing with
- How equipment works
- Particular aesthetic considerations
- Historical origins
- People, their perspectives and their needs

What we do outside of work can have a strong bearing on how we see the world, and each of our perspectives is unique. We can bring these interests into our creative work in a range of ways e.g. by keeping photographs of these interests in sight around the place where we do our creative work. At the start of a project, much in the same way as described for professional interests, we can quickly write down a few things that come to mind from the domain of that interest, that we could apply in this project. Remember, it doesn't have to have direct relevance, as our brains will find the connections for us.

As for professional knowledge, for this technique to work best, we need to have an active engagement in what we are doing, and be able to sort that information. It could be through keeping a photo album, writing a log in a notebook about what we are doing or even writing a blog. At the very least, we should approach our outside interests with a questioning attitude that asks: 'How does what I do here apply to what I do in my work?', as we shall discuss in the next section.

Questioning attitude

Children often ask: 'Why?' a lot. It helps them make sense of the world. As we grow older, that questioning subsides as we become familiar with our surroundings. From a brain-efficiency point of view, this is quite a helpful feature of our mental circuitry, as it allows us to walk down the street without stopping every 10m to question something, like a toddler.

But from a creative point of view, it is important to carry on frequently asking: 'Why?'. Asking why a thing is the way it is will either reveal new information to you, or it will yield more questions to answer. Perhaps a more advanced question than the toddler's is to ask: 'Why isn't a particular process done differently?' Over time we become familiarised to the expectations of how things are done and how to behave, and we need to adopt a questioning attitude in order to remember that things can be done differently.

If you have been questioning the world around you, then you will be able to bring those answers to the creative process. If not, start asking: 'Why?' and see what new insights you uncover.

Put the information into the creative system

All of the steps so far in this section have been on information gathering. To feed the creative system and strengthen both your own creative thinking and the creativity of those around you, think about how you collate and share this information.

The simplest way to share information in a shared working space is to pin it on the wall. Gather your assorted inputs and think about how you can make connections — and therefore new ideas — between the elements in the way you display them. Creating a display achieves three things:

- It helps the individual process the information
- Others can more easily add their own inputs to the problem
- The materials from this project can become inputs for a completely different project which may be happening at the same time

Of course, there are also plenty of ways to collate and display information online. The important factors are making the display clear and easy to read, accessible to as wide an audience as you can and including an element of serendipity, so that someone you never expected might access your online materials and provide the one new insight that is going to spark a new idea.

2.3 The importance of divergent thinking

Earlier in this chapter we saw that an idea was simply a new connection between existing elements. The previous section looked at potential sources of information that could form the basis of new ideas. In this section, we will talk about approaches for actively stimulating new connections. I use the word 'active' to distinguish these steps from the ideas that will probably be stimulated, either consciously or unconsciously, as soon as we start engaging with the source material.

Maximum creative potential exists only towards the start of the design process before many of the parameters have been set. It is at this point that it is worthwhile considering the widest possible set of ideas, however briefly, in order to see if they have any potential. One aim therefore is for the pool of ideas to diverge.

However, while our brains are natural idea generators, as Mary Ann Collins and Theresa Amabile describe in the *Handbook of creativity*[13], the creative process can be restricted when attention is split between the creative process at hand and external factors, such as a looming deadline. Since it is rare for engineers to be working without at least some of their attention directed to external factors, it is useful for us to have some tools at our disposal to help stimulate new connections when the ideas stop flowing. In other words, we need some ways to turn the Kalideascope (Figure 2.3).

Figure 2.3: Techniques for turning the Kalideascope

Draw from a different perspective

This technique was inspired by the title of a book I find endlessly stimulating, *The art of looking sideways*[14]. When doing the initial desk research for a project, we may only look at a new project from a single perspective e.g. a site map or a cross-section. The ideas we initially generate may arise from that frame of reference, but that initial frame of reference may restrict our thinking.

The technique here is to quickly sketch the project from a few different perspectives:

- From above
- From behind
- From inside looking out
- From the neighbour's side of the party wall
- From the perspective of a worm looking up at the building
- From the perspective of a child wondering what on Earth his or her ancestors were thinking
- From *any* position that would force you to see the project in a different light
- Changing vantage point potentially prioritises different elements of the project, reveals new connections between the project and its adjacencies and boundary conditions and alters the importance we place on different aspects

As with the 'What if' technique below, we are turning the Kalideascope to form new connections between the various things we already know about the project to help us catch sight of something new.

Ask: 'What if?'

In this technique, we change the parameters of the design brief in order to widen the range of possible solutions you would consider. The options generated don't need to be plausible, but any one of the ideas might provide a glimpse of how the original problem could be solved differently with a new approach.

The technique is to ask: 'What if?' and then to complete the question with a change to the design brief e.g. What if:

- the entire structure needed to be build by hand?
- the building needed to created from on-site materials?
- we doubled the budget?
- we reduced the budget by 90%?
- all the materials had to be fully recoverable?

Some of these questions open the brief right up, while others restrict the options — but even these can cause divergent thinking. By asking: 'What if?', we are effectively changing the settings of the filter that determines what is permissible, allowing us to broaden the ideas we generate.

The central question that provides a hundred answers

In any design problem there is likely to be a central question that, once answered, dictates how a hundred other questions can be answered. In a city masterplan, the question might be how the flood management system is arranged — the answer to which sets the geometry for the rest of the streets and the infrastructure. In a tall building design, it might be the choice of lateral stability system that has the greatest influence on the form of the tower and everything in it. Equally, the big question could be the construction methodology — the choice of installation method could set the rest of the parameters for the design of a footbridge over a railway.

The technique here then is to look at an existing idea, identify a central question that was answered at an early stage that has dictated the solution space, and to ask: 'What if I asked that question differently?'

I have found that engineers particularly like using this technique when they have already invested considerable time in developing an initial idea. The mental activation energy required to come up with something significantly different can be considerable. Asking what would happen if we changed the key system seems to catalyse fresh thinking.

As ever, the aim is to give the opportunity for new connections to be formed; for the Kaleideascope to yield another idea to contribute to our diverging pool of thinking for later analysis.

Brief disruption

On first reading, the design brief usually looks like a clear set of requirements. By adopting a rigorous, questioning attitude we can easily open up something that is apparently clear-cut, and turn it into something full of ambiguity and possibility.

You can use this technique on your own, but it is more fun to do it with someone else. Go through the brief word-by-word. Stop at every noun, adjective and verb. For each word, check its meaning. Pull out a dictionary and check alternative meanings. Open up a thesaurus, write down synonyms and see what reaction they provoke. Wherever the brief is describing some sort of requirement for the design, ask: 'How will I know when this requirement will be met?'

These questions are intended to deliberately disrupt a straightforward reading of the brief, and turn it into something in which there is more space for exploration and creative possibility. This 'brief disruption technique' is the basis for the 'creative collage' technique described at the end of this chapter.

All four of the techniques in this section on divergent thinking are intended as ways to actively rearrange elements to create new ideas. But as we shall see in the next section, even when our active brain has clocked off, our subconscious is working away.

2.4 Unlocking your subconscious

If we added to our timesheets the time our subconscious spends problem-solving for us, there would be no engineering project delivered to budget. Although we are not aware of it, our subconscious is busily working away on problem-solving — forming new connections and creating ideas — even while our active brain is taking a break. The only time we are really aware of the work of our subconscious is when an idea pops into our consciousness as if from nowhere.

The guidance in this section is to help you manage to unlock the power of your subconscious.

Give it something to chew over

When I ask engineers where they have their best ideas, very few say 'sitting in front of their computers'. It is in the in-between time — walking to the kettle, unlocking their bicycle, gazing into the middle distance from a train window, having a shower. It is in these moments that our conscious brain is able to spot what our subconscious has been working on. For our conscious selves, the ideas appear to pop out of nowhere. The subconscious has been busily working away, we just haven't given ourselves the chance to notice the results.

Knowing that there is subconscious work going on in the background, I have started to deliberately give it something to chew over. This can be as simple as reading the design brief for an upcoming project, or writing down a problem to ponder. I find the timing of this action to be important. I usually like to give myself a creative problem to think about just before I do something that I know is going to require little active mental effort e.g. just before I cycle to a meeting. Almost inevitably, I find that when I have reached my destination I have had at least one idea in response to the problem. Let me emphasise, in this process I am not actively thinking about the problem, it just sits there and bubbles away with apparently very little mental effort.

This powerful creative engine can work on many problems at once. I recently had an intensive period of work in which I needed creative input on multiple projects. While I only actively worked on any given project at one time, I made sure I kept reading over the briefs for the other projects and found that, sure enough, by the time I reached them, I had already had in mind a store of creative ideas in response to these briefs.

The preceding paragraphs are of course based on personal experience, but hopefully you can see how you can develop your own habits for harnessing the creative power of your subconscious. Whatever process you adopt, you will need to beware of the peril of distractions, as we shall see next.

Manage distractions

I took a five-minute break between writing this section and the last. The intention was to give my active brain a rest and to allow my thoughts to assemble for the next. As I stepped outside a storm was blowing and I took a brisk walk across the Floating Harbour in Bristol. As I was wandering, I realised that this experience could be a good way to illustrate the benefit of minimising distractions, which I am now doing. Had I, while wandering across the bridge, checked for messages on my phone, my attention would have been on whatever urgent matter demanded my attention therein. Probably, I would have returned to my screen with no new ideas on how to write this next section and, quite probably, my conscious mind would start thinking about this new invading topic.

As knowledge workers in the digital workplace, engineers are exposed to ever more distractions. According to philosopher Matthew Crawford[12], our attention has become a good that is being exploited for commercial benefit. Everywhere we look, we are beset with distractions — email notifications, headlines, messages etc. that pop up on our screens. The phones in our pockets are engineered to be highly distractive — even when we can't see the screen, the temptation to check our social media feed is sometimes very difficult to resist. Again, a highly engineered experience.

Distractions do two things:

- They give new topics for your subconscious to be working on, instead of the creative problem you were working on
- They give new things for your conscious mind to be actively thinking about, instead of being in the calm state that will allow you to notice the outputs of subconscious thought

In my experience working with engineers, subconscious thought is a very underused creative resource at their disposal, either because they don't give it time, or because they allow themselves to become distracted.

Take a moment to think about where distractions come from and how you can avoid them, for example by:

- turning off all pop-up notifications on screens that you use
- removing or hiding social media apps on your phone, so that they are not so easy to reach for when you crave a distraction
- assigning a portion of your working hours each day when you are not contactable, so that you can work alone with your thoughts

Happy engineers

Researchers in the domain of positive psychology have found evidence of the benefits of positive emotions on our creative abilities. According to Barbara Fredrickson, one of the pioneers of the field:

"Joy, for instance, creates the urge to play, push the limits and be creative...Interest, a phenomenologically distinct positive emotion, creates the urge to explore, take in new information and experiences, and expand in the self in the process".[15]

There are many factors that can affect the emotional state of an engineer during a typical working day. Many of these may be outside your control, but you do have some contro over the impact of stress-raisers that arrive uninvited through your inbox e.g. an email from an angry client that pops up when you are in the middle of doing some creative thinking.

When our stress levels go up, our happiness levels go down and the extent of our creative thought becomes limited. Before starting work on a creative project, think about how you can avoid the negative effects of stress impacting on your idea generation process.

Summary

The subconscious is probably the most powerful creative tool at our disposal. We need to be conscious about how we make the most of it.

2.5 Ideas through conversation

Another common answer I'm given when I ask engineers where they get their ideas is: "in conversation with others". We can probably all relate to an occasion in which we only start to see a solution to a problem when we talk it through with others. At the same time, we can probably relate to a time when we've presented an idea to someone else and their reaction to it has completely quashed it. Worse still, there may be people we work with whom we don't feel we can ever share our ideas.

Whether or not our conversations generate ideas depends on how we conduct them, and the important role that the listener plays, as this section explains.

Stay focused on the speaker

When someone is sharing an idea with you, one of the most supportive things you can do is to listen. Through articulating their idea, they may be clarifying for the first time something they have never properly understood before. As Nancy Kline puts it in her excellent book on the power of conversation, *Time to think*[16]:

"We think we listen, but we don't. We finish each other's sentences, we interrupt each other, we moan together, we fill in the pauses with our own stories, we look at our watches...".

To stay focused on the speaker we can use the catalytic questioning techniques described in Section 2.2.1.
The principles are the same — keep the person speaking, show you are listening and play back to them what you
are hearing.

Build the 'creative system'

According to the 'creative system' model, it is the audience that decides whether or not an idea is accepted and
transmitted to others. If a listener is too quick to reject an idea, that idea may never be shared with others — and its
merits, however unsuitable for the given project, may never be of benefit to other projects.

It is easy to see how a self-reinforcing creative system can be nurtured, in which the diverse range of ideas
generated for one project could provide an even larger range of starting ideas for the next project. Equally, it is easy
to see how the converse could be true: a system in which no one shares their thinking, the feedstock of creative
ideas diminishes and the diversity of thought wanes.

Allow the richness of ideas to emerge and be shared by fully listening to other people's ideas and letting them share
their thinking. The time for idea selection and analysis can come later.

Yes and...

The two previous techniques have presumed someone is sharing an idea they have had with another person. In
contrast, this technique is for two people to use together to generate ideas in conversation, and so could equally
have been located within Section 2.3.

This technique recognises that good ideas sometimes emerge from humourous conversations, where our mind is in
a more relaxed state and the bar for what it is acceptable to say is lowered. It is called 'Yes and...', and is borrowed
from improvisational theatre:

1. Person A poses a problem to be solved.
2. Person B suggests a possible solution.
3. Person A, responds by saying 'yes, and...' and adds an additional element to this solution.
4. Person B, responds to this embellishment by saying 'yes, and...' and further extends the idea.

Usually a simple idea becomes more and more ridiculous, and usually funny. Do this from a few different starting
points for the same problem and you will quickly generate a range of ridiculous ideas, in the midst of which you
might find something quite useful.

Summary

Conversation is a powerful tool for creative thought, but the potential may not be realised if individuals are not given
the opportunity to fully express their thinking. The first two techniques in this section suggest how you can help
others in their thinking. If you can stimulate a culture in your team where people listen to each other, it is more likely
others will help you in your creative thinking.

2.6 Iterative creative thinking

We don't do 'rapid prototyping'

'Rapid prototyping', the process in which ideas are quickly mocked-up and tested, is a popular technique in product
design. 3D printing means life-size prototypes can quickly be produced so that users can easily interact with, and
give feedback on, a realistic product.

I believe this ease of prototyping and user testing is one of the reasons that iterative design is much more
established in product, software and consumer electronic design, than in structural engineering, where scale and
cost make physical prototyping impractical. I believe the tendency in structural engineering is to think of design as
more of a linear process.

But even without the possibility of creating full-scale mock-ups, iterative design approaches are valuable to the
structural engineer. Indeed, they are a necessary response to the 'Designer's paradox', discussed earlier.

If we ignore the 'Designer's paradox' (Section 2.2.1) we risk, at least, disappointment when the client changes their mind about what they want and, much worse, could end up building something which is not fit-for-purpose.

The way to work with the paradox is to keep both the ideas in development and the original brief under constant review. Each new idea provides an opportunity to re-examine the adequacy of the original brief. Each change to the brief requires further development of existing ideas, or creation of new ones.

In practice, this approach to design requires us to keep many things in mind at the same time. It draws heavily on what Leonard Mlodinow describes as 'elastic thinking'[17]:

"Elastic thinking is a nonlinear mode of processing in which multiple threads of thought may be pursued in parallel. Conclusions are reached from the bottom up... in a process too complex to be detailed step-by-step. Lacking the strict top-down direction of analytical thought, and being more emotion-driven, elastic thinking is tailored to integrating diverse information, solving riddles, and finding new approaches to challenging problems."

In practice – the 'creative collage'
That's the theory. In practice, it can be difficult to know how to manage multiple threads of thinking, particularly when working in a group. The following is a technique I have developed to help design groups do just that. I call it 'creative collage'.

The set-up is as follows:

1. Find a large working space which allows the whole group to simultaneously write down an idea. You could use a large sheet of paper on a table around which everyone can stand, or a wall onto which ideas can be written or sticky notes applied.
2. Write a high-level brief in the middle, capturing the essential aims of the project.
3. Write three headings around the brief: 'Questions', 'Information' and 'Ideas'. These headings are points around which to cluster different types of input.

- Around 'Questions', we write down any questions we might have about the brief. Is there anything ambiguous? Is there anything that is missing? Is there any information that would be useful to know?
- Around 'Information', we write down any useful information provided by the brief. We also write down any information we find out in response to questions that have been asked
- Around 'Ideas', we scribble down any ideas that emerge. The ideas can be fully formed, or more likely, fragments of ideas, or characteristics that we feel the ideas might have

The stage is now set. The procedure is roughly as follows:

1. Explain to all participants that:
    a. The aim is to simultaneously develop our understanding of the brief, the ideas we develop and the interplay between them.
    b. This is a nonlinear process, and that they can make their contributions to any part of the diagram at any time.
    c. If an idea jumps to mind, stick it up. If you find yourself blocked because you don't know the answer to something, write it down as a question. If you have some nformation that can inform the process, or that answers a question, write it down under 'Information'.
2. Guide participants to start off by exploring the brief, writing down any information it supplies, and asking any questions that might arise.
3. Invite participants to answer each other's questions, either directly, or by carrying out the necessary research.
4. Encourage participants to write down ideas or fragments of ideas as they emerge.
5. Use string or draw lines to tie together related items.
6. After a period of time, say half an hour, hold a group discussion in which the ideas developed and the brief are looked at together:
    a. Do any of the questions asked or ideas developed suggest any alterations which need to be made to the brief? In which case, update the brief.

    b. Do the ideas proposed meet this updated brief? If not, suggest to participants that these ideas need to be revised.

    c. What happens if elements displayed on the diagram are rearranged and stuck down in a different position — what new connections or relationships emerge?

7. Encourage participants to use good listening techniques and to apply the 'catalytic questioning' style discussed earlier.
8. Continue through cycles of group working and facilitated reflection.

As with all the techniques in this chapter, I've shared with you what participants have found helpful in the workshops that I run. Play with the process and make it work for you.

This process of developing a collage of the brief, questions, information and ideas, ties together the four principles upon which this chapter is based:

1. It allows new connections to be formed between existing elements in the mind, and allows connections to be formed between elements in different people's minds.
2. It provides a structured way to consider what information we draw into the process, and how to make connections between the different bits of information.
3. It provides space and time for our subconscious to make nonlinear connections between all the different elements at play — especially if we leave the process and return to it later.
4. It supports a strong creative system in which lots of people contribute elements which can be the basis of ideas, and in which one person's creative output can be the input for the next person's thinking.

Summary
We need an iterative creative process to make sure we have fully understood and developed the brief. The way the brain sifts through the information that we provide it with is complex and nonlinear. To take into account these two factors, we need a way of working that allows lots of inputs to be viewed at once, in a shared way, and that allows developments in one area of the creative space to influence the other parts.

2.7 Conclusion

The ability to have ideas is an important skill for an engineer — albeit one that is rarely taught. I have written this chapter to provide guidance to enable engineers to develop their own creative practise. We started with four principles:

1. An idea is a new combination of old elements.
2. We can exercise some control over what elements i.e. information, we put into our minds to begin with, and how we stimulate our minds to form new combinations between all these elements.
3. We can prime our subconscious to make the most of its creative power.
4. We can manage the creative system in which we operate, to ensure that we draw upon a wide range of existing elements and that our ideas propagate effectively between the people with whom we are working.

I introduced the model of the Kalideascope to describe the process by which we have ideas. We looked first at how we can gather creative inputs — filling the Kalideascope — acknowledging that some inputs we access at the start of the project, while others we build up over time. We then looked at how we can actively force new connections between these creative inputs — turning the Kalideascope.

We then shifted focus to consider the involuntary role of the subconscious in idea generation and how we can harness its power. Then, shifting the focus from the individual to the group, we looked at the role of dialogue, and in particular listening, in creating a strong creative system in which ideas can flourish. Our final step was to consider the iterative nature of the creative process, and how practically to share inputs, questions and ideas iteratively in a group setting.

At the start of this chapter I said my approach is to offer theoretical models to which individuals can map personal experiences, and to encourage individuals to apply the techniques which emerge from these models reflectively, to see what works for them.

The hard work of skills development is integrating into our daily routines the techniques and approaches we have read about. Good creative work is as much a function of good daily habits as it is a function of the techniques you might use in a design team meeting. Changing habits requires reflective practise. The exercises that follow are intended to help you develop reflective creative practise, as well as to help you practice the techniques described in this chapter.

2.8 Practical

The following exercises are based on 'workwork'. This is something I set for my trainees in my conceptual design training. Workwork is like homework, except that you do it at work in the context of, and in support of, your day-to-day design activities.

Explore the brief

This exercise explores the concept of the 'Designer's paradox'. Each day, for a week, write down any activity you are working on, for which you have been given a brief. It can be a formal design activity, or any task in which you are required to come up with a response to a set of initial requirements. Capture the information in three columns:

- The first column is for a shortened, bullet-point version of the brief you have been given
- The second column is for additional information you have inferred from your understanding of the words that the client has chosen or the context in which you are operating
- The third column can't be filled in straight away — it is for capturing requirements that only become apparent once you have started developing a response to the brief. Entries for the third column may emerge when you present solutions to the client, or when you start working on the project and realise what is and isn't possible

At the end of the week, reflect on how 'unknown' requirements emerged as you did your work, and what steps you can take to account for the 'Designer's paradox' in your design work.

Giving others time to think

Twice in this chapter I have emphasized the importance and power of giving others the opportunity to speak, uninterrupted. This sort of engaged listening supports development and exploration of a brief with a client, and supports the development of ideas through conversation. Practise using the 'catalytic style' described earlier in this chapter in your conversations with clients and colleagues. Keep a note of your attempts to use catalytic style, and ask yourself:

- What do you think the impact was on the person you were speaking with and the quality of your dialogue?
- Which part did you find tricky and what did you think you did well?
- How can you use this approach to support your design work?

Boost your curiosity for the built environment

Asking questions is a powerful way to open up a brief and develop more divergent thinking. However, it is easy for familiarity to dampen our curiosity. Artists are often able to see the familiar from a different perspective, providing new insight and the opportunity to access deeper meaning to the context in which they are operating.

The aim of this exercise is to re-prime your curiosity about the sort of built environment contexts with which you are probably familiar. You may choose to do this exercise in a room on your own if you don't want anyone to see you — or you can do it in your office and see if other people will join in.

Close your eyes for a minute and see if you can put your mind at rest for a few breaths. Having cleared your mental desk, you are ready for the exercise.

Open your eyes, scan the room, and let them settle on the first object that catches your eye. Sit there and look at the object for a while. Now walk up to it and look at it more closely. Notice how it appears to change, more detail coming into focus, as you approach it. If you can, touch the object — how does it feel, what is its texture and temperature and does it feel different to how it looks?

Now, turn around and see what next catches your eye. Approach it, inspect it, investigate it, ask questions about it. Continue this exercise for as long as it is enjoyable.

When you have finished, return to a seated position in the room, and look back over the room you have investigated. How different does it look now that you have spent time really looking at it, and what have you learnt about the space you are in?

You can conduct this exercise whenever you have a few moments to spare. Rather than reaching for your phone and filling downtime with whatever distraction your portable device has to offer, use the time to nourish your curiosity and allow the possibility of discovering something new and beautiful that may just be the seed of a great idea on your next project.

'Action-learning' for building creative skills

This exercise takes an 'action-learning' approach to developing creative skills. Spend a couple of minutes thinking about and writing down the answers to the following questions:

- Defining the goal — what kind of creative thinker do you want to be, how will it support your work, what will you be able to do?
- Analysing the current situation — how would you rate your current creative thinking skills, how do you use creative thinking in your work at the moment?
- Identifying the barriers — what challenges do you need to overcome in order to become the creative thinker you want to be?
- Getting started — which challenge will you address first and what is the first step you can take to addressing this challenge?
- Finding your support — who can support you in overcoming this challenge to becoming a more creative thinker, how will they support you, what could you offer in return?
- Tracking progress — how will you know when you are making progress towards meeting your goal, what will be different, how will it feel? How might others notice?
- Time for review — when will you review your progress, who can offer you regular support, what questions do you regularly want to ask of yourself to help you stay on course?

Gavin Knowles
University of Bath

3 Sketching

3.1 The importance of drawing

Engineering is a creative practice and we need to communicate this in different forms, so the old adage of 'A picture is worth a thousand words' rings true for communication in the construction profession. Engineers are thought of as the 'numbers people' but really great engineers have a much wider skill set and this includes communication through drawing and sketching. The built environment is constructed by interpreting drawings into reality so, as a communication medium, it's a great skill set to acquire and hold on to throughout a career in engineering. Hand sketching really is the basis of understanding and communication in the design world — many issues are solved in design meetings with a simple pencil and paper sketch. These sketches are very basic but tell the story. This chapter provides tips and tricks for engineers to be more confident in their sketching, and encourages them to invest time in practicing this important skill. The guidance is supported with a series of workshops to practise the skills necessary for each sketching technique.

For thousands of years, civilizations have communicated through the medium of drawing — think cave paintings depicting animals and hunting, or early Egyptian paintings (Figure 3.1). These were ways to record and pass on

Figure 3.1: Egyptian wall painting

knowledge or tell stories. In our modern world we still communicate using drawing, or photographs — which are essentially digitised drawings.

3.2 Types of sketching

This section explains the ways in which different types of sketches are used as an informal and formal communication tool. The exploration of these types is useful to understand the level of complexity that a sketch needs to convey at different stages of the design process.

3.2.1 Concept sketches

These can be very quick and are used as an immediate way of explaining and getting thoughts and lines onto paper (Figure 3.2). They are probably completed with a single line pencil (HB or similar). They intentionally lack detail, but start the conversation between the design team about form. These concept ideas may identify the way the structure might want to look or feel. Generally, the image will be accompanied with notes to show the intent of the main elements, and perhaps some questions for the different design team disciplines.

Figure 3.2: Early development sketches to engineering competition work

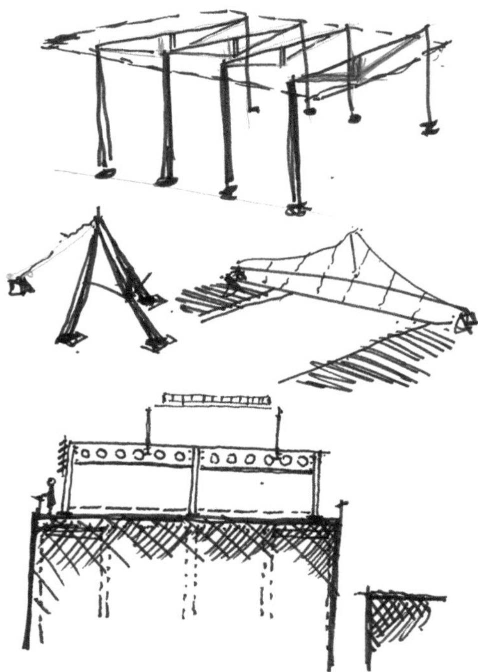

To practice your concept sketch technique see Workshops 1 and 2 (p. 40).

3.2.2 Sketching in meetings (coordination sketches)

An important sketch skill to develop is to be able to draw ideas quickly within the format of a design team meeting. Good communicators often draw something while explaining how the detail might work. This is used a lot when it comes to solving a complex issue or resolving how something might be put together. This initial sketch can then be offered up to the group, and others might want to draw on top of this and show their understanding of the problem (Figure 3.3). In this way, ideas can be readily resolved and there is also a record of the resolved problem. These sketches are quickly produced, and often in a single line pen or pencil. They are free flowing and don't go into great detail, and are a quick response to put an idea out to the design team. It's a good idea for different team members to have different coloured pens to help define inputs.

The skills to develop this should be based around quick drawings of a minute or so and practise verbalising what you are drawing as you draw. This hand, eye and language skill is essential for this type of drawing.

Figure 3.3: Typical coordination sketch — highlighting the coordination of the foundation arrangement around existing drainage

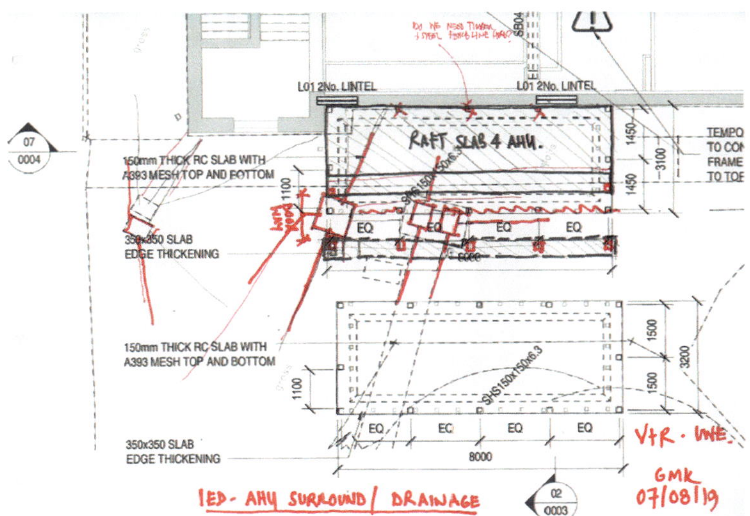

To practice your sketching in meetings see Workshop 3 (p. 41).

3.2.3 Sequence assumed in design (sketches for the contractor)

As designers of structures and the built environment, engineers are responsible for providing an intent of the steps involved for the safe construction of the proposed structure. This is formulated as a series of sketches showing the sequence of the construction process. Think of this as drawing the individual steps or instructions of any day to day task e.g. how to load and turn the dishwasher on, or putting a set of ingredients together to make a cake. If we put this into the context of the construction of buildings, we might want to show the contractor how we expect them to introduce a steel stability frame into an existing building. Figure 3.4 shows this on a project, and you can see it is supported with text information on critical operations, and reads like a storyboard (see also Figures 10.2 and 10.6).

Figure 3.4: Typical storyboarding

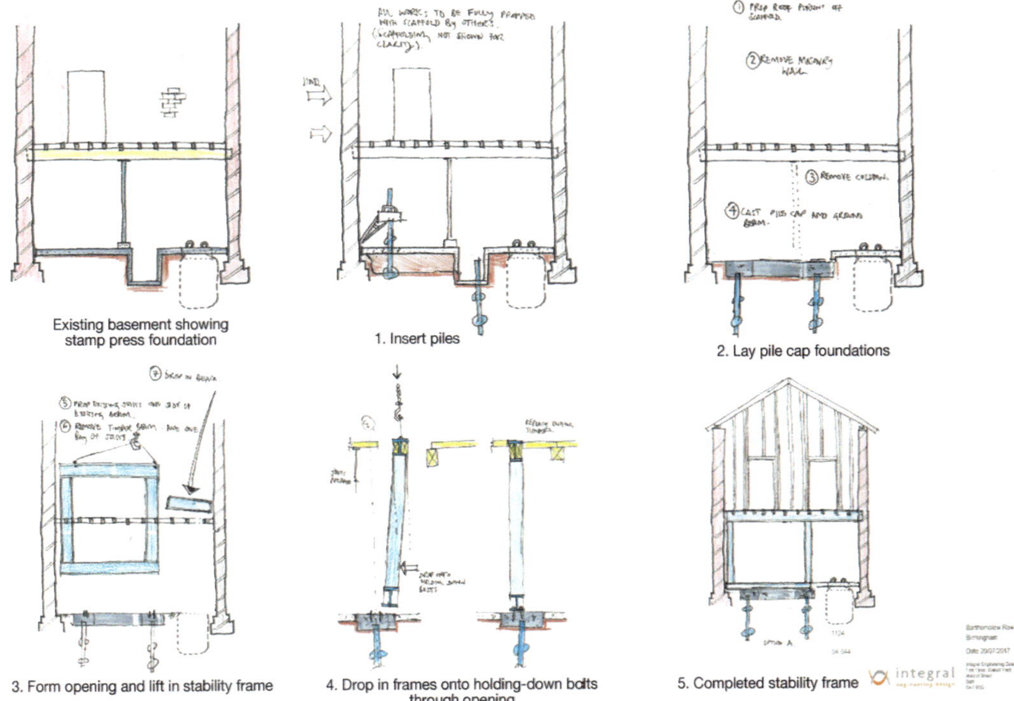

Figure 3.5 shows a steel connection broken down into it's component parts. This is put together as a series of drawings which allows the fabricator to interpret how it can be welded together.

Figure 3.5: Typical storyboarding to aid steel fabricator

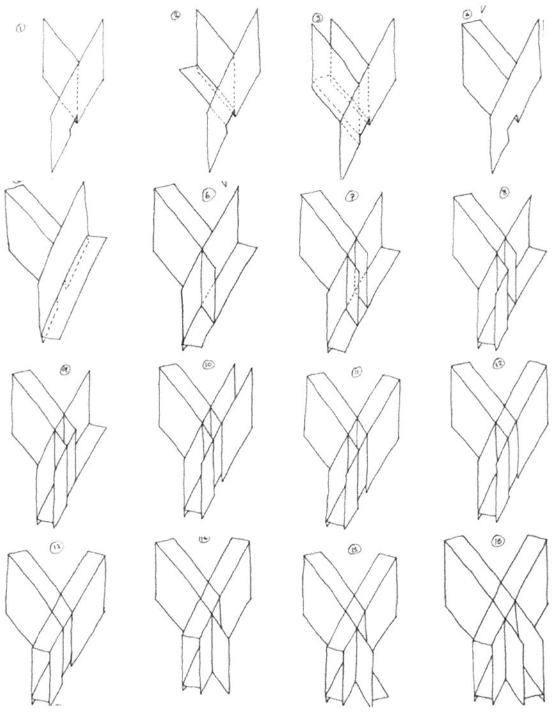

To practice your sequence drawings see Workshop 4 (p. 41).

3.2.4 On-site sketching

Site visits are a great place to practise your sketching. A sketch produced on-site is a useful way to record what you have seen and capture dimensions and arrangements. Add text to show issues and concerns that you need to take away with you to review (Figure 3.6). Sketch it in elevation and plan if necessary. Take lots of photos of the area in question as this is a good aide-memoire when you are back at your desk. Remember this style of sketching is quick

Figure 3.6: Site sketch detail recording existing roof and wall connection

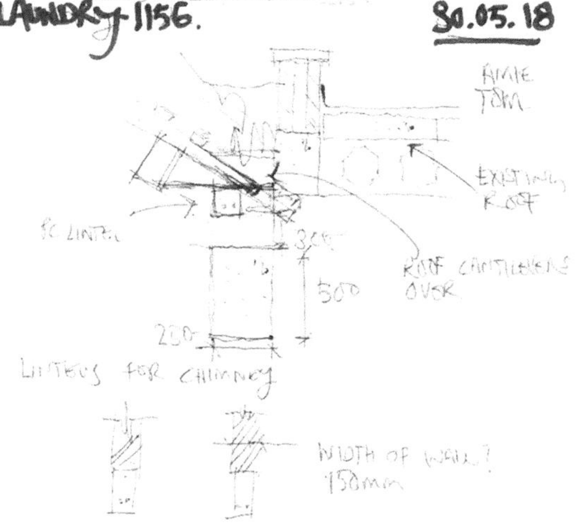

and very rough. The key is recording the details for use back in the office. Don't get lost in the detail and overwork these sketches — they are a record only.

To practice your on-site sketching see Workshop 5 (p. 41).

3.2.5 Details sketching

Often, we need to concentrate on a localised area to communicate how a piece of structure needs to be layered or constructed. This can take the form of a small section or detail. This type of sketch is drawn at a large scale, say 1:5 or 1:2, depending on what we are trying to show. Figure 3.7 shows an elevation of a timber infill wall, with the section on the right-hand side showing in detail how the wall is to be constructed, with its layering and fixing positions.

Figure 3.7: Detailed timber build-up with connection

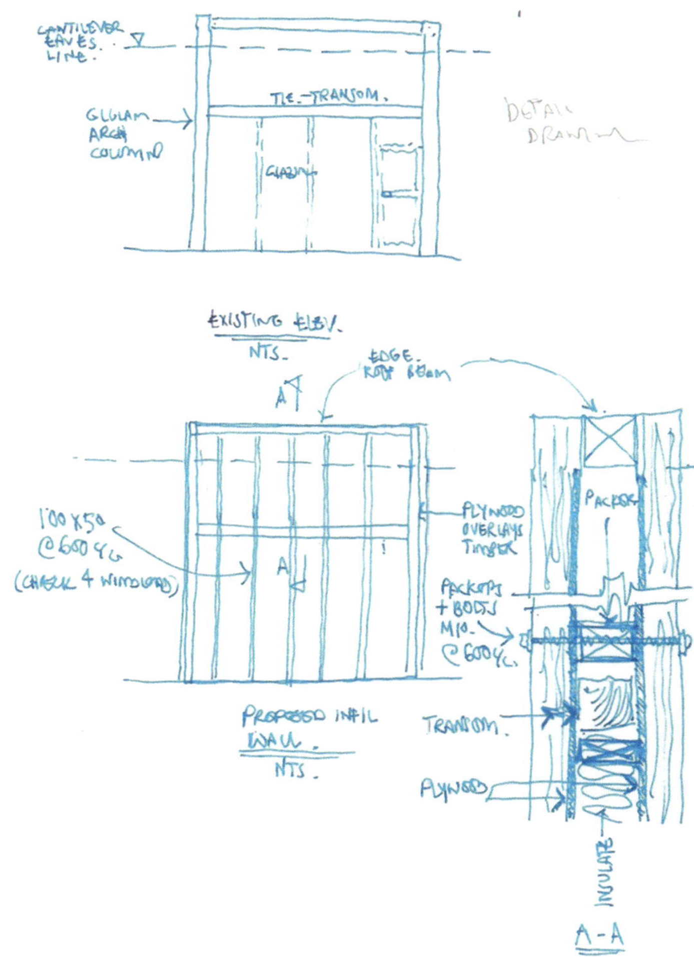

To practice your details sketching see Workshop 5 (p. 41).

3.2.6 Sketching for reports/competition work

The sketch used for a final report or competition work should be more refined, and you may look to use colour and shading to add context. The image is still loose in its form, but time is spent to make it look more like the finished article. Figure 3.8 allows a prospective client to understand what the space may feel like, and how the structure will integrate with the architecture. Note how the shadows are used to show off the light nature of the space, due to the long central roof light. The sketch is in perspective to allude to the size of the space.

Figure 3.8: Detailed timber build-up with connection

Figure 3.9 shows a new floor intervention into an existing building. This sketch was used in a structural report to show the client how the space could be used. The person gives an indication of scale, and the key point here was to show that the floor could only fit between the trusses to give usable space.

Figure 3.9: Mezzanine intervention

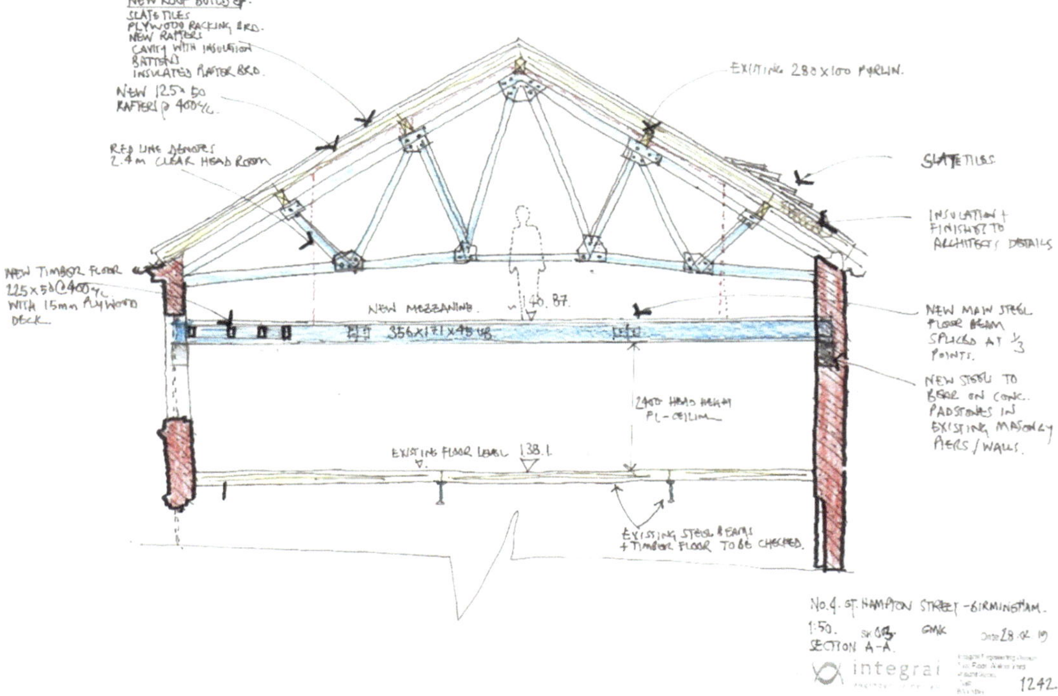

To practice your sketching for reports see Workshop 5 (p. 41).

3.2.7 Sketching to solve a problem

Putting pen to paper, by definition, is problem solving. This type of sketching is very rough, and not exactly to scale, but can be done on-site, in a meeting or simply at your desk when you need to see how an arrangement will come together. Maybe it's to explain to the design team how a piece of structure will integrate with the architecture and services (Fig. 3.6).

3.2.8 Sketching from other disciplines

Other disciplines are good areas to explore. Art is another go-to place for sketch inspiration. An artist never jumps into the finished article, they explore form and shape through sketching and that is what we as engineers are trying to develop with our early sketches. By sketching, the artist allows themselves to explore the form, make mistakes and realise the pitfalls in their drawing, correcting them on subsequent sketches to produce the end result. As engineers, we should use this idea to make sketches fluid and use them as a discovery tool and proof of concept. Drawing the objects from different directions is a good way of finding out how things go together. In doing this we can attempt to consider how we want to position elements on a page to allow an easy understanding of the drawing. The examples in Figure 3.10 are completed quickly and show how early sketches become the basis for the final work. Note how the series of sketches are drawn from different angles to fully understand the form — this process develops a better understanding of the subject.

Figure 3.10: Life drawing

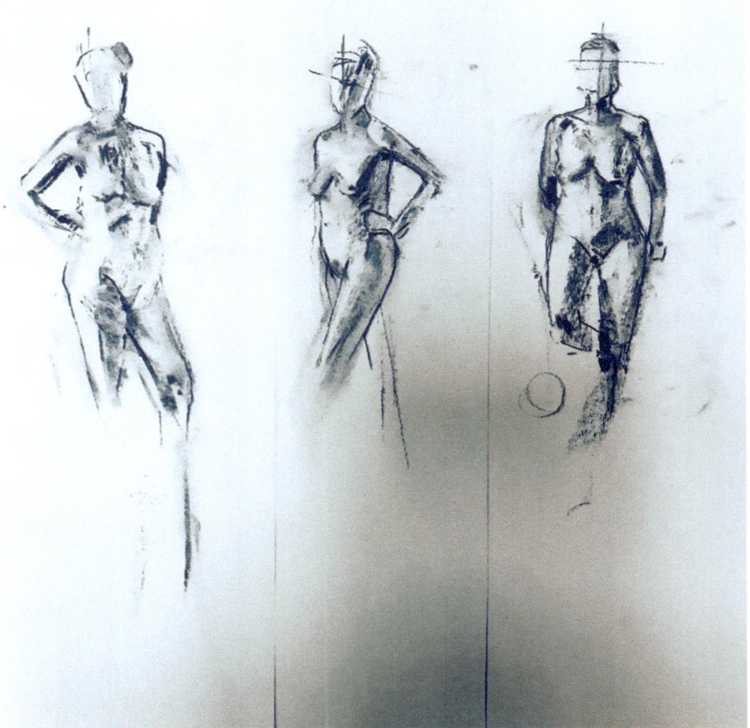

Life drawing is a good way to practise developing lines and speed of drawing, and develops a sense of iteration. Having a few attempts is easy if you are producing sketches quickly. We should take this artistic dynamic and apply it to our engineering sketch drawings. This ability to produce quick drawings, that are not complete in any way, is a very useful technique in meetings where the sketch is being used to explore an idea.

3.2.9 Create a drawing habit

The key to getting better at anything is practise, practise, and practise. There has long been an established theory that 10,000 hours of practise will make you an expert. Don't worry — you won't need this much to become proficient, but it does help to practise a skill and don't worry if you don't get it first time. The joy of sketching things a few times often helps with the development of the design. The important thing is that it communicates the intention. Note in Figure 3.11, adding thick pen outline really defines the sketch.

Figure 3.11: One-minute sketch challenge

Practicing can be difficult to fit into our busy lives. How many of us have signed up to a gym with the intention of getting fit? Those first few sessions seem easy, but a few weeks in and our lazy side kicks in. How do we create a drawing habit which you won't drop?

Make your daily drawing habit attainable by using simple warm-up sessions at regular times. Maybe make a pact with yourself that you will doodle when you are waiting for your computer to start up in the morning? Could you sketch when the kettle is boiling for your morning coffee? Use the simple warm-up exercises in Figure 3.12 as a guide. Try reproducing these five shapes as neatly as you can in order to warm-up your hand-eye coordination. Devise your own warm-ups and try consistently to do them before you start or end the day.

Figure 3.12: Warm-up exercises

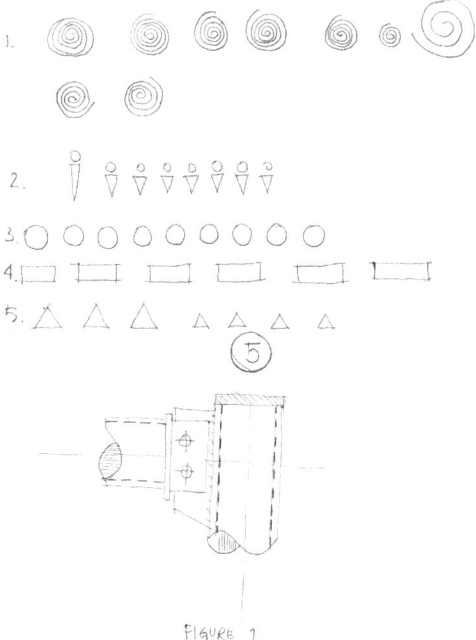

Another way to make time for your sketching is to try a sketch a day. Put a sketch book next to your bed and draw from memory just before you go to sleep — it beats staring at your phone!

The best way to really give yourself a window of opportunity is to start a sketch club. Getting a group of friends/colleagues together in a consistent place and time allows you to sketch freely knowing you've made time for it in your week. You can theme the different meetings and each bring in some working drawings to share with the group. Find an area at your place of work or study to display the sketch club's work. This will be helpful in promoting the club and expanding its membership. A sketch club should be a place where an informal and comfortable environment is developed, and sharing skills should become an essential part of this learning group.

3.3 Tools for drawing

The principal tools for sketching and drawing can be kept very simple which makes it accessible to all, and allows the skill to be very mobile (Figure 3.13). This is especially important when you need to sketch 'in the field'. An A5 or similar sketchbook is really good to have with you all the time so that you can record ideas. You can capture details of buildings, and how elements of the building are working. Also draw objects to help develop your eye for detail (Section 3.6).

Figure 3.13: Tools for drawing

If you don't have a sketch book to hand, you can easily make one by putting together a series of folded A4 sheets of paper, card covers and bulldog clips to retain the pages (Figures 3.14–3.15). Add graph paper, isometric paper and coloured paper, as well as plain paper. The beauty of these sketch books is you can remove and add sketches as needed. You may want to make one for each project you work on.

Figure 3.14: Different paper options

Figure 3.15: Complete sketch book

3.3.1 Paper and pencil
Start all drawings with a pencil, standard paper and tracing paper. Using pencil as a beginner means you can erase errors, but don't rely on this — we want you to become proficient in using pen.

3.3.2 Line types
The style or thickness of the line you draw is important to add depth and meaning to the sketch. Figure 3.16 and Table 3.1 show the basic line types and their uses. Note the use of pen thickness (weight) used to differentiate sections or outlines compared to typical display lines. Have a thicker black pen as part of your sketching tool kit.

Figure 3.16: Line types

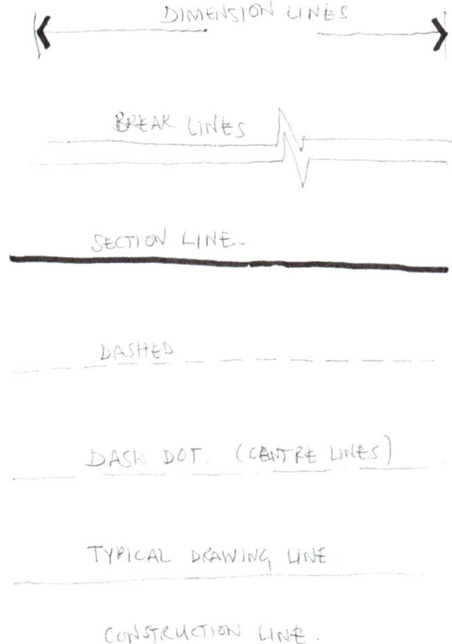

Figure 3.17: A building edge section showing line type thickness and style

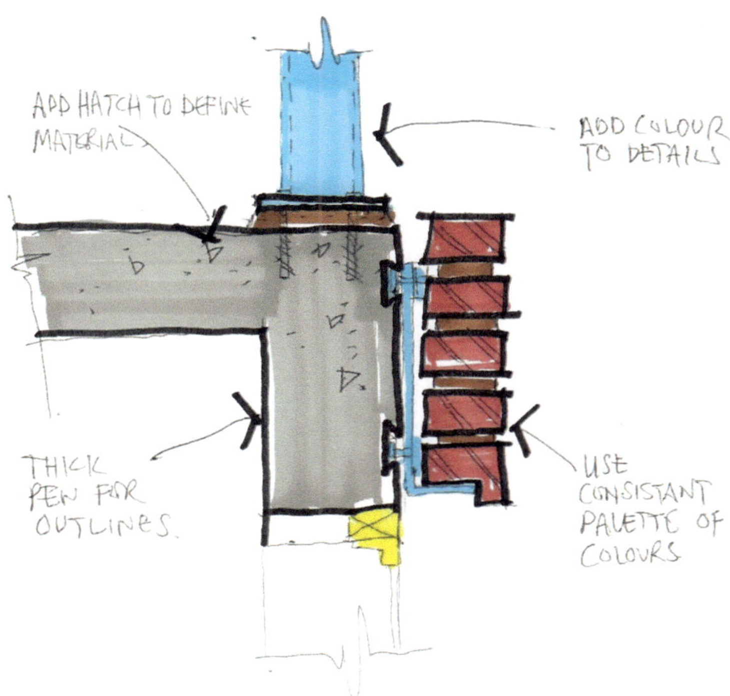

Line types give the sketch depth and definition and allow the viewer to distinguish what the sketch is trying to represent. Note all lines can be made with a fineline black pen. Figure 3.17 shows a standard suite of line types used in a sketch drawing. Using these different marks really brings a sketch alive, and quickly transforms what would have been a very flat sketch into something that looks a lot more like a professional line drawing.

Table 3.1: Line types and their meaning

Line type	Description
Dimension lines	Used to clarify dimensions. Often, we want to communicate specific dimensions to the reader of the sketch, these could be critical to the layout or fit of the object drawn
Break lines	Helpful to reduce the sketch to show the critical information, and enable it to fit on one page
Section line	When a section is cut, a thicker line is used
Dashed	Represents a building element that is screened or hidden under what you are sketching in the foreground
Dashed dot	Allows you to represent a centreline of an element, or it can be used for gridlines. Gridlines are useful as a wayfinder on sketches to relate to architectural or other engineering plans
Typical drawing line	Used to define any common line in the sketch
Construction line	Used to set out a sketch, and can be a lightweight line that is traced-over or blocked-in with a more solid line when the sketch is defined and you're happy with the scale and proportion. It can also be used to show an elevation of something in the background, to give context

3.3.3 Hatching

To create highlighted character and depth to a sketch we can use simple hatching or shading. Figure 3.18 shows different grading of shade created with a pen.

Figure 3.18: Shading

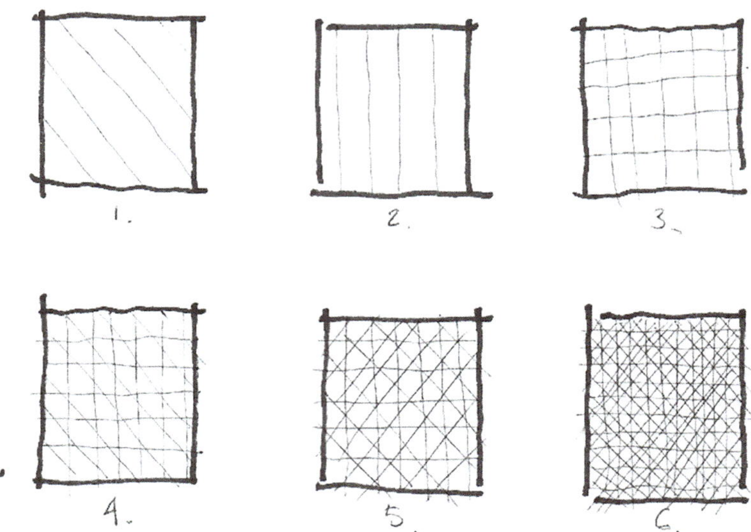

Decide on the light source and use this hatching technique to add shade to your image. Areas in total darkness should have an almost solid fill. Elements in the foreground are highlighted if they are filled more solidly. This will give depth to the sketch.

In engineering detailing, we often use the hatch standard in Figure 3.19 to delineate material type.

Figure 3.19: Typical engineering hatch styles

3.3.4 Touchscreen and stylus

CAD is at the forefront of engineering design, and the majority of drawn information coming out of the modern design office is in the form of 3D models, drawn with sophisticated software. This form of drafting language has its place in the modern construction world, although it does take time for the user to learn these programs and become proficient with the technology. It may seem a little odd to devote a section in this concept design book to digital sketching, but the hand sketch is currently going through a renaissance period and there is a way to develop this, using the computer or tablet as a tool. Figure 3.20 shows how the computer can help in developing a sketch.

Figure 3.20: Developing a sketch from a digital image

3.4 Top tips and practical guide to drawing

Drawing lines

The easiest way to draw a straight line is to connect two dots on a vertical axis. You should make a mark where you want to start and finish the line and then, while looking at the end dot, make a positive continuous movement. You should, with practise, be able to create a good positive line. Make sure you give consistent pressure through the pencil. To prove this is an effective way to draw a straight line, try doing the same task but draw on the horizontal axis — you will find that the results are not as effective.

Figure 3.21 shows vertical line practise – note as you start to gain confidence with this you can lose the dots at either end. When joining lines to form corners or edges (Figure 3.22) don't hold the line short — always overrun your line. This will give the impression of a closed corner, and adds to the look of the sketch.

Figure 3.21: Making positive lines

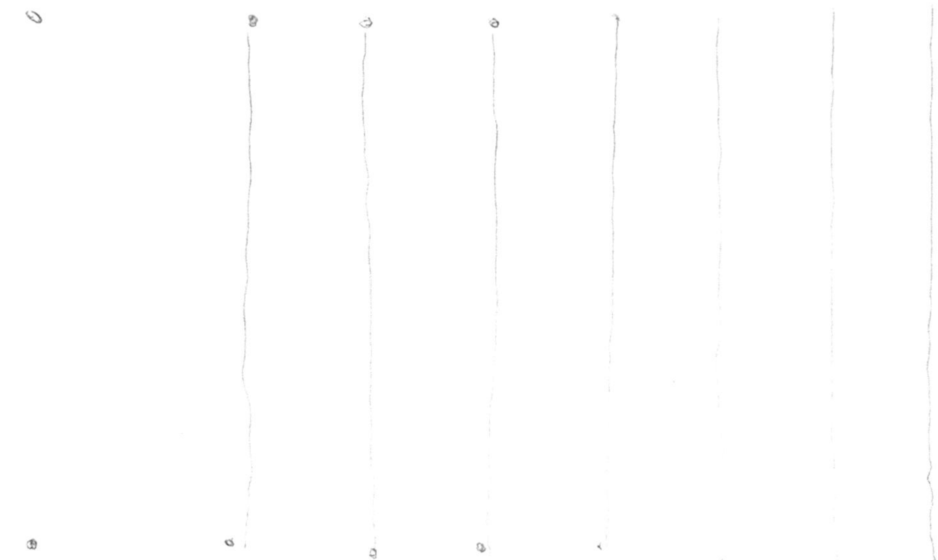

Figure 3.22: Typical corner line treatment

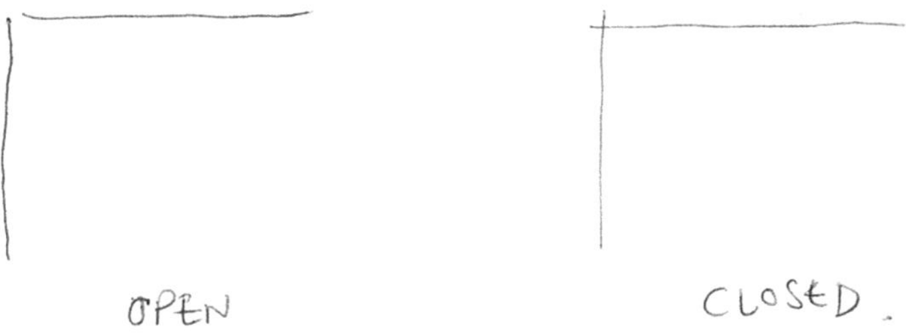

If you stick to just using a pencil and rubber you can get into a habit of being a bit lazy with your approach to making definitive lines. As you get more confident, start your work with a black fineline pen, this will train you to be more accurate with your line, and commit to putting permanent lines on the page. It also speeds the sketch process up.

Drawing curves
Curves are easiest if laid out initially with three dots on the page, denoting the start, finishing point and the depth of the curve/arch. Join the dots, in a similar style to the straight line, keeping your eye on the dot you are travelling toward with the pen (Figure 3.23).

Sketch over photos
A quick and easy way to create a sketch of an existing subject is to print a photo to the correct size for the paper and then overlay tracing paper to copy the main outlines of the subject. This instantly puts the drawing elements into relative scale with each other. You can then remove from the photo and add detail. The image will not be at a true scale, but this is not always necessary. This technique is often helpful when recording any site issues. If time is limited, a photo can be taken and then used back at your desk as a basis for your site sketch.

Figure 3.23: Drawing curves

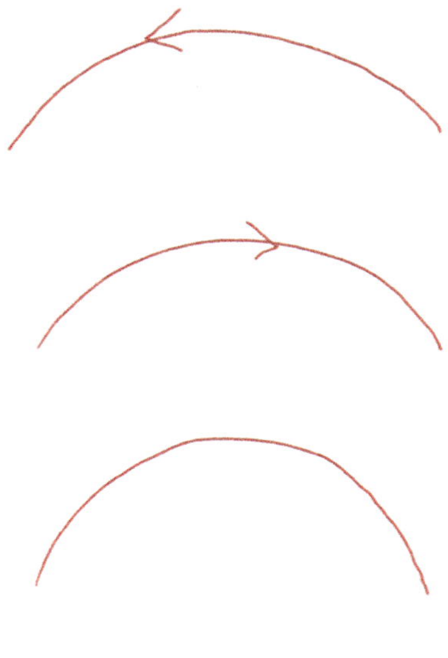

Sketch over drawings

A simple way to create a scaled plan or elevation is to take the architectural drawings and overlay tracing paper. If you don't have tracing paper, put a thin piece of paper over the original and hold both up to a window, this will serve as a natural light box — unless of course you have access to a light box (typically these pieces of kit were thrown out with the drawing boards with the onset of CAD).

3.5 Interviews with the professionals

Simon Hatcher – Hatcher Prichard Architects

In response to a busy life, I sketch to relax and take time out. It forces me to switch off and concentrate on really looking at what's directly in front of me. I guess I subconsciously tailor a sketch to the time I have available. Up to an hour forces me to dive right in, and concentrate predominately on form and shade, using a thick clutch pencil on either textured watercolour paper, or smooth Bristol board. I always draw what I see without preconceptions — I set out key construction lines, then build up form and shade, constantly asking myself questions like whether a given point is darker or lighter than its neighbour. If I have longer, say on holiday, I'll draw a pencil or pen line drawing and apply watercolour washes to add tone and colour, then finish with detail (Figure 3.24). A tip is that I'll tend to do a pencil sketch alongside a watercolour and use it to roughly and quickly work out my own questions.

Figure 3.24: Drawing with watercolour washes

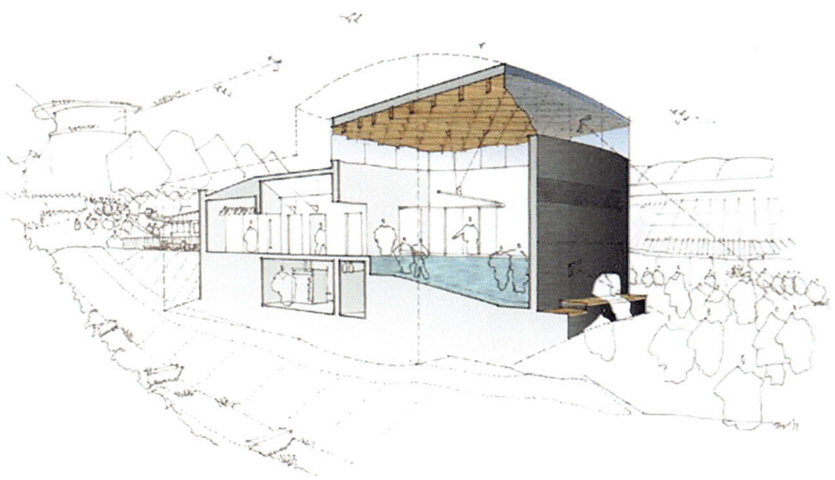

Andy Jarvis – E3

I normally sketch on plain loose A4 white paper, and always have a stack next to my laptop on my desk. I also have A3 plain paper, and an A3 tracing pad for when I want to create sketch overlays. I like to use clean, unused paper, and sometimes use a window or door as a light box, I like there to be nothing on the back, as it can be a distraction to what I'm sketching. I find that a new clean sheet is a good way of developing a fresh idea, or enhancing a developing idea, which can sometimes lead to a rather big pile of scribbles.

I tend to use mainly black pens (Figure 3.25), the Pilot hi-tech points are good, 0.5 and 0.7, and then have a V Sign pen for the thicker lines. I was always taught to use three pen thicknesses in my sketching, although now I normally use the 0.5 and the V sign. I then sometime add notes using a red or blue Hi-tech point, and the red is certainly used a lot in marking-up drawings. However, I still think monochrome is clearest and simplest, as long as there is nothing in the background to hide the linework.

Figure 3.25: Drawing mark-ups (1)

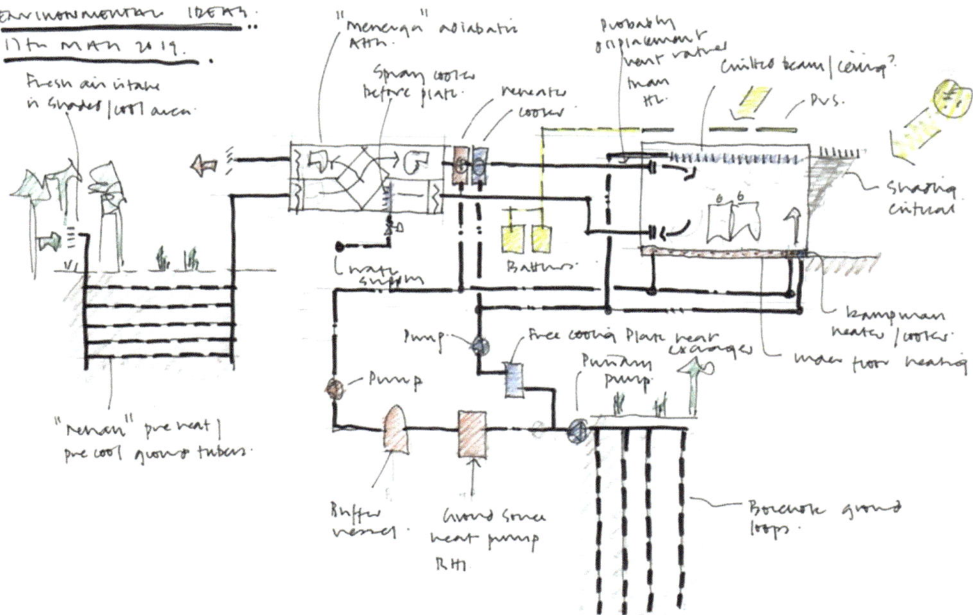

I also have a 0.9 Pentel propelling pencil. I tend to use this for straight line work with a flat scale, normally as construction lines, or when I need lines to a dimension. I then normally over ink with pens to suit. It's also handy having a pencil in design team meetings, as I think that is more polite when overmarking and commenting on another person's sketch, especially if it's in their book. I never like to use ink in someone else's notebook — I was once slapped down by an architect for using black ink in her book. She always used red ink!

I also have an A4 lined Black n' Red, and I use this as my normal day book. I keep notes, sketches, comments, records of conversations etc. (Figure 3.26). I try to keep my book in order— random stuff gets put in at the back.

I think my sketching is primarily about capturing ideas, and then either using them to:

- develop the sketch into a detail and into a drawing
- help someone understand how the principles of systems and spaces work
- describe how something can be installed or put together

In teaching, I like the principle of two people developing a sketch or idea together on the same piece of paper — it's always a useful way to record a discussion or decision. I find that it's also a way to give my mind logic and order, and helps me with processes. I will always sketch as I'm thinking.

Figure 3.26: Notebook drawings

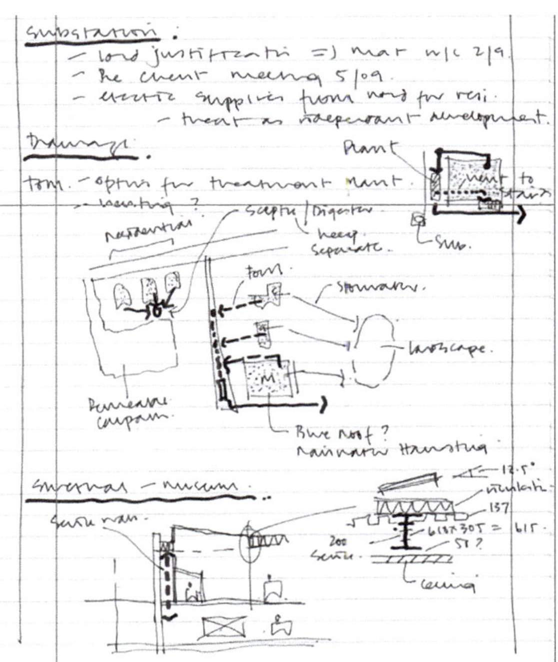

I often mark-up over photos, and mark-up arrangements on existing buildings from Google Earth, or over photos of installations or spaces that already exist (Figure 3.27). This is normally done on printed sheets, using one of my permanent markers. I have a 4-pack of Lumocolors next to me as I type. Normally these would be used in an email conversation, and sometimes will be a response to receiving a photo of something on site. I also mark-up drawings — again I see this as a useful means of communicating (Figure 3.28).

Figure 3.27: Photo mark-ups

Figure 3.28: Drawing mark-ups (2)

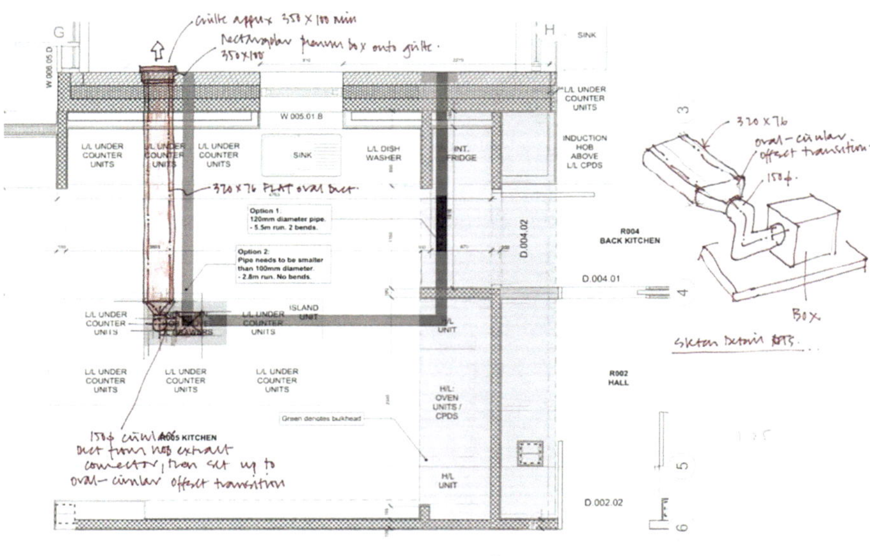

3.6 Workshops

Workshop 1

The one-minute challenge: Take a stopwatch, or your phone timer, and put a countdown on for one minute.
Start the timer and draw the object in front of you. In Figure 3.11, I have tried to capture the outline of the subject
and you can see it is very fluid but describes the basic shape.

The five-minute challenge: Again, time yourself drawing the same subject. You have longer so potentially you can use
line to define the elements better, which will draw the eye to the subject. Remember that this is still a quick sketch to
define the main outline, but you have time to add details.

The ten-minute challenge: This time limit will allow you to explore the shape further, and perhaps add colour to the
image. The line work can be more controlled and potentially you should look to get a hold on approximate scale.

Workshop 2

One-line drawing — Take an object and sketch it in one go, don't lift the pen from the page. This workshop helps
build confidence in line marking and being fluid in your motions. You will naturally find it is a really quick way to draw
and produce some interesting results. It also teaches that it's fine to make errors when sketching, and that accuracy
isn't everything (Figure 3.29).

Figure 3.29: One-line drawing

Workshop 3

For this exercise set up the video on a phone or tablet. Then draw one of the '100 things to draw in one minute' (Section 3.7). Record yourself talking and drawing at the same time. The film element to this exercise will put you under a bit of pressure to recreate the experience of a design team meeting with other team members watching over your work. Try to talk through the salient points of the sketch, and how they connect together and think about the scale of the elements when drawing them.

Workshop 4

Draw the stages of folding a sheet of paper to make a paper aeroplane. Label each stage and use a dashed line type to show hidden folded layers of paper. Write the description of the stages next to each image (Figure 3.30).

Figure 3.30: Paper aeroplane workshop

Workshop 5

Go to a building nearby and pick out a detail. This could be a steelwork connection or a window detail. Sketch the detail and the area local to that detail, and use a measuring tape to record sizes of elements and critical heights or levels.

3.7 Practical: 100 things to draw in one minute

Choose to draw these initially as plan or elevations, and then at your next one-minute attempt draw them as 3-dimensional objects.

Table 3.2: 100 things to draw in one minute

Flowers	Aeroplane	Exhibition stand	Umbrella	Teapot
Wardrobe	Lunchbox	Football stadium	Bolt	Apple
Car	Mobile phone	Beach house	Screwdriver	Hole punch
Bike	Your home	Beach ball	Hammer	Pineapple
Bus	Shed	Cup	Scissors	Headphones

Table 3.2: *Continued*

Tree	Digger	Keys	Coins	Bus stop
Table	Shoes	Window	Swimming pool	Railway station
Mosque	Can of drink	Tap	Tea cup	Horse
Dog	Kettle	Book shelf	Shirt on hanger	Washing machine
Castle	Knife	Toilet	Aubergine	Light bulb
Hand	Fork	Saw	Road signage	Your dinner
Crane	Sledge	Screw	Door	Fence
Telephone	Shopping trolley	Sandwich	Milk bottle	Handle
Computer	Suitcase	Pens	Tent	Charging cable
Insect	Train	Pair of glasses	Statue/sculpture	Bridge
Watering can	Person	Stapler	Bench	Hat
Chest of drawers	Alarm clock	Group of people	Mouse	Watch
Skyscraper	Chair	Open fields	Stairs	Pylon
Church	Lamp	High Street	Music instrument	Scaffolding
Cat	Hair dryer	Cityscape	Half an egg shell	Route to work

Drawing various objects can help develop a sense of scale and proportion. Practise drawing these objects and give yourself more time to play with the sketch techniques you have learnt in this chapter. Use colour and hatch to create shape/form and materiality.

3.8 Further information and inspiration

Books
- Ching, F.D.K. and Juroszek, S.P. *Design drawing (3rd edition)*. Hoboken, NJ: John Wiley, 2019
- Ching, F.D.K., Onouye, B.S. and Zuberbuhler, D. *Building structures illustrated: patterns, systems, and design (2nd edition)*. Hoboken, NJ: John Wiley, 2014
- Farthing, S. and Davey, P. *The sketchbooks of Nicholas Grimshaw*. London: Royal Academy of Arts, 2009
- Hunt, A. *Tony Hunt's sketchbook*. Oxford: Architectural Press, 1999
- Hunt, A. *Tony Hunt's second sketchbook*. Oxford: Architectural Press, 2003
- Slade, R. *Sketching for engineers and architects*. Abingdon: Routledge, 2016
- Zell, M. *The architectural drawing course: the hand drawing techniques every architect should know*. London: Thames & Hudson, 2017

Websites
- Drawing Gym. Available at: https://www.ucl.ac.uk/drawing-gym [Accessed: March 2020]
  Showcasing engineering sketch drawings and techniques
- ExpeditionWorkshed YouTube channel. Available at: https://www.youtube.com/user/ExpeditionWorkshed [Accessed: March 2020]
  Sketching for engineers
- Sketchmob. Available at: http://sketchmob.co.uk [Accessed: March 2020]
  Organised sketching meet-ups in the London area

James Norman
University of Bristol

4 Communication

In Chapter 3, we looked in depth at how to use drawing to create ideas, solve complex problems and communicate. However, while our final ideas may be predominantly encapsulated on drawings, we also need to be able to communicate verbally and in writing.

As an author of this book you may imagine that I have always been confident at writing; I am an artist sculpting words into fantastic shapes; by the age of nine I had written my first novel.

You would, of course, be wrong – instead at age 7, I could read LEGO Technic instructions, but not about Jennifer Yellow-hat (a popular learning-to-read series from the 1970s); at secondary school, I was required to attend 'special lessons' for my writing; at GCSE, I got Bs in English, a subject I worked hard at but found very difficult; at A-Level, I almost did English Literature, a subject I loved but struggled with, but chose instead to do Chemistry, a subject which I enjoyed and I knew I would get a high mark in; at university, my tutor suggested I be tested for dyslexia; for my teaching qualification, I had to submit my case study nine times, because I couldn't capture in words what I was doing in the classroom; I have had multiple rejections of book proposals.

I find words really hard — I have battled with them, and lost many times, but I have persevered.

The reason I am opening up to you on this is because the way to get better is to 'have a go'. I now try and write regularly. I try to spend at least an hour a week typing away. I have gone away on writing retreats, and churned out words which only my wife has seen. I have written blog posts that have never seen the light of day. I have written a book which no one wants to publish. All of this experience has helped me improve. I know my style is too loose and conversational. The Man Booker award is not coming my way any time soon, but I am learning and hopefully improving.

The other thing I have done is read — fiction, non-fiction, magazines, graphic novels, zines, poetry. Choose your own adventure. I have tried to explore a wide variety of sources, looking for inspiration and trying to learn.

You won't get better by not doing — I'm sorry, but it just isn't going to happen.

In a nutshell, this chapter has just one point — 'practise' (Figure 4.1):

- If you want to be better at speaking in public, you need to *practise* (and get honest feedback)
- If you want to be better at writing carefully worded emails, you need to *practise* (and get honest feedback)
- If you want to write project reports that delight your client, you need to *practise* (and get honest feedback)
- If you want to stand up in front of a client and pitch for a job and win, you need to *practise* (and get honest feedback)

The rest of this chapter is just some friendly advice on a few different areas of communication that I have picked up along the way.

Figure 4.1: The importance of practise

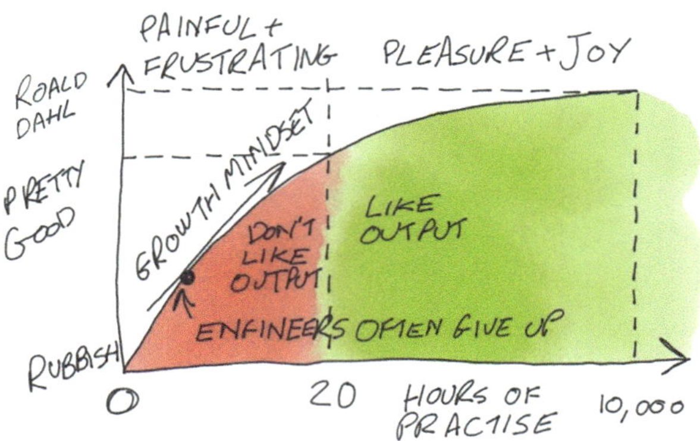

4.1 Learn to tell stories in a way other people can understand

I am an engineering educator. I have spent years teaching people how to design buildings, and how to communicate that design. We train engineers to describe difficult technical challenges and describe how to solve them. If you are a chartered civil or structural engineer, just think about your chartership submission. Did you focus on some large challenging problems and describe how you solved them, maybe focusing on a number of possible solutions before deciding on the 'best' one? I know I certainly did. Of course this is brilliant. Engineers love talking shop. The problem is other people don't care and, even if they did care, they don't understand. If the client wants a large open space which requires a complex 3D post-tensioned tensegrity solution which has pushed your ingenuity to it's very limits, they don't care. They just want to know if they can have it or not. As engineers, we need to learn how to talk about our role in a way that people are interested in. If we have made things which seemed impossible, possible, people care. If we have saved money, or even better the environment, people care. If we have a great narrative around our facade design, people care. I am not suggesting that our engineering skills are not important, they are, but practising communicating with people in a way that they can understand is also important.

Activity
As I said repeatedly in the introduction to this chapter, the way to get better is to have a go. This chapter will include a series of recommended activities to try at home, at the pub, in the supermarket or even at work.

Your first challenge is to tell people about your job in a way that leaves them excited or inspired. To think of a narrative that engages with them emotionally. To leave them wanting to know more. Next time someone asks you what you do, have an answer ready. Then switch it up. Try to never use the same answer twice. Examples include:

- I design schools for children in underprivileged areas
- I have brought abandoned buildings back to life
- I have transformed a local community by being involved in an arts project

What we do is *amazing*. We need to learn how to celebrate the big picture of what we do, as well as the technical detail.

4.2 Hitting the mark — making your communication intentionally targeted

When writing, you need to tailor what you are saying to the audience, if you want them to hear. Before you start there are a number of important questions you should ask:

Who is your report for?
Is it for an architect, client, members of the general public, end users, planners or someone else? This question feels so fundamentally obvious, yet when we pick up a report and read it, can we tell who it was aimed at? And are we able to adapt our approach depending on the answer? Also, who do we ask to review our report? If it's aimed at

the general public, is it better to have a technical or non-technical member of staff review it? Look around your office — you have a number of different people with different perspectives and understanding. Is it always the director or project lead who should review what you have written, or could someone else review it?

Why are you writing it?
The obvious answer is because you have to, but maybe you don't. So why are you doing it? I remember writing reports for design team meetings — they weren't strictly necessary, but we wanted to impress a couple of members of the team, so we did it anyway. Maybe we are writing a short report to influence a decision, to solve a problem or to enable a decision to be made. I remember clearly writing a report on basement waterproofing options for the client, so that they could actively choose which option to proceed with. It was a big decision and it was important that they were consulted and understood the impact of that decision, so we wrote a short report with lots of clear diagrams. On the same project, we had to agree the loading on the floor slab of a large library. Was the space a reading room, library or classroom? So we created a report, explained the implications and constraints of some of these choices, and facilitated the client in choosing a solution.

Should it be formal or informal?
The answer feels obvious, yet formal reports can be so dry and boring that people switch off by the third sentence. If we want people to access it, we need to make it accessible. Of course, this is not a simple 'yes or no' choice. There is a scale of formality. It should certainly be respectful, written in the third person (you are writing on behalf of your employer), and we should avoid offending people, but at the same time it should be joyful to read, or at least not a trudge to get through. Of course, we should always use the correct terminology and avoid colloquial language and slang, whether we are writing for a corporate bank or a community farm, we should keep our language professional, *innit?*!

Long or short?
Again, this feels obvious. Short. Surely short — and I agree. Every report I have written could be improved by being shorter, but there is a danger of being *too* short. Let's remember that our written output is probably the main item of work that our clients sees (and understands). Those lever arch files of calculations (or gigabytes of PDFs) due to go to Building Control won't be seen by the client — all they see is the certificate to say that Building Control has approved the design. We should aim to make our written work short, but not so short that the client is left wondering what you are doing with the £100,000 fee they are paying you.

Is it structured?
As your report gets longer, the structure becomes more important. We want to avoid too many forward references (such as: 'See Chapter 12 for a typical Stage 2 structure'), while also creating a sensible narrative. As we noted back in Chapter 1, many of the decisions we make will occur concurrently (the choice of structural material will affect the grids, which affects the choice of material, so when we choose one, we actually choose both), but a report must be, by definition, linear. Therefore, we need to decide the order, so that it makes the most sense. For very large reports this can take a long time and some degree of standardisation is helpful. Writing out the content and the different dependencies, and creating a logical structure is very helpful. I typically do this in two stages — first I write down everything I need to say and group the content by topic, then I write down the topics in a logical order, and try to write sections in the order they appear. Often this requires a few iterations. Figure 4.2 is an example considering the content from Chapter 6.

Part of a larger report or stand-alone?
It is unlikely that this decision is in your gift, but it changes what you include. If the first third of our report is repetition of someone else's, at some point people will stop reading. It is important we know what everyone else is saying, and therefore what we want to add.

Illustrated or just text?
In my mind this is another obvious choice. It must be illustrated as a picture is worth a thousand words, although some clients may not want this. You need to know what they are expecting and why.

Your 'house style' or someone else's?
This is outside of your control, but is obviously important to know.

Figure 4.2: Planning out the content of Chapter 6

Technical or non-technical?
This comes back to: 'Who is your target audience, what do they need to know and how best can it be explained to them?'

Baffling or clear?
Do we even need to ask? So many of our reports are baffling to others. Very occasionally this is a good thing (we are persuading the client of just how clever we are) — but generally it's not.

Information, question or answer?
Some reports are for information e.g. a site progress report. Others supply an answer to a question e.g. a Stage 3 or 4 report, which summarises the design as you go to tender. Others are posing questions. What type of waterproofing is required? Is this really the design load you want? Would you prefer a building of steel or timber? The purpose will depend on your approach. If we are asking a question, it is important we give as much information as is necessary to the client, so that they can make an informed choice. If it's an answer we may not want to go into e.g. the technical specification of the waterproof concrete, we may just need to explain the degree of waterproofing provided.

Who is this section for?
Many longer reports will have different sections which are intended for different people. The client may not read the full report, they may only read a project summary. The architect may focus more on the structural framing, less on the ground, while the quantity surveyor may be trying to work out how much everything costs. You may need to ask all of these questions many times, especially for longer reports.

Brevity?
It is hard to say well in 300 words what you can say in 1000. Why do you think this chapter is so verbose? It's because making it shorter is really challenging. It requires multiple edits. Questioning and re-questioning. And yet we often neglect these parts of our report. I think back to the 300-word summaries I used to write, and I know they were the worst part of my report. How do you say anything in just 300 words? I also know that this is probably the only section that everyone reads.

Activity

1. On your next report, before you write a word, for each section consider the above questions and make a note of the answers. Now attempt to write each section for the intenced audience, using appropriate language.
2. Choose one section of your report, ideally the section that you think most people will read (hint, this will be the first words at the start of the report). Remove exactly 30% of the words. Use the word count facility, then work out how many words need to go. Now, ask a non-technical member of staff or a friend to read both the original and the shorter version, and find out which they prefer and why. Now do it again. Take out another 30%. You will almost certainly need a blank sheet of paper, a lot of time and some serious ingenuity, but have a go.
3. Pass the report to someone else and give them the list of questions but not your answers. Ask them to look at three or four sections, and guess at the answer to the questions for that section. Do they agree with you? If not, why not and how could you improve?

4.3 Emails

Another form of communication is email. We all send tens if not hundreds every day. We are well practised, but how often do we really think about what we are saying and doing?

Activity

1. Put a one-minute timer on, and find a good email in your inbox.

   Think about what makes it good — I imagine it has three features:

   - It is clear
   - It is concise
   - You know exactly what you and others need to do

2. Put a one-minute timer on and find a bad email in your inbox.

   Think about what makes it bad? I imagine the answer is one thing — it is so long, you stopped reading after the third sentence. You never reached the list of actions buried part way through the fifteenth paragraph and, even if you had found them, you wouldn't really know what they were asking for.

In my opinion you should send an email for three reasons:

1. To send information requested e.g. a set of drawings, prior to a design team meeting.
2. To ask a specific question/action e.g. can you confirm the dimension from grid line D–E or can you send me the latest version of your model?
3. To confirm a conversation e.g. following our discussion, please find attached a sketch outlining the revised connection detail.

Everyone's life would be better if emails were short, clear and you knew what they wanted. Conversely, no one enjoys getting very long emails, which explain in vast detail exactly what they are thinking and why. I really think the telephone is much better for this.

Two don'ts:

1. Send an email without re-reading, especially who it's going to. We have all hit 'Reply all' when we really meant 'Reply to sender', and have regretted the comments which are now being shared. It is important to check. If you are new and the email is important, ask someone else to read it before you send it. If you have been around for a long time and the email is important, ask someone else to read it before you send it
2. Send an email when you are cross/annoyed. Sleep on it. No email that you write today will be worse if you sleep on it and send it tomorrow. Ever.

We all get too many emails. A good work habit to get into is not to send internal email if possible. Have a chat instead. Or, write your question on a paper aeroplane and send it over, or on a Post-It and deliver it with a cup of tea.

A few thoughts on actions:

- Be clear about what you want
- Set a reasonable deadline
- Set a time limit e.g. please spend no more than five minutes doing this

The last one is especially important, because we might think this is a five-minute job, but it may take three days. If we knew it would take three days, we might not want it done in the first place, or we might think of another, quicker way to do it.

Finally, if you have to write 15 paragraphs of text, consider putting the action at the top first and explain the lengthy rationale afterwards.

Activity
Go through your outbox and find the longest email you have sent this week. Now rewrite the email in three sentences with one action. What is missing? How important is this? Is there any other way you could communicate the missing information (a drawing, telephone call, report)?

4.4 Minutes

I know a secret. Learning to write minutes will get you a job (possibly).

A few years ago, when I worked in industry, we were tendering on a framework with a large reputable university. As part of the tendering process, we had to attend a day's workshop where tenderers were placed in groups and discussed problems. As you can imagine, a lot of people wanted to be heard. As a junior member of staff, I felt intimidated and outranked by most of the people present. But then I realised no one wanted to take the minutes. So, I volunteered. Slowly but surely, I was able to direct the conversation, not in an aggressive way but by asking people to repeat comments, making sure other members of the group got to give ideas, avoiding the conversation being dominated by two or three voices. Low and behold, we were included on the framework.

I tell this story to my students who go for interview days where they are expected to work in teams. I explain the quiet authority taking the minutes gives you and, in my experience, they have also got the job.

Of course, taking minutes is about more than this — they are the formal record of the meeting. If you ever have the opportunity to take minutes, I recommend you do it — that way you can ensure all the conversations that are important to you are well documented.

Activity
Start taking minutes at every meeting you attend, even if someone else is doing it, and keep your own minutes for your personal records. Practise this skill at every meeting. Not only will they be an invaluable record for future meetings and conversations, but over time people will notice and, at some point, you will be asked to take them.

4.5 Connect

The previous advice is generally most useful for written communication, although it is also useful for spoken communication. The next two pieces of advice are more useful for spoken communication.

Connecting to your audience is an essential part of presenting as well as writing.

I remember pitching to a client. I walked into the room and the client was bored. Their heads were down, they seemed to be taking little interest in what we were saying. That is until I told a small joke. A joke so unfunny and specific to the situation, that it is not worth repeating. But a joke all the same. Suddenly the heads of the interviewers lifted. They started giving eye contact. They started nodding to what we were saying. Smiling as we answered their questions. The difference in body language was palpable. We won the job.

When you meet with clients and other members of the design team, what they really want to know is whether or not they can work with you. To persuade them that they can, requires you to make a human connection. This can be achieved in a number of ways — through laughter, shared experience or shared emotion.

I told my story at the start of the chapter because I wanted to connect with you emotionally. I could have increased its impact by explaining a minor detail, like the smell of the carpet I remember sitting on age seven, as I struggled to read the books, my friends have moved on from years before. In this situation that would have been inappropriate, or perhaps an oversharing of emotion, which doesn't add anything else to the story.

I slipped a small joke into the sentence on different types of written word. I wonder if you spotted it. Whether you dared to smile or laugh when you read it.

I used these devices because I know that if you feel some level of human connection with me, you are more likely to trust what I say. You are more likely to take the advice on board. You are more likely to have read up to this point in the chapter.

When you present, I suggest you try and connect with people. That you are open and honest.

Activity
Unlike writing, finding opportunities to present is more difficult. However, it is also an activity which I believe improves with practise, so you may want to try the following:

- Volunteer to speak at one of your office lunches
- If you don't have office lunches set one up and as the owner of the event give a short (two-minute) introduction and add yourself onto the list of presenters
- Offer to give talks at schools
- Join a book club
- If you have small children make up your own stories and tell them to them
- Tell stories around the fire
- Go to an improv workshop
- Join an amateur acting group
- Give a talk for the ICE or IStructE on your latest project
- Apply to give a TEDx talk
- Seek any and every opportunity to present, and if none exists start the event yourself

4.6 Listen

It seems important in a chapter on communication to talk about listening. The art of listening is one that is seldom well-practised. I believe if you really want to build a relationship with your client or architect or acoustician, it is important to make the time to listen well. This may well not be in the meeting, it might be over a coffee or pint afterwards, but taking time out to really listen to them will do wonders to build a long-term relationship. Practise listening well — if you are not sure what I mean by listening well, I suggest you search the internet and see what others have to say, as I am no expert.

Activity
Practise listening well. Practise with your friends and loved ones. Practise with your colleagues and clients. Practise with the caretaker at the school you are working on. Practise with the rebar-fixer you meet on site.

4.7 Feedback

Getting feedback on our communication can be difficult, but it is also really helpful. Practise will get you so far, but unless we get feedback it is very hard to really improve.

Reflect
The first person who can give you feedback is yourself. Taking time out to reflect on what you do is valuable — plus the feedback you give yourself will hopefully be easy to accept. I suggest you don't reflect on recent work but take

an old report and review it as if someone else had written it. What has worked well? What could have been better or clearer? What would you change if you were doing it again? What could you take and use on other projects? Reflecting on work is really helpful. Be constructive and don't be too critical. Look for the easy wins, the items that you can change quickly and easily without too much effort. Don't ask how could this be perfect? Ask what one thing could I have changed to make it better?

Ask for feedback
At some point you will want others to give you feedback. To receive feedback well and improve there are a few tips:

- Don't take it personally — try to see your work as separate to yourself, so when you receive feedback it is not a criticism of you, but of the work
- Try and ask for feedback when you have emotional capacity. Working on a job and creating an extensive report to a deadline is exciting and exhausting. I remember dreaming about the projects I was working on. However, if you ask for feedback (not just a review) of your work the day afterwards, when you have been up late, working hard, you may not be in the best emotional state to accept the feedback. Why not wait? Ask for feedback a month or two later, when you are invested intellectually and emotionally elsewhere
- Learn to absorb the impact of the criticism

A short story. I mentioned that I had to write my teaching case study nine times. What I didn't tell you is that the first review was brutal. It was brutal for a number of reasons but the biggest was this — I thought I had produced the perfect report. I had recently won a teaching award and thought I was pretty marvellous, so when I was told that I had to start again it was like a blow. I felt myself crumple. I was furious at first (though only internally), then deflated. Finally, I accepted that maybe, just maybe, the reviewer knew more than me. I picked myself up and tried again. I could have stayed cross. I could have given up. They are normal responses. But I wouldn't have improved. I wrote that case study nine times, and by the ninth I had learnt a lot. I had genuinely improved. I can see that now.

4.8 Suggested further reading

There are so many great books, blogs and talks on this subject. I really feel you should look at them and not this chapter, although I have waited for the last paragraph to tell you so. If you want to discover more about communication, I have enjoyed the following:

- Buster, B. *Do listen: understand what's really being said. Find a new way forward*. London: The Do Book Company, 2018
- Buster, B. *Do story: how to tell your story so the world listens*. London: The Do Book Company, 2018
- Heath, C. and Heath, D. *Made to stick: why some ideas take hold and others come unstuck*. London: Arrow, 2008
- Munroe, R. *Thing explainer: complicated stuff in simple words*. London: John Murray, 2015
- Trott, D. *One plus one equals three: a masterclass in creative thinking*. London: Pan Books, 2016

However, I also suggest you do your own searching and find resources that help you find your own unique voice.

James Norman
University of Bristol

Rachael De'Ath
University of Bristol and Arup

5 Developing the brief — "You want me to design what?"

The very first stage of working on a project is defining the brief — what is it the client wants to achieve with their project? This is set out in the RIBA Plan of Work 2020[2] Stages 0 and 1 where:

- Stage 0 is the strategic definition
- Stage 1 is the preparation and briefing

As an engineer, the point at which you become involved in every project will vary greatly. On some projects, you will be involved from the start and possibly, very rarely, even before the architect. On others, you won't get involved until much later, and much of what is referred to as 'concept design' will have already happened. As a practising engineer who enjoys the creative part of the design process, it is better to be involved in the design process as early as possible, ideally working in collaboration with architects and other engineering disciplines who want to create something incredible.

I can think of examples when I have had to set the brief. Take for example the HIVE project, where the University of Bath wanted a facade testing facility (Figure 5.1). The project was to provide a building with no external walls where facades could then be tested to assess their long-term performance. It included a variety of different spaces, such as a flooding cell, a blast cell, some double-height cells with further testing capacity and several single-height cells for long-term testing. As the engineer, we were appointed to lead the project, which meant working to the client brief[18]. However, the client didn't have a complete technical specification to work to, so we were required to write it. Working collaboratively with the client we developed a design brief, outlining the technical requirements for the number and size of each different space. It was only much later, once the project was in the detailed design phase, that we employed an architect to help us detail many of the facade connections.

While this example is not typical, being involved in setting the brief is far more common than you might think. From deciding the load for different spaces, to choosing how much of an existing building to use, we are often required to make decisions that could be considered to be 'setting the brief'. Sometimes this is done explicitly, but often it is done implicitly.

For example, a university recently built new teaching spaces. The teaching space was for practical work and could have been defined as either classroom or laboratory space. By choosing to take loading for one situation or the other, we are effectively limiting (or not) the activities that take place in the space. The loading decision may not seem to be part of the brief, but by choosing certain allowable limits on load does determine what a space can be used for, not just when the building is built, but for the rest of the building's life (which is another big decision when considering the brief), unless future strengthening works are considered.

Figure 5.1: The HIVE Project, University of Bath

In this chapter we will explore what a brief is and how we check that we have completed it successfully. Sometimes, our involvement with this will be integral to the project; on other projects we will only have peripheral involvement, if at all. However, every engineer should take a step back on every project and ask: 'What are we trying to achieve and could we be doing this better?'

One of the most important jobs of an engineer is to ask the right questions.

5.1 What is a brief?

A project brief should be the final stage of a process in which a client has defined their requirements. It should state the client's needs and outline how these are to be achieved. Project briefs often continue to develop during the concept design stage, as additional information is gained and further consultation with the client takes place.

Often the project brief which you will be given will be ill-defined and vague. To ensure that the project which is ultimately delivered is what the client actually intended, it is important that the whole design team, including you as the engineer, helps to develop the brief. At the beginning of a project, clients don't always know what they want or need. They may think they know but, with the help of a team of specialists, this can be clarified and defined.

Sometimes the engineer's role is to help develop the brief by adding requirements and information. However, just as important is making sure the brief does not overstate the client's intentions.

The following are two conflicting views on adding to the brief:

- When you are employed on a project, you may want to add your own elements to the brief e.g. a particular material, aesthetic, or approach to design. You may want to make the design low-carbon, even if that isn't the client's desire. That's OK. It's good to have values that you bring to a project, as long as you are clear about these from the outset. You may even want to take it further, refusing to work on projects that don't match with your own values. This is a big decision, but you should also know what you would and wouldn't do before working on a project. Over time, you will become known for your values, and people will choose you because they want you on their project
- Conversely, sometimes you may need to reduce the ambition of other members of the design team (or your own) who are looking to achieve something with the project that the client hasn't asked for e.g. a level of sustainability the client doesn't want to achieve (I am often encouraging clients to use rammed earth, for example), a level of

aesthetic beyond the client requirements or a level of temperature control beyond the client's needs. We are all guilty sometimes of adding to the brief, and it is good for the design team collectively to ensure the brief captures the client's requirements, but also that we don't add our own desires

Either way, sometimes we may need to ask questions, and check the brief really is achieving what it is supposed to.

The RIBA has broken projects down into a number of clearly defined stages to help everyone as they go through the often long and complex process of designing, constructing and using a building. At the start of this eight-stage process is the strategic definition. Following this is the brief definition (Stage 1). The RIBA has clearly defined the core tasks of Stage 1 as:

- Prepare project brief including project outcomes and sustainability outcomes, quality aspirations and spatial requirements
- Undertake feasibility studies
- Agree project budget
- Source site information including site surveys
- Prepare project programme
- Prepare project execution plan

Setting the brief requires us to do all of these. It would be useful to break them down further:

5.1.1 Project outcomes
Project outcomes include a wide variety of different items. Some are easily definable e.g.: what is the building purpose (is it a school, hospital, residential)? How many units are required and what size should they be? For example, for a residential development, the number of units and bedrooms for each unit are key considerations.

Others are harder define e.g.: the impact of the design on the client's brand, or the type of learning environment we are creating in a university. These are, however, just as important.

A good question to ask at this early stage is: What are the main priorities of the client?, as these can vary hugely from project to project e.g.:

- for a commercial project, the main priority may be to maximise the net lettable area, while keeping cost down
- for a school, the main priorities may be ensuring the project is delivered on time, as the school cannot open after the start of the new academic year, ensuring adequate space can be provided, and that the project can be built safely on a live site
- for an art gallery, the main priorities may be acoustics, light, humidity and fire suppression

In each case, the priority is something different and, for other types of project, a different list of main priorities may well arise. It is important, from the outset, to try to define not just the requirements, but which ones are key for the client, and which ones are flexible.

These may change during the design process. If you have tried to buy a house, you may well have experienced this. We often start with some set criteria (local school, number of becrooms, garden), but when we find a house that fits our criteria and step inside, we often feel like it's not right. The criteria didn't capture what we were looking for. And when we finally step into the house we fall in love with (and go on to buy it) we find that the spare room we absolutely had to have is no longer so important, that there are lots of good schools in the area, and a small back yard is really a lot less work than the large garden we thought we wanted.

So, defining 'project outcomes' is a little like nailing jelly to a wall!

5.1.2 Sustainability outcomes
Many projects today have sustainability outcomes within the brief. These can sometimes be a condition of planning permission or funding requirements. Typically, these are measured in some way, using certification methods such as BREEAM[19], LEED[20] or Passivhaus[21]. Sometimes they are less defined, and other ways of defining and considering

sustainability will be discussed. RIBA Plan of Work 2020 now includes a substantial section on sustainability considerations for all eight stages and I strongly recommend that you review these.

Generalised scoring systems are not always a guarantee of sustainability and can, at times, lead to perverse outcomes. As an example of this, school projects receive points for installing bike racks. You may think there is nothing wrong with bike racks, but if you have been to a primary school recently you will have noticed the plethora not of bikes, but of scooters lining the walls. Sadly, bike racks do not offer a practical storage solution for scooters, and installation of scooters racks won't score points, even though scooting and cycling to school are equally sustainable (note this was the case five years ago, when I was in practice, but hopefully things have moved on and scooter racks are now given the BREEAM points they deserve). The client is left with a quandary — do they install bike racks and gain the points, or install scooter racks and forgo the points? Or do they install the bike rack, gain the points and then replace them soon after with scooter racks? As we can see from this simple example, generalised scoring systems can be overly rigid and may not cover the needs of the client.

Vague notions of sustainability can be equally tricky, especially if the client and the design team have a very different understanding of sustainability. Sustainability to some means:

- Designing to building regulations
- Wind turbine and/or solar
- Re-use of materials or a whole building
- Natural material and/or natural ventilation
- Highly efficient materials and plant which minimise energy usage

All of these approaches can be considered sustainable, but they are clearly very different. It would seem a conversation might solve this, but it may not be that easy as it is not always possible to articulate, especially for non-technical staff.

One approach may be to ask for the client to give one or two examples of the type of sustainable building they would like, and why they think they are sustainable.

Another approach is to outline a number of different approaches and give local examples that the client can look at. They may not be able to look around, but just looking from the outside will help them get a feel for the different approaches to sustainability. Other approaches include:

Generalised scoring systems
While generalised scoring systems may not be the best approach, taking the time to do a quick review of the different categories and seeing if they apply and if they make good sense can be a helpful way to spot opportunities, even if you decide against formal certification.

Calculating embodied carbon and energy
As a minimum, engineers should be considering the embodied CO_2 and energy in their buildings. Roger Plank has written a short guide on how to carry out the calculation[22]. If we create 3D models (possibly BIM ones) of our structure and carefully select different layers for different components, calculating the volume of each material is relatively straightforward. We can use these to start to estimate the embodied CO_2 and energy of our options, which should help inform the solutions we choose to progress later on.

Optimised design
Engineers have a terrible habit of overestimating — we want to be sure that our building is safe. This is, of course, understandable. Inefficient buildings not only cost more money to build, they also require more energy and CO_2 to construct. While we would not advocate intentionally under-designing your structure, we would suggest that you work hard to ensure that the building is working hard e.g. 95–100% member capacities, and that if serviceability limit states govern the design, you consider seriously the impact that has, and ensure the loadings assumed are not over-conservative. For more information on this topic, we recommend you monitor the MEICON website[23] which is looking at optimising structural design and the associated cost, energy and CO_2 savings that accompany this.

5.1.3 Quality aspirations

Quality is a wide and varied topic, which should be covered at this stage. It includes quality for safe design (designing to codes correctly, quality assurance, ensuring what is specified is built, or that changes achieve the same outcome), but it also includes other types of quality. Some of these are definable and measurable — from air quality to the acoustic performance of a space. Others are less tangible, such as how a space looks or feels. Some sit somewhere between the tangible and the intangible — the acoustics of a concert hall, the light in an art gallery. These are, however, equally important. At the brief definition stage, we need to highlight the quality aspirations for the project. This can be achieved in a number of ways — from highlighting exemplars, to providing prescribed standards that need to be met. It is, however, important to be explicit about what you are trying to achieve, and why, as the conceptual design will be tested against this brief in the next stage.

5.1.4 Spatial requirements

When designing a building, it is of course important to think about the spaces that we are providing. How big, how many, how they feel, the inter-relationships between different spaces and required adjacencies.

Some requirements can be given as simple numbers, such as the number of apartments/bedrooms. Others will be deterministic, but based on legislation or best practice e.g. the number of toilets. Others will be based on exemplars — we want the acoustics of *this* space to sound like *this* existing space, or the facade to look and feel like *this* existing building. Other requirements will be more abstract — I want a building which delights. These are often the hardest to define in any measurable terms, but carry significant weight with the client. Thinking carefully about how you define your spatial requirements and how you communicate them to the client is of upmost importance. It may be more than just the words and numbers you use. On one project I worked on, an element of the facade was described as being like a giant Kit-Kat wrapper, in that it was shiny but also a little crinkled — I remember when the scaffold was removed, the client exclaiming that 'the facade looks like a giant Kit-Kat wrapper!' Perfect. Brief delivered. We provided exactly what they wanted. Or did we? We always thought a giant Kit-Kat wrapper would be a good thing, but the client was not so sure. Maybe a more abstract (and less shiny) description would have brought the client more joy?

5.1.5 Develop the initial project brief

It is easy to think that, once we have the project outcomes, sustainability outcomes, quality aspirations and spatial requirements, these will be fixed. This is simply not the case. The brief will develop with the project.

We can't assume that the initial project brief is achievable within the defined budget (Section 5.2.1). We can't assume that we have got the brief right first time and that the client will not, at some later date, change their mind about what they want. While this might be the case, it isn't always. The brief should be seen as a changing document that reflects the project's current status. However, this should be articulated honestly and clearly so that everyone can see how the brief is developing.

Once the initial project brief has been delivered it may be prudent to consider whether:

- the client expectations and project constraints match
- the client can afford what they want
- the site has adequate space for the project, both in the final condition but also during construction
- the ground is suitable

If the answer to any of these is no, further work is required on the brief before you carry out significant further work.

Also note during the development of the initial project brief stage, it is often helpful to ascertain how much your client knows. In construction, clients can vary greatly in their experience, from developers to completely novice clients who are embarking on their first ever project. You will be much better positioned to help your client if you understand the limits of their knowledge and experience. It is also helpful to know how flexible your client is. Do they have a very fixed idea about what they want and what that should look like, or are they open to new ideas and suggestions, based on your own experience, which might provide them with a better solution?

The RIBA stages suggest that the brief development occurs in Stage 1 (and therefore before the conceptual design process in Stage 2). At the end of this stage the brief should be captured and clearly articulated. The reality is that the conceptual design and brief will continue to develop in tandem in Stage 2 (RIBA Plan of Work 2020 refers to 'feasibility studies' as a way of separating brief development from conceptual design). However, at the end of Stage 2, the brief should be articulated and agreed again between the design team and the client. This then forms a reference point for the project going forward. It would be nice if, from here on, the brief didn't change, but the reality of the situation is that the brief will alter. Nevertheless, this 'written and agreed brief' acts as a good reference point for all future changes.

5.2 Beyond the brief

Having reviewed the project outcomes, sustainability outcomes, quality aspirations and spatial requirements, we should now be in a position to define the brief. But you'll notice the list doesn't stop there. And so we thought it might be helpful to add a bit more information to some of the other tasks. Site investigation is covered in detail in Chapter 7. We suggest that you refer to a project management resource for the implementation plan and programme, but we thought it might be helpful to discuss the budget, the feasibility study and some other aspects which directly impact the brief, but were not listed by the RIBA.

5.2.1 Project budget

The project budget is not ours to define. The client should have some idea of how much they have to spend and, if not, we may suggest they find a quantity surveyor who can assist them. We can, however, come up with some very broad-brush advice. Based on years of experience, we may suggest something along the lines of the following:

- **'Cheap as chips'** ($£1500/m^2$) — very hard to achieve and budget will control all decisions
- **Budget** ($£2000/m^2$) — the normal target for projects and is often what the client has in mind. There is some small margin for adding design flourishes, but ultimately the building will be utilitarian and may require value engineering (a phrase I hate as it rarely offers value and doesn't involve any real engineering) to reduce the cost
- **Deluxe** ($£3000/m^2$) — if the client is willing to pay more, they will have far greater freedom over the design and can achieve more. Note, however, that just because a design is more expensive does not mean it is better. Often the constraint of the project cost forces engineers and architects to collaborate and problem solve in a way that creates a better overall design
- **'Starchitect'** ($£6000+/m^2$) — if you are working with one of many incredible architects then be prepared (and ensure the client is prepared) for the cost. We can all think of examples of such projects where this is the case

Project fees are a complex issue and the above values should be used as a very loose guide to help estimate the overall project cost. These figures are approx. at 2020 and an allowance for inflation should also be included.

In addition to the project cost, it is incredibly important that you get paid for the work. While there are no hard and fast rules to fees, Table 5.1 is a good indicator of fee levels. On most projects where the architect is the lead

Table 5.1: Approximate fee levels for different size projects

	Structural engineer's fee	Architect's fee	Total design team fee
Large project (>£50m) with a lot of repetition in detail	Fee <1%	Fee <3%	Fee <6%
Medium/large project with some repetition	1% < Fee < 1.5%	3% < Fee < 4.5%	Fee <10%
Small project or project with a large amount of reuse/ refurbishment	Fee >2%	On small or highly complex projects the architects fee and the total fee is hard to define, as you may be the only one working on the project, or there may be a host of professionals	
Large basement	Add +0.1–0.3% for additional complexity		

designer, the structural engineer's fees are typically one third of the architects. In our experience, this leads to a happy balance in terms of the team.

A 10% contingency of the build cost should also be included plus tax where appropriate.

The final cost = the build cost + fees + contingency + tax.

These numbers are useful — not to tell the client what their building will cost, but to help understand the compatibility of the brief and associated budget (note the budget will typically be for everything, not just the build cost, which is why it is important to have an estimate for all elements, including fees, contingency and tax).

It is worth noting that there is an assumption that the larger the project, the smaller the percentage of the fee. However, experience shows that while this is true up to a point, when a project becomes very large (typically too large for one person to hold the whole project in their head), the cost of designing a project actually starts to rise again (Figure 5.2). This is due to the extra expense of management and coordination that is required on these very large projects.

Figure 5.2: Fee/project size

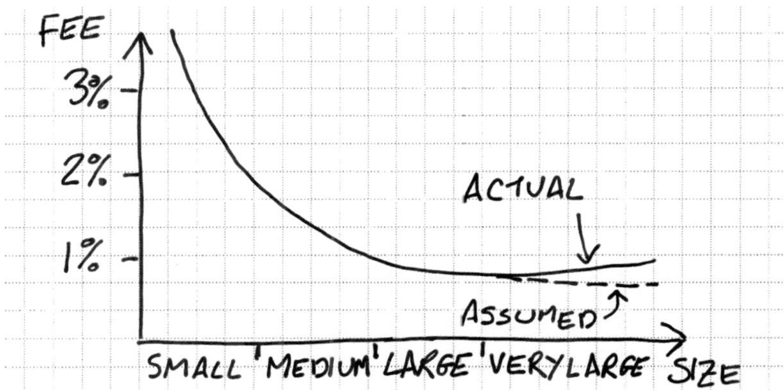

5.2.2 Other considerations

There will invariably be a number of other considerations on any construction project. These may include programme and site/resource availability. As well as client constraints, such as those listed, you should also consider external constraints which may be imposed, such as statutory authority requirements, planning authority and heritage bodies.

Be prepared for unexpected requirements. Briefs are always broad and often vague, and it is impossible to determine an exhaustive list of requirements.

Keep asking questions:

- Is there a planned site?
- Are there any existing buildings on the site?
- What is the site access like?
- What sort of contract will the project be working to?
- What next?

As you progress as an engineer, you will find that you learn to ask better and better questions and that these questions lead to better solutions (or avoid ill-conceived ones). Do not stop asking questions. Learning to ask the right question is:

- an art form and, like all art forms, you will get better with practise
- a skill and like all skills you will get better with practise
- an essential part of your job as an engineer and like all parts of your job you will get better with practise

One way to learn to ask better questions is to take note when other people ask good questions. If you are in a client meeting at the very start of the project, listen out for good questions. Write them down. Reflect on why they were helpful in that situation, and what they helped to unlock.

5.2.3 Undertake feasibility studies — 'The importance of playing'

It is almost impossible to set a project brief without 'prototyping'. I use the word prototyping intentionally as the early part of the design process is fluid, messy and involves asking many questions. We often fixate on the 'what works'. But it is equally instructive to ask: 'What doesn't work?' If an early idea is not right, why not? This is where we find that our simple numerical tests of the brief (cost, plan area, number of rooms etc.) are woefully inadequate at defining the extent of the project. You will be able to draw several solutions which fulfil these requirements but are not fit for purpose. It may be down to aesthetics, or the spatial organisation, or the lack of natural light. Many of these may not seem to be engineering concerns, and in one sense that is right, but to come up with a solution that works as an engineering solution alone, misses the point of the complexity of building projects. This process of prototyping will help to develop the brief rapidly (similar to 'rapid prototyping' in manufacturing), and will help solidify the preferred outcome.

From my own experience, one of my favourite day's work was physically prototyping the roof of the Lee Sixth Form Centre. We sat with the architect and used paper and scissors to play with a variety of roof options. The final solution was clean, elegant and simple, but we arrived there through playing around with a variety of possibilities. The project has gone on to win several prizes.

Figure 5.3: Lee Sixth Form Centre

This process is clearly articulated in the book *Yes is more* by Bjarke Ingels[24], which demonstrates how, through play, a team develops and defines a project brief, which it is then able to deliver. This form of rational articulation of the design process looks simple, but is anything but.

It is also important to highlight that these feasibility studies should not be seen as the concept design. They are a tool for developing the brief. The concept design, which starts once the brief is defined and agreed by the client, should be a response to the finalised brief, while acknowledging that the brief may change. It is important to make both yourself and your client aware of this fact, as it is easy to build emotional attachment to a design, which may not get built. Therefore, explaining that the purpose of the feasibility study is to create a brief at the start of the process helps everyone to let go of it at the end of the stage.

5.3 When might you need to develop a brief?

Engineers are not always required to develop the brief — often this is the role of the architect. We would suggest that our role in brief development may be one of three options:

1. **You are leading the concept design** — occasionally the engineer is asked to lead the design. This may be because the job is small, or believed to be more technical. Typical examples include bridges, which are normally carried out by bridge specialists, but sometimes we get asked to design elevated walkways which straddle the building/civil split. We may also lead on technical facilities such as university laboratories or rail projects where, as engineers, we are best placed to understand all of the technical constraints. In these cases, it is essential we develop the brief, working collaboratively with the client, to ensure they understand exactly what they are receiving.
2. **You are co-creating the brief** — there are times when the architect asks the engineer to help develop the design from the very early stages. The architect's role will be to develop the brief, but we can choose to be an active part of this process. This can occur on any project, but is more likely on a competition or team bid where collaborative working starts at the very earliest stages. It is important to respect the architects and to work with them in the process. Just as we are experts in structural design, so they are the experts in brief development.
3. **Implementing an existing brief** — the brief appears set, but our role as a professional is to gently test that brief. Throughout the concept design process, we need to ask appropriate questions, to test the brief and ensure that it is delivering the best and most appropriate solution to the client. We should remember that others, the client and architect for instance, will have put considerable work into the brief, and therefore we need to approach our questioning with caution. You will need to apply a degree of business intelligence as you want to work with both of them again. Therefore, we need to ask:

   - What impact will this have on the overall design?
   - What is the impact on the architect and/or client?

   It may well be that we are 'right' in our questioning of the brief but, if it has little effect on the project and will have a negative impact on our relationship with the client or architect, it could be wise to hold back. However, there are times when the brief leads to a major issue which in your professional judgement leads to an inappropriate, costly or, even more importantly, unsafe design. In these situations, it is important to challenge the brief, regardless of the impact on our relationship with others. However, experience tells us that, even in these situations, there are good and bad ways to go about these conversations.

When we work on a brief, whether leading, co-creating or simply implementing, there are both implicit and explicit brief documents. A typical example of this is the loading report. An engineer may produce this long after the brief is set, yet this document may well implicitly define the use of the building when opened but also, more importantly, in the future. Current usage is often defined in the brief, but while future flexibility and possible usage may never be discussed, we implicitly agree it by giving load requirements for different floors. Even if we don't think we are involved in defining the brief, on almost every project we are, even if only in an implicit way. We should of course not see this as a reason to overdesign a project, but as a reason to think flexibly and communicate well, so that the client knows exactly what they are getting, not just on the first day, but for the life of the project.

5.4 Brief development — taking an ill-defined brief and making sense of it

Alex Wright's excellent paper on 'Design education'[25] outlines an iterative approach to brief development, which involves coming up with a brief or question, developing a solution and then critiquing the solution to update the brief. Often, we make the mistake of updating the solution without first reviewing the brief. He then goes on to outline several ways to approach coming up with a solution which can be broadly placed into three subsets:

1. **Typology** — most buildings have a type, whether houses, schools, libraries or theatres. Typology takes what makes a school a school and applies it to the new problem. This can be a powerful approach, but there are also inherent pitfalls as existing problems, or more likely missed opportunities, remain. However, much building design is undertaken on this basis, whether explicitly through reference to other projects or implicitly through the knowledge of the designer, as they put pencil to paper.
2. **Determinism** — determinism carries out design through solving a set of numerical problems which lead to an answer. A simple example is the design of a timber building where every item works to 600mm, 1200mm, 2400mm and 4800mm dimensional systems, based on the standard sizes of available boarding material. This can lead to efficient and cost effective design, but does not necessarily create space for creativity (although many would argue that constraints force creative solutions to occur). It is worth noting that the briefs given in the IStructE Chartered Membership exam are all of a deterministic nature.

3. **Abstraction** — the stereotype of architecture is the abstract approach. They pluck an iris from the site and decide that the building will be developed to replicate this beautiful flower. It can lead to a beautiful story of development and a remarkable project, but it can also lead to a painful and ill-conceived project. This is by far the hardest approach to get right, and is often the most challenging to those coming from an engineering background.

These are used as different starting points for iterating the solution, and are then further subdivided into 14 different approaches.

As previously mentioned, this early solution should then be critiqued, pushed and pulled, questioned from all sides — not to directly change the solution but to update and better understand the brief. This brief development process will start with some major changes (number of storeys, material choices etc.), but with each iteration less should change and a clear brief definition should emerge.

Sometimes the solution will diverge, each time you go around the iteration you will see more problems. This suggests two things:

1. That the iteration starting point was wrong.
2. That rather than start again you have already developed several questions and answers which should help to determine a new starting brief.

In understanding the brief, we need to not consider the project in isolation. In Chapter 7, we will understand the technical constraints of the site, but here we will consider the impact of the site and of people on our brief development.

The site

In Paolo Belardis' *Why architects still draw*[26], the author writes two essays — one on the importance of drawing and the other on the 'survey'. It is this second essay which is so important, as it outlines the depth of information that can be gained from being present in the location of the building and surveying it with our eyes, ears and other senses. To be clear, by 'survey' he does not mean a topographic survey. While we detail in Chapter 7 the importance of technical survey data, there is another level of information which can be collected from being on site. From the existing buildings and building style locally, to the ambience and scale of the space. Taking a step back, there are also key questions around site access, local industry and the possibility of reusing and recycling local materials. While there is no specific checklist of questions to ask on a site, it is important to spend time there, exploring the possibilities at the very early stages of a project. You may think we are in danger of straying into the realm of architecture, but many important opportunities may be missed if you don't explore the site. As an engineer, it is likely that you will see different things which may impact the project. The different perspectives that different members of the design team will bring are all helpful.

In exploring the site, it is important to highlight two other considerations — health and safety, and taking your time. Your place of work should have a policy on site visits, and before you go it is important to carry out a risk assessment. We would not recommend visiting site alone, and if you must visit alone then ensure to check-in regularly so that if you don't check-in, people can send someone to check you are OK. Do not go into existing buildings if you are not sure of their condition, both in terms of structural safety and in terms on other risks, such as asbestos.

When on site, take your time. Ensure you take more clothing layers than you think you need, a flask of hot drink and some chocolate. Getting cold and hungry will encourage you to rush your time on site. The longer you spend on site thinking, observing, feeling, reflecting, the more you will get from it. Take some work with you. Do some initial concept drawings on location. But don't rush. If you have bothered to take the trip to site, you may as well make the most of it.

The people

You will, of course, want to spend time with the client. Understand their aspirations and get their feedback on the brief as it develops. However, there will be other people who can also provide a wealth of information. We have all met site caretakers who have worked on the site for 50 years, whether at a school or community hall, and can tell you the complete history of the building e.g. the problems that were encountered when constructing the extension in the 1980s due to unforeseen ground conditions. These people are a valuable source of information. They are also an end user, and while ultimately you are delivering a project for the client, if the maintenance is intelligently considered the building will be more efficient and productive. When on site, take time to sit down and have a cup of tea with the people you meet, and really listen to what they have to say.

5.5 Using the brief to select the solution at Stage 2

It is quite possible that, despite developing the design to fulfil the brief, you find that you have more than one possible solution. At some point, you will need to decide how best to proceed. There are two ways to do this:

1. **Single criteria** — it may well be that at this point there is a single criteria that is used to select the design. This could be the cost, or simply the client's preference. It may be that the priorities outlined in Section 5.1.1 provide a single criteria. This assumes that all other criteria are met. For example, for a school there may be two solutions that provide the right number of classrooms, and the criteria is which of these solutions is most likely to be completed before the start of the new academic year. Single criteria can be powerful and effective, but they also may miss a number of subtle points e.g. if the project cost of two options is similar, but the cheaper option has considerably less prefabrication, it may seem that the cheapest option is the best, but when the risk of delays on site due to inclement weather is considered, the small cost difference may appear less important and the advantages of the prefabricated construction (less time on site, reduced risk of delays, reduced risk of accidents and reduced risk of poor finishes) may outweigh this cost.
2. **Multi-criteria** — it may be that there are multiple criteria to be considered. In such cases a decision matrix can be used. This creates a list of criteria, gives them a weighting or preference, and then scores the options to see which one scores highest. While this appears a rational approach, there are many vagaries. How do we score quantitatively and qualitatively in a way that we can assess them objectively against each other? For example, one category may be cost, where the cheapest scores 10 and others score in relation to how much more expensive they are (even this is hard to define). For sustainability, it is much harder to create a ranking. A timber frame may score higher than concrete, but by how much and how do we define this? If we need thermal mass the concrete may lead to a more sustainable solution, so on what basis do we give scores? Additionally, we have all worked on projects where the scores say one thing but the decision is not based on these but on something else e.g. the client prefers the aesthetic of the other solution. There are, in effect, some criteria that trump all others regardless of our scoring matrix.

The solution is to be sensible, rational and flexible. Neither single or multi-criteria approaches are wrong, but what we are really trying to do is find the solution that the client is most happy with. If the client genuinely likes both solutions and is looking for support in the final decision, it may be that the matrix helps them to rationalize the decision and feel confident that they have made the right one. It may be that, like a child in a sweet shop choosing between different chocolate bars, there are no bad choices, so the decision becomes them being satisfied that they have made the best choice. At the same time, it may be that the cheapest, or quickest, or biggest solution, while on paper appears the best, actually does not leave the design team or the client happy, and that another perfectly good solution offers other benefits not considered or counted in the process. It is by articulating these things that we are able to make a rational account of the solution that we have reached.

5.5.1 Measuring the success of your solution

What is success? Can we measure it? Do we need to be careful about placing too much emphasis on numerical measures and not enough on other measures?

A family of five wants a six-bedroom sustainable house for £400k. What is success? Is a standard masonry house with six bedrooms and costing £400k more successful than a four-bedroom house made from reclaimed timber from a local source and reusing existing foundations for £350k? Does it matter that we only provided four bedrooms? Did they really need the six requested? Maybe they did, they might have an elderly relative or be planning to foster, but maybe they didn't. Is being below budget good or bad? What if it is too far below budget? Would a £600 tent with six bedrooms really be OK? How do we measure successful outcomes in sustainability? Is it enough to say the building is reusing materials or do we need to do a BREEAM assessment, and what happens if the BREEAM leads to an outcome that seems perverse? What if the family love the timber house and don't like the masonry one even though numerically it doesn't fit the brief? Do we just change the brief, but then we could build a masonry house with four bedrooms for £300k? With a new brief, all the options could be reconsidered.

A simple design raises so many questions, so we need to make sense of all of them. I am not sure this can be done by simply scoring different aspects out of ten. We inevitably bias the scoring consciously or unconsciously, to ensure the design we (the client and design team) like goes forward to the next stage. So why not be honest about it? That is what is so helpful about the book *Yes is more*[24] — in simple terms it states why the solution is successful. Often the success of a solution cannot be measured with simple numerics, but instead should be articulated in other ways.

One way of approaching this is to use reflective practise to understand what success looks like. This can be in the form of a dialogue that captures the changes in both the scheme and the brief, and makes explicit the success that is achieved. This may feel uncomfortable for many engineers, as articulating reflection is not something we have been formally trained in. It may feel vague and ill-defined. It may raise questions about why this approach? At this point, it is useful to go back through our notes and sketches (it is important to ensure you articulate and file this information as you go), and make sense of what we have done. Doing this well is a valuable skill and one that good architects have crafted over many years. It is only by engaging with this process and developing our own approach that this is made possible. This, of course, does not mean that numerical measures of success are jettisoned, but that we consider them in the richer dataset that is the narrative around the project development.

5.6 Having a go

For the six briefs in Section 5.6.1 try going through the following process:

1. Decide on three simple requirements.
2. Spend five minutes sketching a solution.
3. Spend five minutes reviewing the solution. What works well? What doesn't work? What is included in the design that wasn't included in your original requirements?
4. Repeat Steps 1–3 but with seven requirements.
5. Repeat Steps 1–3 and revise the seven requirements.
6. One more time repeat Steps 1–3, now reduce the requirements back to three to five.
7. Finalise the brief and present the solution along with the clear requirements. Remember that requirements can be abstract, typological, or deterministic. They can be words, equations or pictures.

Before you start, choose a location for each brief. Choose somewhere you are familiar with, and have a look around, maybe not physically, but at least using Google Maps. When looking at the briefs, the solution will vary hugely, depending on location e.g. the public toilet — locating it in a woodland or in a city square will make a huge difference to the approach you take.

The process should take no more than an hour, so just the right amount of time to carry out in teams over lunch at work. Make a time, pick a brief, pick a site and then have a go.

5.6.1 Six ill-defined and poorly articulated briefs which need developing

1. Family home for five

- A family are looking to build a new house on their current plot before demolishing their existing house. They are keen to embrace a self-sufficient lifestyle as much as possible but have limited funds.

2. A car showroom

- A car dealership is looking for a new showroom. They have clearly defined style guides which ensure every showroom looks and feels the same.

3. A public toilet

- Mixed, different sexes, gender-neutral, disabled — you decide.

4. A school

- This could be a new school or an extension to a local primary school. If an extension, think about what would happen if the school was asked to add a new class to every year group (seven classes in total) — how would they achieve this and what other problems would this throw up?

5. A nightclub

- Decadent or downmarket — what makes a nightclub successful and how can this be captured in a design?

6. A sustainable office development

- Imagine your work is relocating — where would you like to work and why? Try and design this ideal office. It can be new build or a refurbishment.

5.7 Brief development — an example

Design a walkway through a woodland at a wildlife park, which will become a new visitor attraction.

Step 1: Three simple requirements

- Safe for users
- View of wildlife
- Accessible for as many people as possible

Step 2: sketch a solution (Figure 5.4)

Figure 5.4: Woodland walkway: initial sketch

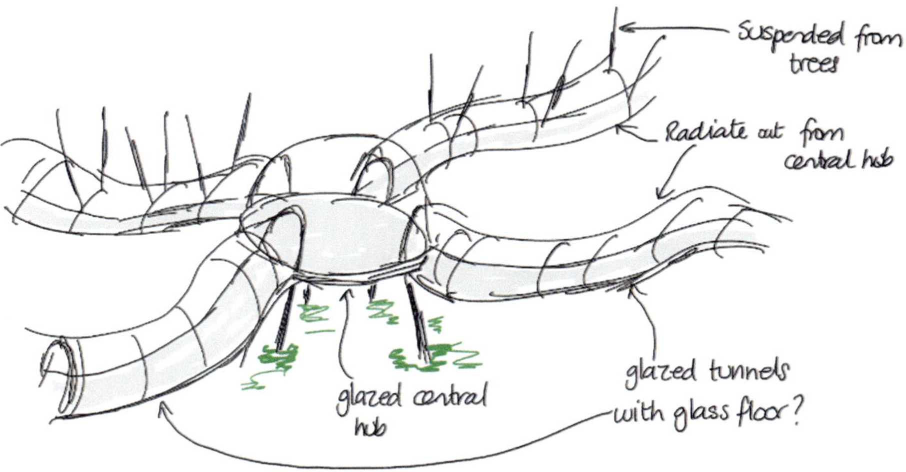

Step 3: Review solution
What works well?

- Good visibility of surroundings due to glazing
- Visitors safe and protected from wildlife and weather

What doesn't work well?

- Heavy, brittle structure suspended from trees — too much movement if you could support it at all
- How do you build it?
- Not clear how you initially get in, up in the trees

What is included in the design which wasn't in the original brief?

- Complex, sculptural structural form
- High-cost solution
- Protection from weather

Step 4: Seven simple requirements

You can use your review in Step 3 to help you develop your new requirements. I considered what didn't work well to expand the requirements to address some of these issues:

- Safe for users
- View of wildlife
- Accessible for as many people as possible
- Sympathetic to woodland setting
- Buildable in short programme
- Sustainable materials — conservation ethos of client
- Step back in time

Step 5: Sketch new solution (Figure 5.5)

Figure 5.5: Woodland walkway: new solution

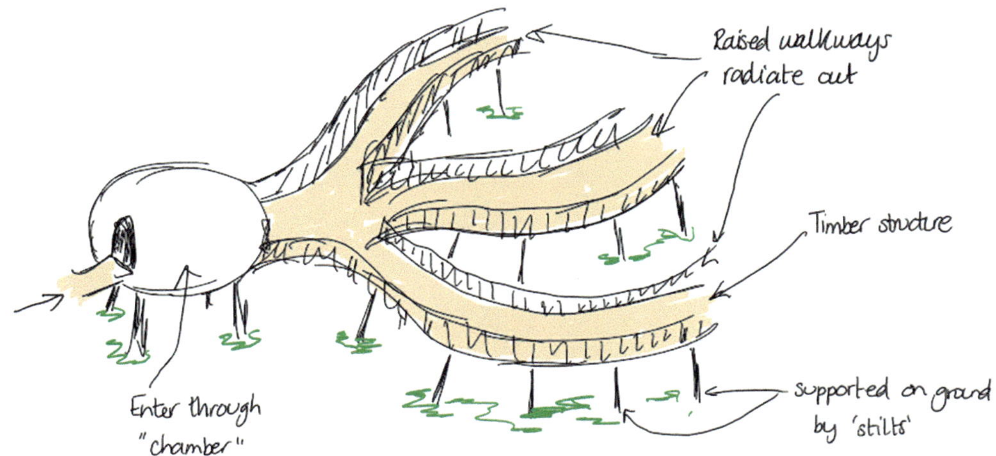

Step 6: Review solution

What works well?

- Use of timber as a material fits with woodland setting
- Support on stilts gives views but more structurally feasible
- Can source timber from sustainable sources

What doesn't work well?

- Are visitors safe? Can animals climb stilts?
- No weather protection — does this matter? Closer to nature?
- Needs foundations in ancient woodland

What is included in the design that wasn't in the original brief?

- Multiple routes/paths to follow

Step 7: Revise seven simple requirements

- Experience nature/natural habitat
- Safe for users
- View of wildlife
- Fit in with rest of centre
- Tell a story of the woodland

- Sympathetic to woodland setting
- Buildable with relatively low-tech techniques

Step 8: Sketch another solution (Figure 5.6)

Figure 5.6: Woodland walkway: second solution

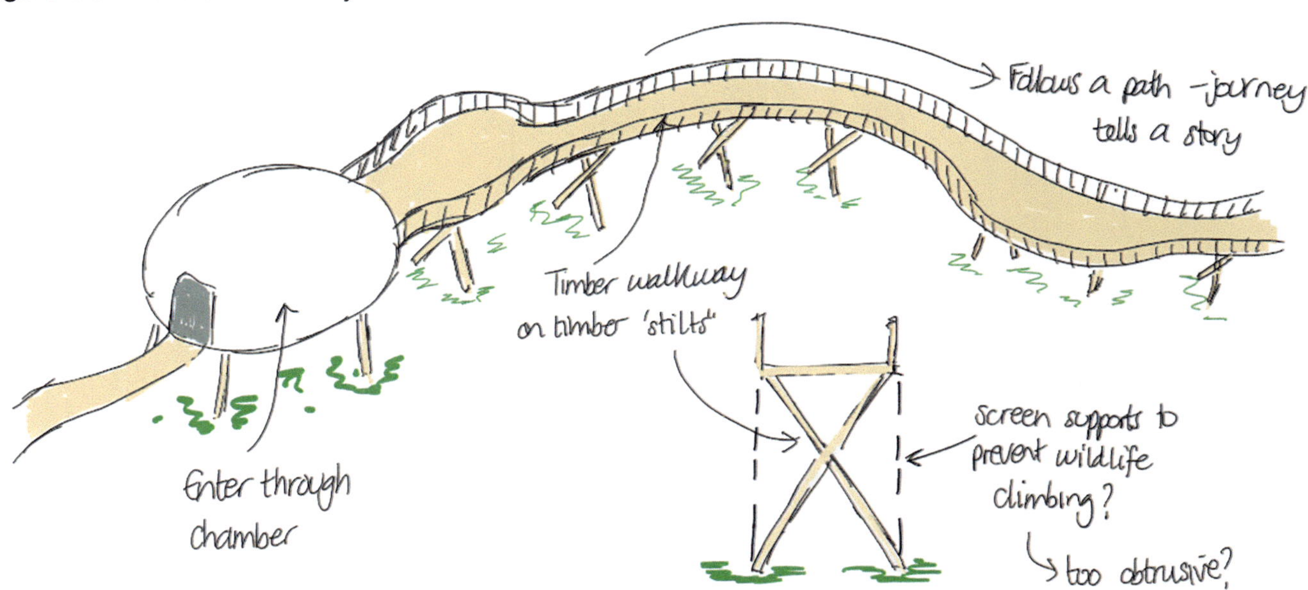

Step 9: Review solution
What works well?

- Path/route though woods — a journey tells the story
- Still natural materials — timber in woodland
- Supporting stilts can be screened from animals — prevent climbing

What doesn't work well?

- No weather protection — is this OK?
- Inclined stilts difficult to build in woodland
- Horizontal component of force at foundations — bigger foundations in wood

What is included in the design which wasn't in the original brief?

- Wider section viewing area for wildlife?

Step 10: Reduce simple requirements (three to five)

- Safe for users
- View of wildlife
- Sympathetic to woodland setting
- Tell a story of the woodland
- Experience nature

James Norman
University of Bristol

6 Questions we must ask

I have a confession to make — I hope that it is five years from now and that you are holding this book in your hand wondering why it's so obsolete, why the solutions that you want to explore are not covered and why only outmoded and old techniques are included?

I have worked in structural design since 2000 and in these 20 years very little has really changed — the predominant form of large construction is still steel and concrete. In fact, the design of buildings now is very similar to that of 40 years ago. If you were to back-analyse a building from the 1980s you would use much the same formulas, and there would be much the same methods of construction. The biggest difference would be in the accessibility of the information, and the degree of structural survey you would be required to do. The point is simply this — that we are stuck[27]. We know that we cannot carry on designing buildings in the same way as we have in the past, but we also can't see a way forward — that we continue to design buildings out of the same materials and with the same methodologies as we have for the last 40 plus years.

In Richard Rogers' book *Architecture: a modern view*[28], he writes:

"… beyond the exploitation and injustice, which is so central a feature of our civilization, looms the prospect of ecological disaster. Our predicament now is that the means for our emancipation threaten our very existence and the existence of other species."

He wrote these words in 1990, and it is hard to see what change has occurred since then. Biodiversity is down, CO_2 emissions are up. The climate is changing. We know that there are huge challenges ahead, yet there is little evidence of significant action. I had a very depressing meeting with a large client recently who was very blunt that the projects they do are in the 'mainstream' — and the mainstream today is exactly the same as it was when Richard Rogers wrote those words thirty years ago.

So, what should we do? How do we move forward? I really don't feel I can answer these questions with answers, but instead I think we should answer them with questions. Questions that inform the way we work and what we choose to do differently going forward. I would like to suggest three types of questions:

1. What if? These questions ask us to imagine the future and think about how we might design for it.
2. How much? These questions challenge us to count the cost, not just economically but in a much broader sense.
3. Where do we start? These questions are offered as a way to shift our design paradigm, and challenge us to design differently.

6.1 What if?

I love 'What if?' questions. I spent much of my childhood asking them. I would sit beside my father watching sport, trying to explore the limits of the rules and boundaries of the game, and asking questions like what if someone takes an indirect free kick, it touches no other player and goes in the players own goal — would it be a goal? My long-suffering father would always take time to consider my questions and debate the outcome. As engineers, asking 'what if' questions is essential. It helps us imagine new ways forward. I would like to ask some really big 'what if' questions. I hope in answering them you might see new ways forward in the design of buildings. I will break them into three components: materials, loads and 'other'. I will attempt to add my thoughts, but please don't see these as the answer — more a further description of the question.

6.1.1 Materials

With the world committing to being carbon neutral by 2050, and a number of businesses, organisations and governments promising to be carbon neutral by 2030, we need to ask some really big questions about materials. For much of the last decade, the focus has been very much on the amount of energy in usage but, as this comes down, the embodied energy and CO_2 becomes a larger concern. If we are to be carbon neutral in ten years' time, we need to change the way we use materials. Here are some great questions to ask:

What if cement is banned by 2030?

Cement production produced 8% of global CO_2 emissions in 2016[29], and concrete is the second most used commodity after water[30]. But what would we do if we could no longer use concrete at all? What then? Every building I have designed, whether steel, masonry, timber, rammed earth or straw bale has included concrete somewhere. This is a huge question, a massive challenge, but inevitably we will need to think of alternatives going forward. Another related question could be what if we have no more usable virgin sand or aggregate (a very real challenge for certain parts of the world)? How would this affect our approach to design? We might also ask why don't we use stone? Stone is stronger than concrete, has not been processed in the same way, so has much lower embodied energy and embodied CO_2. However, the use of stone is now generally limited to conservation projects and expensive floor finishes!

What if energy-intensive materials are taxed so they cost ten times more than they do now?

This question is subtly different to the previous one. Whether using steel or concrete as designers, if the cost of materials increase ten-fold, the drivers would change. The desire to use less would suddenly trump the desire to build quickly or simply. We would value every kilogram of material and try to lose as much as possible. Yet we can design like this now. We know how (to a certain degree), but the cost of innovation remains a barrier.

What if your project was the last project to use this particular material?

Imagine that every beam, column and slab that you design will be the last to be made. Would it change the way we designed? I imagine I would consider re-use much more seriously. I imagine I would value every beam and slab and consider, if and when my building is disassembled, how they could be used again. Designing for re-use is already a consideration in design — but how many buildings are really re-usable? Yes, there are some excellent examples (such as the Cork House that was shortlisted for the 2019 Stirling Prize), but is every high-rise designed with re-use in mind?

What if all metals had to be recycled — that we could no longer use raw materials?

This invites more questions. How do we reduce our material usage? How do we ensure our material is recyclable and not just some of it, but all of it? How do we ensure that we don't lose material quality, that our high-grade steels are recycled into high-grade steels, our high-grade aluminium is recycled into high-grade aluminium? And can we avoid melting everything down and starting again? Can we use as much as possible in its current state?

What if our timber usage is limited to guarantee the world's forest is growing?

Similarly for timber — how do we ensure that our usage is sustainable? We know that the world forest is reducing, not increasing[31], but we also know that European forests are growing (by looking at the 'annual increment') and that we can certify timber to ensure that it is well-managed and sustainable. But what if we stopped using steel and concrete and moved across to timber? Do we have the capacity for that level of expansion? And bearing in mind that trees take 35–45 years to grow to a harvestable size[32], do we have a forestry plan to compliment any significant increase in usage?

What if designing for re-use was not just a 'nice to have' but was a legal requirement?

We have mentioned re-use already — that we *should* think about it — but what if we *had* to think about it? What if it was a legal requirement? What if buildings were chosen every year at random and had to be dismantled and reconstructed, and failure to do so would be a criminal offence? Would we approach the design differently?

What if we were less worried about preserving our buildings, and more worried about preserving the environment?
Alongside the question of re-use, there is another tension that we must face; that of the preservation of our built environment (and especially our historic structures), and that of preserving our environment. One of the large contributors to greenhouse gas is the heating of buildings. If we were to wrap all our buildings in insulation, their energy performance would improve — especially if we re-glaze them at the same time, but the identity of the building, street and maybe even city would be altered. There are no easy answers to these questions, but better to have a frank debate and forge a way forward, than to do nothing.

What if we couldn't build on greenfield sites? What if we had to re-use existing buildings and what if all new build was limited to enhancing these existing buildings?
These are radical questions, but look around the city or town where you work — would it really be as disastrous as we first think? Yes, it would require us to think radically. Yes, it would require us to design differently. Yes, the engineer's role (along with the rest of the design team) may become different, but would that be a problem? I often think that the best value an engineer can bring is to prove to the client that they need to do nothing (at least structurally) to use their building in the way they want.

6.1.2 Loads
We can reconsider the loads we design for. Historically, we have assumed that the higher the load, the greater the level of safety. If we overdesign, we are really making everything safer — and it's hard to argue I want a cheaper building rather than a safer one! But overdesign comes at a cost, not just financial, but also in terms of embodied energy and CO_2, so now there is a more complex problem to solve. How do I make my building safe, affordable *and* sustainable? 'Safe' is not a simple definable binary question. The question really should be is my building safe *enough*? While this is a rich and complex question, one of the big areas engineers could challenge is around the loads that they design for. We could ask some of the following questions:

What if I designed only for the load for this specific use?
Over the course of my career, the loads that I designed for have gone up. When I started designing in 2000, we designed for the current use, but then we started to worry about the future. What if the use changes? What if we decide to put partitions everywhere? What if the space becomes a gym with lots of partitions? Design loads have crept up. While the desire for future flexibility is a noble one, we need to weigh these questions against the environmental cost of overdesign.

What if I designed only for the temporary load case — and then provided a capacity?
This is a more radical question. Most designers start with the final constructed load case and later consider the temporary loads. But what would happen if we turned it around? What would happen if we said that to build this concrete frame, it needs to be this strong (allowing for back-propping etc.) and then worked out what the floor capacity was? This may do two things — we may find that the capacity in the final condition is enough, and we may change the design approach, making it more efficient.

What if I assumed an intelligent user?
By 'intelligent user', I mean a user who can apply some engineering thought, or at least follow set guidelines. Often our designs are reductive, assuming a worst case e.g. the load patterns we apply, but what would happen if we advised the user not to stack all their heavy items in alternating bays? Or to limit the load? Or to only place heavy storage items near columns, significantly reducing bending stresses? Or even if we created a map of what loads could go where? Our current approach is the opposite — any load, anywhere, in any configuration (but we will assume the worst).

What if I assumed a feedback system?
Of course, we don't have to have an intelligent user to achieve this outcome — we could also create feedback systems. If the limit on our slab design is a set deflection, we could measure the deflection, and when the slab deflection gets close (say within 10%), the client is sent an amber warning, and when the deflection limit is exceeded a red warning is sent, using panels similar to those used for fire alarms. We could also measure vibration, stress, strain and all sorts of things. Could we then be more relaxed about the loads, especially SLS loads, maybe applying ψ_1 and ψ_2 reduction factors for the live loads?

6.1.3 Other 'What if?' questions

Some of these will seem obvious and sensible, others more 'out there' but we need to think differently if we are going to behave differently.

What if the structure defined the finishes?

This may seem like a strange question. But how often do the finishes define the structure? I would suggest often if not always. Deflection criteria are often driven by the needs of the finishes, especially around the perimeter where low tolerance curtain walling systems demand such small deflections that we are almost trying to make our structure *infinitely* stiff. What would happen if we instead flipped it, so that all deflections would be limited to span/150, and that the finishes had to work with this?

What if we could upgrade our building?

What would happen if we could upgrade our structure? We already can, and there are a variety of techniques that we use when retrofitting existing buildings, but what if we designed these in at the start? What if we designed for just $1.5kN/m^2$ of live load but had a clear strategy for upgrading if necessary? Or if we designed for construction loads as our baseline, but had a strategy for upgrading if necessary? Of course, you might ask whether any client would actually upgrade? Maybe rather than assume they won't, and therefore designing for a greater load to start with, we should assume they will, and create sensible and achievable ways of doing this.

What if we designed for 100 years (or 200 years or 400 years)?

You may think that this question is out of step with everything else I am asking — that longer design lives lead to higher loads — and you would be partly right. This evokes so many good questions e.g. if I design my building for 100 years of corrosion protection (rather than 50), how much difference does it really make to my concrete cover and does it fundamentally change the diameter of bars in my beam, or just the location? What about the design for stability — the stability of a building is only a very small proportion of the overall structure, but if it were to fail the whole building fails, so what would happen if we design for 100 years, especially for the ULS? We may be surprised by how little the sizes increase. Durability is of course one of the biggest questions here. It is all very well to design for this stress or that, but are our buildings durable? Will they still be standing in 400 years' time? If not, why not, and how can we change this?

What if we (engineers) didn't design the building — what if they were designed by other people?

Everyone comes to design with a set of values and preconceptions and this influences the output. This will come from a number of different influences including our engineering education. Approaching the design as someone else can be helpful. I had a painful lesson in this as a young designer. We were designing a theatre and had provided full access at ground floor, but were struggling to make the balcony accessible. I was discussing the challenges with the architect and asked was it necessary for both floors to be accessible? What I haven't mentioned up to now is that the architect had broken their back some years before, and often used a wheelchair. They asked the simple question "why should you have access to both floors when I don't?" I was both humbled and stumped. I learnt a very valuable lesson that day.

What if we designed for risk?

I have saved the biggest question for last. Most engineers don't design for risk — they apply factors of safety, then assume that everything is determinant, rather than really understanding the statistical significance of what they do. But do we really consider questions like: Will this fail in a ductile or brittle manner? Am I using the upper or lower bound theory of design? What is the statistical distribution of my load? How much do I trust the contractor to build what I have designed (a question we ask when designing masonry but no other material)? What is the probability distribution of my material (after all, not everything is normally distributed, despite our frequent assumptions to the contrary)? These are huge and complex questions, but if we really understood risk we could design better. Rather than assume that a beam at 99% capacity was OK, and a beam at 101% capacity will definitely fail (which it almost certainly won't), we could understand what is really going on and assess what the risk really is.

There are so many other 'What if?' questions you could ask. I also believe that we need a paradigm shift, that we cannot carry on doing things the way we do them, and that by posing these 'What if?' questions we can imagine different ways to answer the questions we face. This was recently reinforced to me by two artists I've been talking to who are asking: 'What if we moved to Mars?'[33] What was interesting about answering this question was it made you reimagine everything from food to water to waste. Much of this reimagining can be applied here on Earth, right now. We don't need to move to Mars to use less and save more.

6.2 How much?

'How much?' is not a new question. I imagine since people have been building, they have been asking this question. We do though need to broaden the question. It shouldn't just be how much money, but: How much CO_2? How much energy? How much waste? How much damage to the local ecology?

I think we may be surprised by some of the answers. I would love to think when an engineer designs a concept they would, as a matter of course, present how much CO_2 and embodied energy is required in each option e.g. if we take a design and set it as a benchmark, we can ask some good questions, such as:

- If we double the grids and half the spans, what is the change in embodied energy and CO_2?
- If we reduce the load, what is the embodied energy and CO_2?
- If we relax the deflection criteria, what is the embodied energy and CO_2?
- If we relax vibration criteria, what is the embodied energy and CO_2?

I think answers to these are really important because they enable the client to make an informed choice e.g. they may agree to double the number of columns but not to change the vibration criteria. This information shouldn't be hard to produce — a 3D CAD model will give volumes of material, and experience will help us estimate other factors, such as percentage of reinforcement. This information would be really powerful.

If I want a six-storey concrete frame, 40m × 20m, there is a baseline for slab depths, based on fire resistance. We can then, on the basis of the wet weight of concrete, calculate the maximum span in the temporary condition before we need to increase the floor depth. This in turn defines a structure which has a load capacity in the final condition. At this point, we can start saying how much more CO_2, and how much more embodied energy to increase the load, or span, or stiffness. I hope you can see just how powerful this approach could be, and also how it might help the client to make informed decisions. We may ask them to reduce the loads. Maybe the saving for this type of construction is minimal, but until we do the sums we don't know.

6.3 Where do we start?

Design is seen as an iterative process. We start at a point and iterate towards a solution. In mathematics, some equations cannot be solved by iteration. If you start in the wrong place, you will never find the solution, with your solution either shooting off to infinity or converging on a different solution, rather than the one you are looking for. So it is with design.

In Chapter 8 we will discuss the six decisions that need to be made to develop a concept, and how, if we decide one thing, we often decide many at the same time e.g. if we decide to use a 10m grid, this instantly excludes a number of structural options. We may feel we can then iterate from this point. But I don't believe we will ever reach a number of solutions from this starting point.

In design we often use typology — the replication of a type — to design, but typology also locks us into solutions. An open-plan office is a type. We need columns at a certain grid for the type to work.

What would happen if we changed the starting point e.g. if we started with the material, and stated our structure must be timber? If we really understood the materiality of timber and all its nuances and quirks. What would happen then? We might end up with:

- low-rise rather than high-rise
- flexible structures rather than stiff
- light structures rather than heavy
- exposed structure rather than hidden
- fire prevention rather than resistance

- a smaller structure but more regular
- a very different structure than we first assumed

Likewise, what if we started by assuming that we would re-use everything on site, and that all structures would remain? It might completely change our approach to the project. If we are designing an office, we might start asking questions like: Do we all need a desk? Do we all need to come to work every day? Do we need a large open-plan space? Do we need to store anything, or can we go completely paperless?

Finally, what if we started by stating that the embodied CO_2 of the structure would be less than 99kg CO_2/m^2? This might feel like a reach right now, but it is already possible under the right conditions. What would happen if, rather than starting with materials, or a set grid, or a certain type of frame, we started with this aim? What if we let this drive all the other decisions we make? Not only could we achieve it, but it could (and I would argue should) become normal. In fact, and this really is radical, maybe we should aim for even less embodied CO_2. If 99kg CO_2/m^2 becomes normal, we should go a step further and aim for 50kg instead and, you know what, if that's where we're going to end up anyway, why not try and go straight there! That really would be both exciting and radical.

Often, one of the biggest challenges an engineer faces is that by the time we are involved, the really important questions have already been answered, and we are faced with the question: Can you make this out of timber? and the answer is: "No, not without starting again". This is where relationships become important. As we build relationships and gently describe why the building cannot be converted from a concrete frame to a timber one, we also point to a future where we work together again, and we outline the types of questions that we should ask on the very first day.

Of course, none of this is easy. If it was, we would have done it already. But I sincerely believe that the future is very bright for outstanding engineers, that we will need to face up to these challenges with intelligent solutions, and that the best place to start is by asking the right questions.

I really do hope that over time this book becomes obsolete — cr at least out of date. That we move forward, coming up with new solutions, using both old and new materials. That the world of design changes, and that the buildings of the next decade will be radically different to the buildings of the last four decades.

Isobel Lloyd
University of Bristol

7 Geotechnical decisions

7.1 Introduction

Many engineers think that geotechnics is a 'dark art', but it is more like a good detective story, in which the clues may be limited and contradictory, but from which logical deductions can be made. The first stage of any geotechnical design is the collection of these 'clues', followed by the construction of a ground model — a 3D representation of what is in the ground beneath the proposed building, including the key material properties of:

- strength
- compressibility
- density
- plasticity
- angle of friction
- permeability

Once the ground model is taking shape, it is possible to start thinking about how the building will interact with the ground e.g. a deep basement beside a river will raise concerns about waterproofing, buoyancy and uplift, whereas for a car park at grade, the concerns are likely to be surface water flooding and filled ground.

An example for an industrial estate site in St Phillips Marsh in Bristol is given in italics for each stage — collecting the data, interpreting the key material properties and suggesting quick designs.

The ground model is *never* finalised. The good engineer will be revising the ground model throughout the design and during construction, as more 'clues' are uncovered, which may or may not support the original assumptions. A wise engineer once told me that "every site holds one more secret and that it is up to us to find it" — this has proved a useful adage. Despite our best efforts, the greatest uncertainties lie in the ground and, for this reason, it is recommended that a cautious design approach is adopted to everything in the ground, until data proves otherwise.

As engineers, any problem can be overcome, but if they can be avoided or minimised then the overall project will be easier and cheaper to deliver. When considering the viability of any project, there are multiple aspects of the ground conditions that need to be taken into account. The obvious one is the depth to a suitable bearing stratum for the structural loads, followed by it's compressibility. Equally significant to the cost of the project might be:

- disposal or treatment of contaminated land
- previous land uses, including old foundations
- instability either from extractive processes or landslips
- viability of soakaways or attenuation ponds
- flooding of excavations
- lateral variability across the site

Other factors may also affect the design e.g. if high levels of radon are expected, that would influence the structural solution and push the project towards suspended floors and ventilation.

The design of a site investigation (SI) is not covered in this book because it is assumed that it will be undertaken once concept design is complete. An SI carried out after concept design can place strength tests under buildings, permeability tests in soakaway areas, CBRs under parking etc. An early site investigation risks being wasteful, by either testing in irrelevant areas or by not doing enough tests. It is difficult to explain that you have wasted the client's money or that you need to go back for a second investigation.

7.2 Desk study

7.2.1 Background research

The desk study is the first source of clues. This stage should not be skimped on. It might be much more fun to sketch exciting images of the above-ground structure, but it may not be feasible to deliver the architect's/client's dream on the proposed site at a reasonable cost. Time spent establishing the constraints on these dreams, before they become requirements, will often save pain for all later in the project. For example, in many UK cities with high land values, the most economical solution is to provide a single-storey basement, because the excavation removes contamination and soft alluvial strata, to ensure the foundations are on competent strata. The addition of a second level of basement, however, will often be below the groundwater level and require excavation of rock, which adds significantly to the cost of the project.

Useful sources of desk study information can be collected from internet sources (Table 7.1). Unfortunately, in the UK, the Environment Agency's 'What's In Your Backyard?' site is being decommissioned gradually, so it is necessary to search for the individual elements. A list of keywords is provided as the website addresses are likely to change. There are companies that will compile this factual information, such as Envirocheck and Groundsure. The types of information considered in this stage of preliminary information gathering are given in the following example:

A search of the site at St Phillips Marsh in Bristol reveals that in the area there:

- *is 1 in 100 (or greater) annual probability of river flooding*
- *are high levels of radon found within 1km of the site*
- *are no coal workings*
- *are records of Unexploded Ordnance around the site*

The historical maps indicate that:

- *the area was used for lime kilns, brick and tile works prior to 1883*
- *in 1903 there was a foundry, corporation depot and railway sidings with an engine shed*
- *in 1951 there was a varnish and paint factory*
- *in 1955 there was an addition of stone works, petrol tanks and a refuse heap to the east of the site, with continuing industrial uses up to the present day*

Common contaminants related to historical uses are shown Table 7.2.

If contamination is anticipated, it is recommended that a specialist is included in the team and appropriate testing undertaken as early as possible in the project. At concept design stage, without the benefit of any ground investigation, it is necessary to make a best guess on the levels of contamination and extent of remediation that might be required. It can be deduced from the history of this site that it is likely there will be toxic heavy metals, oils, tar and solvents close to the ground surface, so a specialist should be involved in the project.

Starting a risk register is a good way to track the risks that have been identified and record what extra testing, design or construction is going to be incorporated into the final design. Examples of the things that would be included for this site are provided in Table 7.3.

Table 7.1: Sources of site data

Aspects for investigation	Information	Website	Notes
Underlying ground conditions	Geological online maps, memoirs, lexicon, borehole logs	http://www.bgs.ac.uk/ http://mapapps2.bgs.ac.uk/geoindex/home.html	British Geological Survey BGS GeoIndex
Man-made voids	Coal mining records, mining and quarrying	http://mapapps2.bgs.ac.uk/coalauthority/home.html http://www.bgs.ac.uk/geoindex/	Coal Authority BGS GeoIndex coal
Historic uses which might result in contamination, excavation, pollution	Topographic and historic maps	https://www.old-maps.co.uk/index.html#/ http://maps.nls.uk/ https://digimap.edina.ac.uk	Old Maps National Library of Scotland (UK wide) Historical maps
	Street maps	https://www.streetmap.co.uk	Any street map of the vicinity may have names indicating previous uses or conditions
Risk of flooding	Long-term flood risk/flood map for planning	http://apps.environment-agency.gov.uk/wiyby/default.aspx https://flood-map-for-planning.service.gov.uk/	Environment Agency
Constraints on land use	AONB, SSSI, RAMSAR, nature reserves, etc.	http://www.magic.gov.uk	MAGIC
	Aquifer designations	http://www.magic.gov.uk	MAGIC
	Groundwater vulnerability, nitrate vulnerable zones, drinking water protected areas, safeguard zones, landfill sites, pollution, etc.	https://www.gov.uk/government/organisations/environment-agency http://apps.environment-agency.gov.uk/wiyby https://naturalresources.wales/?lang = en	Environment Agency (England only) Natural Resources Wales (Wales only) Scottish Environment Protection Agency (Scotland only) EA's What's In Your Backyard?
Changes to the landform	Aerial photography (past and present)	https://www.historicengland.org.uk/images-books/archive/collections/photographs/ https://support.google.com/earth/answer/148094?hl=en	Historic England Google Earth
Hazards to excavation	UXO (Unexploded ordnance)	https://www.zetica.com/uxb_downloads.htm	Zetica
Radon	Radon risk	https://www.phe-protectionservices.org.uk/radon	Radon
Planning portal	Local site investigations	https://www.planningportal.co.uk	Local authority for site/planning

Table 7.2: Common contaminants from historical site uses

Historical uses	Probable contaminants
Metal works, foundries and metal mines	Toxic heavy metals e.g. cadmium, lead, arsenic, mercury
Gasworks	Coal, tars, coke, Blue Billy (ferrocyanide), sulfur, sulfides
Railways	Coal and coke (also anything that has been transported), asbestos
Tanneries	Chromium
20th century buildings	Asbestos
Agriculture	Pesticides and unrecorded anthrax pits
Paint works	Lead, chromium, cadmium and zinc, VOCs, chloroform, ethylbenzene, solvents

Table 7.3: Elements for inclusion in a risk register for St Phillips Marsh

Potential hazard	Risk	Possible actions to investigate	Source
Radon	Human health	Check and suspend floors/vent voids	Desk study
Unexploded ordnance	Contractors especially piling	Magnetometer survey	Desk study
Contamination	Health risk to contractors, end users and water course	Employ specialist	Desk study
Flooding	To underground services and structures (permanent and during construction)	Check with LA, EA, flood resilience measures	Desk study
Buried obstructions	Extra cost to remove or design around them	Site investigation	Desk study
Mine workings	Ground collapse, shafts and methane	Mining report/site investigation	Desk study
Uneven site	From the walkover survey it was observed to be largely flat but there were spoil heaps. It is unlikely boreholes were drilled on these	Topographical survey	Cross-section
Flooding of basement. Groundwater level is assumed at 0.9m below ground level	Flooding, uplift, flotation of drains	Install fast response piezometers as close to river as possible	Cross-section
Variable thickness of made ground	Foundations	Site investigation	Concept design
Variable thickness of gravels	Pile foundations	Site investigation boreholes	Concept design

7.2.2 British Geological Survey (BGS) information
BGS maps

The BGS website is the main UK source of geological and geotechnical information, and is constantly being improved with new layers of information. At the time of writing, 3D models are being developed for the website, but are incomplete.

The logical starting point is the solid geology map[34], which is predominantly rock but may include engineering soils (Figure 7.1).

Figure 7.1: Extract of solid geology from BGS website

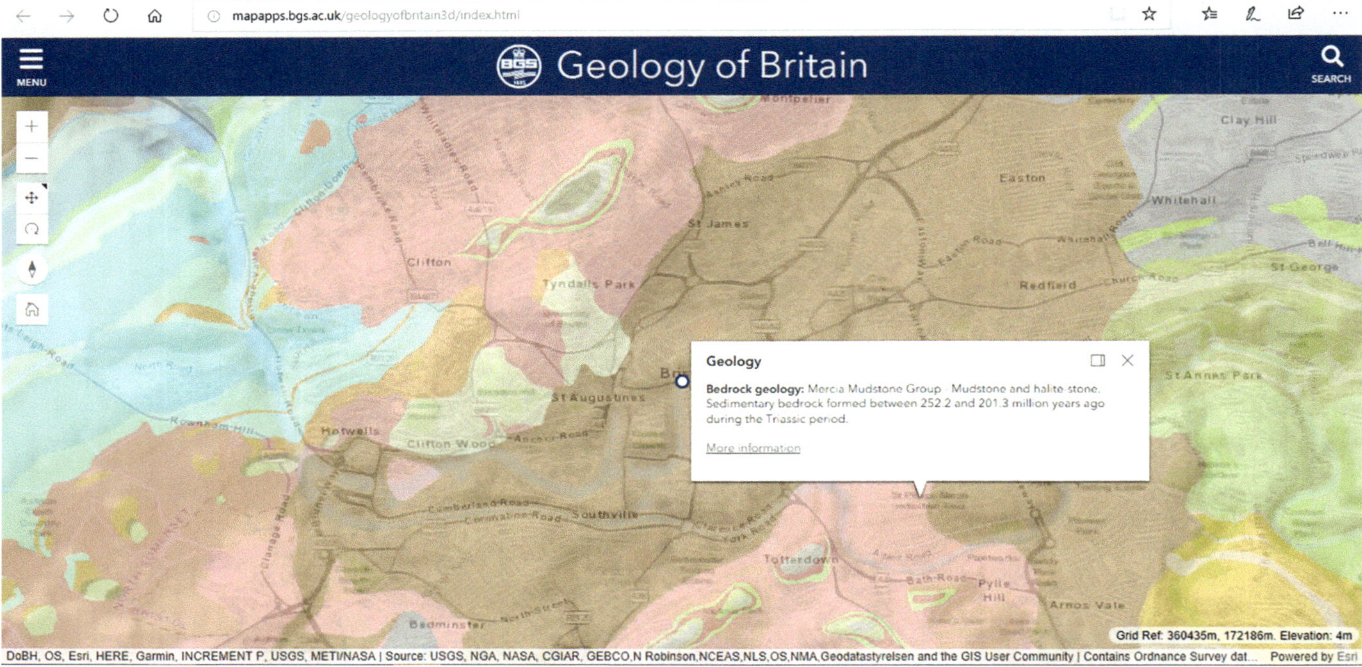

Note that solid geology has been selected for this view. By clicking on the site, a window opens with information about the underlying solid geology. In this example, the St Philips Marsh in Bristol is on Mercia Mudstone Group deposits, which is around 200 million years old. Clicking on 'More information' links to the BGS Lexicon which provides information on:

- depositional environment
- the strata above and below
- previous names

The latter is particularly useful when searching for references about the rock type. In this example, the Mercia Mudstone Group was previously known as 'Keuper Marl', which was the subject of a CIRIA Report[35].

The next step is to add the layer showing superficial deposits (select 'Combined'), which have been deposited in the last two million years, and are likely to be unconsolidated, loose and compressible soils (Figure 7.2).

The superficial layer indicates that there are tidal flats deposits of clay and silt across much of the area. At this stage, we have no idea of their thickness, but they are less than 20,000 years old so we can expect them to be soft and compressible.

Additional information is available from the BGS website within the BGS 'Civils Bundle' (Section 7.2.4). At this scale, it is useful indicative information, but is not site specific, and does not replace interpretation of the data.

Figure 7.2: Extract of geology from BGS website including superficial layers

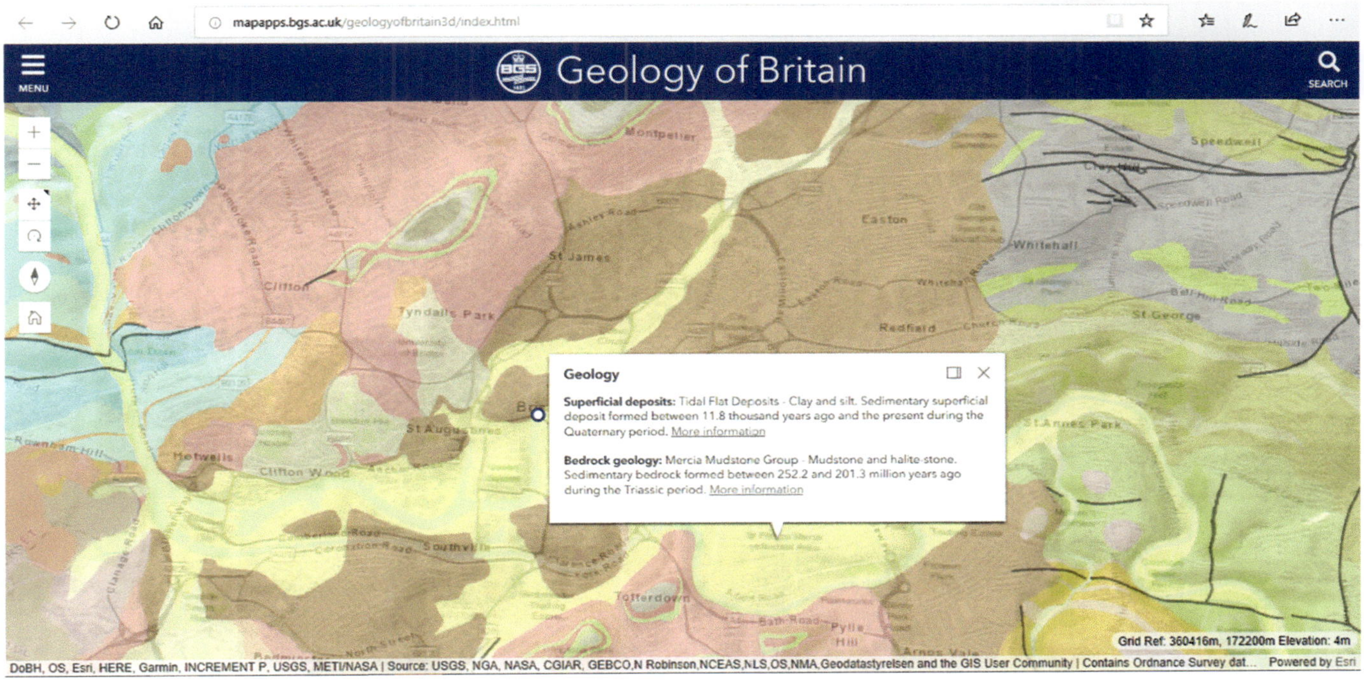

BGS boreholes

If a site-specific ground investigation has not been undertaken, which is commonly the case at concept design stage, the best available borehole data is, again, on the BGS website[36]. This will probably be the only source of clues as to the thicknesses of strata. Unfortunately, the quality of data is highly variable as much of it predates BS 5930, first published in 1981[37] or even CP 2001 from 1957[38]. Many of the early boreholes were drilled for local water supplies and were hand-written in copperplate script. At first glance, these are difficult to interpret but in the absence of any other data, the effort is usually worthwhile.

There are four types of material found at the ground surface:

- Made-ground
- Rock
- Granular soil
- Cohesive soil

Rock typically has an undrained shear strength greater than $0.6MN/m^2$ and cannot easily be excavated. The definition of the soil/rock boundary has varied and in BS 5930:2015[39] was lower at $300kN/m^2$. Made-ground, by its very origin, can include anything — generally being domestic waste, inert waste or clinker, but has been found to include complete double decker buses and even human corpses! At concept design stage, it is only viable to look for a competent stratum beneath the made-ground. Soil can be divided into either:

- Granular (coarse-grained) soils
- Cohesive (fine-grained) soils

The important difference between these is that cohesive soils stick together when wet and can be rolled into a thread that supports the soil's own weight, so they are said to have cohesion and plasticity (may change volume with moisture content). Granular soils do not have these properties and are particulate in their behaviour. A clay fraction as low as 35% will dominate the behaviour of a soil. All soils can be excavated by a backhoe excavator.

Soil description is like a code which has to be deciphered — each word means something very precise e.g. a 'soft grey CLAY with a little sand' without knowing the code suggests something that is a bit 'squidgy' and fine-grained. Between 1981 and 2015, it was actually a shorthand description for a fine-grained soil with an undrained shear strength of 20–40kN/m^2, with the majority of the material having a grain size of less than 0.002mm diameter, with less than 5% by weight of the grains between 0.06–3mm diameter. The use of capitals for the clay, shows that it was the main constituent and therefore it would behave as a cohesive soil. The previous standard for soil description, CP 2001, had descriptive terms but did not include correlations to values of strength. Since 2015, BS 5930 no longer requires inclusion of strength in the field description.

Table 7.4 gives some useful guidance to break the code, giving what the descriptive words actually mean numerically, and also their physical behaviour.

Table 7.4: Grain size

Description	Range of grain diameter (mm)	Typical attributes
Cobbles	>60	Granular soils, high permeability
Gravel	2–60	Granular soils, high permeability >10^{-1}m/s
Sand	0.06–2	Granular soils, moderate permeability 10^{-5}m/s–10^{1}m/s
Silt	0.02–0.06	Can behave either with or without cohesion, and are often very sensitive to water content. Mostly described as a 'fine-grained cohesive material of low permeability'
Clay	<0.02	Fine-grained cohesive soils, low permeability <10^{-7}m/s, strength range from stiff to very soft, high to low compressibility

Generally, for foundation design, it is most important to establish a strength profile. Combining tables from BS 5930:1981, Table 7.5 summarises the commonly used strength descriptions for cohesive soils.

Table 7.5: Cohesive soil strength

Description	Field test	Undrained shear strength (kN/m^2)
Very soft	Exudes between fingers when squeezed in hand	<20
Soft	Moulded by light finger pressure	20–40
Firm	Can be moulded by strong finger pressure	40–75
Stiff	Cannot be moulded by fingers. Can be indented by thumb	75–150
Very stiff	Can be indented by thumbnail	>150

Density of granular soils can be estimated from N values, recorded on the borehole log. Table 7.6 shows typical angle of shearing resistance and unit weight.

With these three tables alone, it is possible to start forming a ground model for any site and start attributing some approximate engineering properties to the ground beneath. Unfortunately, soils are rarely composed of a single grain size, but if one grain size is in capitals then it will dominate the behaviour of the soil. If two soil grain sizes are in capitals then they are present in equal amounts, which is commonly seen e.g. SILT:CLAY.

Table 7.6: Density of granular soils

Description	SPT N value	Ease of excavation with a spade	Angle of shearing resistance ϕ (°)	Bulk unit weight γ (kN/m^3)	Dry unit weight γ_{dry} (kN/m^3)
Very loose	0–4	Very easy to excavate	25–28	<16	<14
Loose	4–10	Fairly easy	28–30	16–18	14–16
Medium dense	10–30	Difficult to penetrate with spade or crow-bar	30–36	18–19	16–17
Dense	30–50	Requires a pick for excavation	36–41	19–21	17–19
Very dense	>50	Difficult to excavate with a pick	41–43		

The same principles of words meaning a range of numbers also applies to rock descriptions, certainly since the publication of BS 5930:1981, based on Brown[40], and refined in BS EN ISO 14689[41]. The important tables for engineers are those for rock strength and bedding thickness (Table 7.7).

Table 7.7: Rock strength based on BS 5930:1981

Description of strength	Unconfined compressive strength (MN/m^2)	Field identification
Extremely weak	0.60–1.0	Can be indented by thumbnail
Very weak	1–5	Crumbles under firm blows with point of geological hammer. Can be peeled by a pocket-knife
Weak	5–25	Can be peeled by a pocket-knife with difficulty. Shallow indentation made with a firm blow of geological hammer
Medium strong	25–50	Cannot be scraped with a knife. Can be fractured with a single blow of a geological hammer
Strong	50–100	Requires more than one blow of geological hammer to fracture
Very strong	100–250	Requires many blows of geological hammer to fracture
Extremely strong	>250	Can only be chipped with geological hammer

Degree of weathering may also be important, but it usually coincides with a reduction in strength. Examples of the classifications of weathering are given for chalk and Mercia Mudstone in BS 8004[42].

Two important points to note:

- The unconfined compressive strength test, to measure the strength of rock, uses the *diameter* of the Mohr circle, whereas undrained shear strength of cohesive soil uses the *radius*
- These strengths are measured on lumps of intact rock, and may not represent the strength of the rock mass

There is a continuous spectrum of strength, from very stiff soils with an undrained shear strength above 150kN/m^2, to extremely weak rock with an unconfined compressive strength of 600kN/m^2 or undrained shear strength of 300kN/m^2.

Prior to 1981, there were considerable discrepancies in the meaning of the strength descriptions (Table 7.8).

Table 7.8: Various strength classifications for intact rock

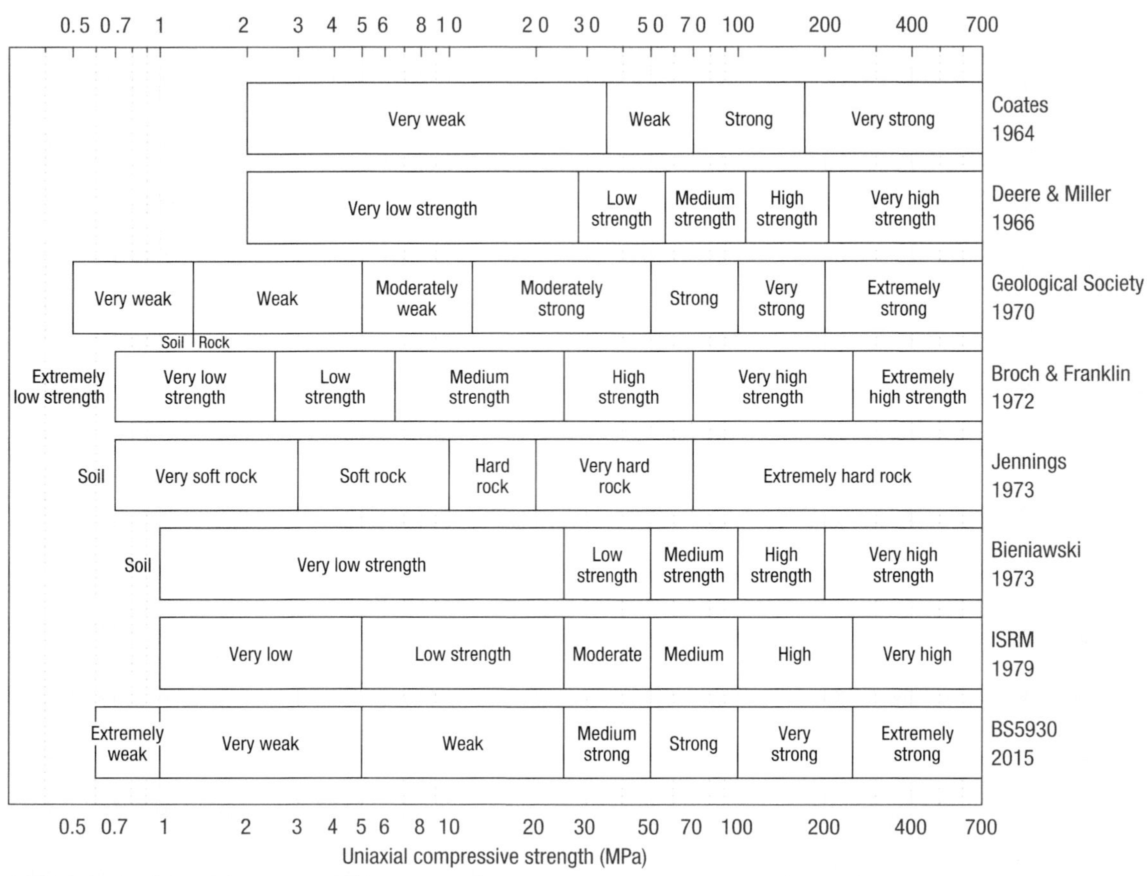

Note: Derived/adapted from Bieniawski[43].

Planar discontinuities in rock are commonly due to joints or bedding in sedimentary rocks — joints are fractures which can form in any rock, generally due to stress changes, and often form in parallel sets. Bedding discontinuities form in sedimentary rocks as deposition alters, so they also appear as parallel sets.

Discontinuities are important to engineers, as the more there are, the lower the strength and the more compressible the rock tends to be, the easier it is to excavate, and in all likelihood the higher the permeability of the rock mass (Table 7.9).

Table 7.9: Bedding description based on BS EN ISO 14689

Bedding description	Discontinuity spacing description	Spacing (mm)
Very thickly bedded	Very wide	>2000
Thickly bedded	Wide	2000–600
Medium bedded	Medium	600–200
Thinly bedded	Lose	200–60
Very thinly bedded	Very close	60–20
Thickly laminated	Extremely close	20–6
Thinly laminated		<6

Figures 7.3–7.5 are three examples of boreholes from within the St Phillips Marsh in Bristol. These were selected to show the range of detail that might be available.

Figure 7.3: BGS Borehole ST67/23

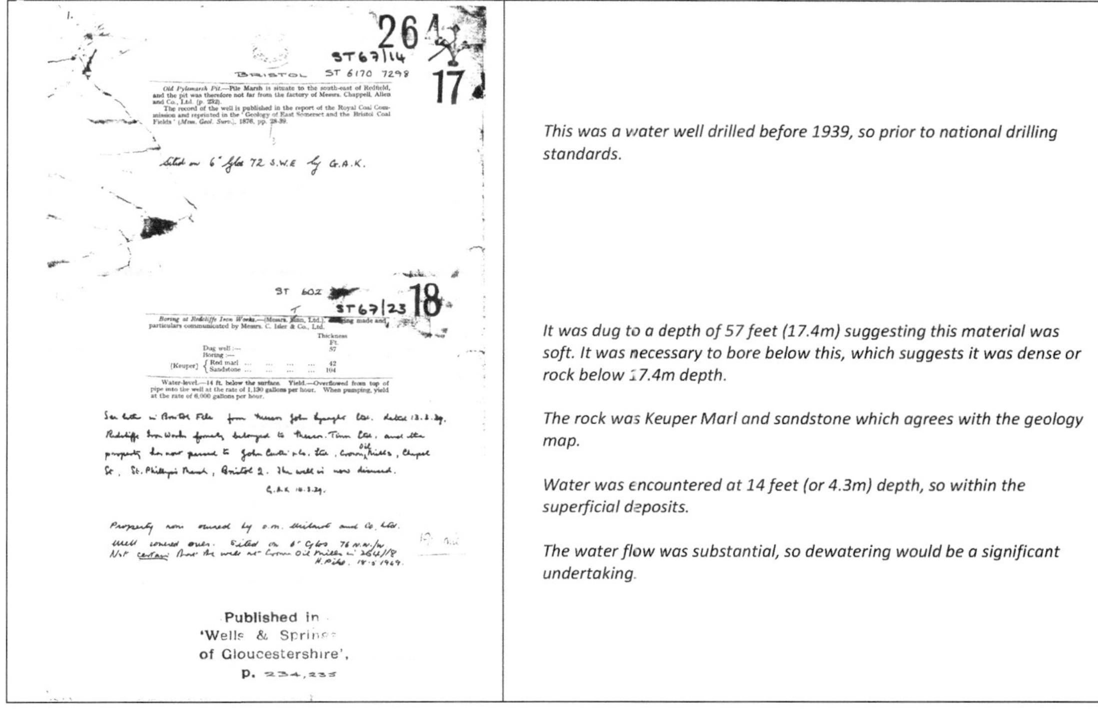

This was a water well drilled before 1939, so prior to national drilling standards.

It was dug to a depth of 57 feet (17.4m) suggesting this material was soft. It was necessary to bore below this, which suggests it was dense or rock below 17.4m depth.

The rock was Keuper Marl and sandstone which agrees with the geology map.

Water was encountered at 14 feet (or 4.3m) depth, so within the superficial deposits.

The water flow was substantial, so dewatering would be a significant undertaking.

Figure 7.4: BGS Borehole 3

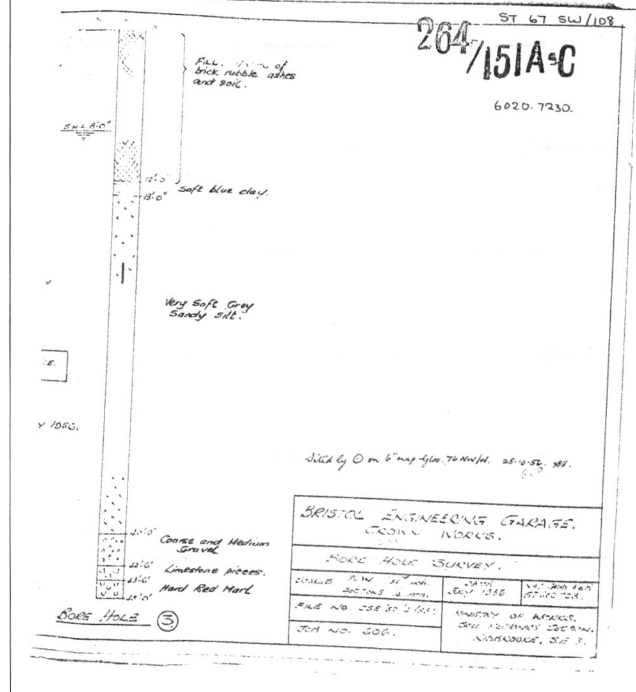

The borehole was drilled in 1956, after the draft CP 2001 was in circulation.

Made ground is present to a depth of 3.7m and comprises brick rubble and ash.

Between 3.7m and 12.2m is soft or very soft clay or sandy silt.

The descriptions may indicate that these soft layers can be easily moulded in fingers, but this is not correlated to a range of undrained shear strength

This was drilled after the publication of CP2001 so the descriptions should indicate that these soft layers can be easily moulded in fingers, but this is not correlated to a range of undrained shear strength

There is gravel and limestone pieces. The log does not tell us the drilling method used so we need to consider that they may be large enough to obstruct cfa piling.

The top of red marl was observed at 13.3m depth.

At this point in the concept design, there is sufficient information to suggest that any building will require piling, with support only beginning about 14m below ground level. Standing water was seen at 2.4m, but we have no idea how reliable this reading is. There is rubble and ash, which we would have expected from the desk study but confirms our need for contamination testing.

Figure 7.5: BGS Borehole 1

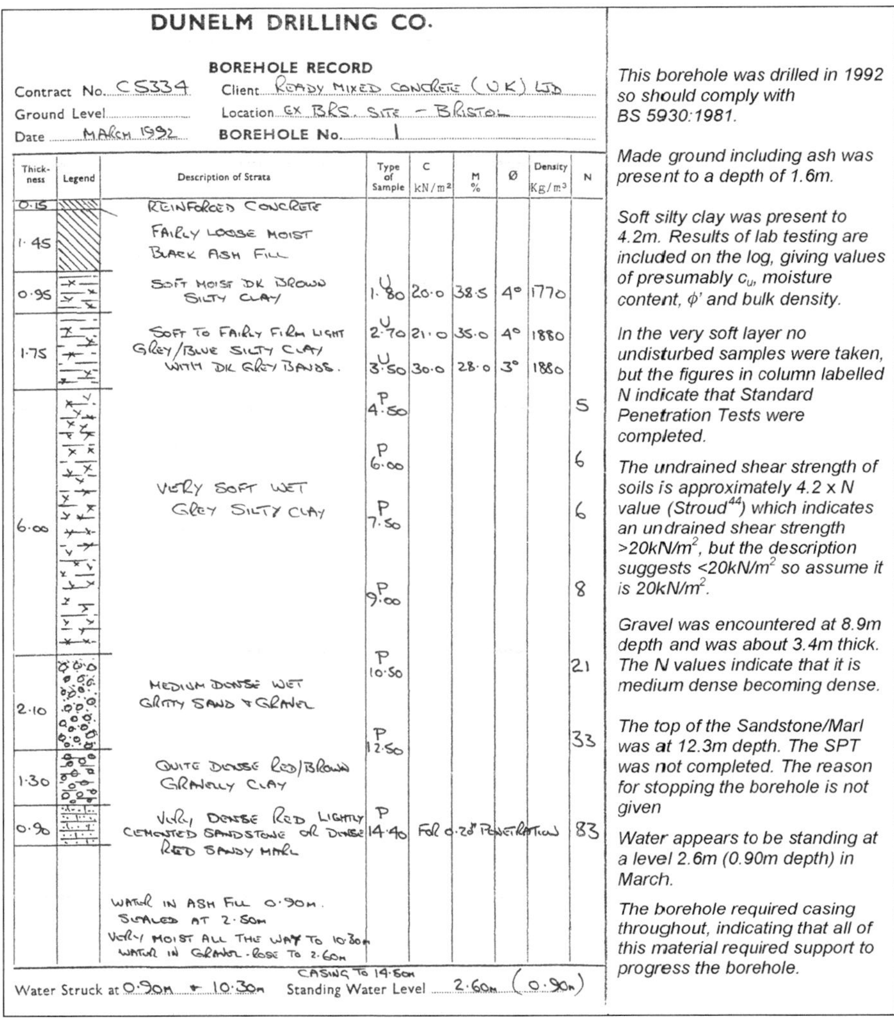

The borehole information in all three examples is broadly similar, so even historic data is useful. Made-ground can be expected to be between 1.45–3.7m thick. It is described as largely being of brick, which fits the presence of brickworks dating back to before 1883. We have to assume that this is contaminated. This overlies soft, presumably alluvial clay based on the geology map, with a thin layer of cobbles at the base. The top of the Mercia Mudstone Group was found variously at depths of 12.3, 13.3 and 17.4m.

7.2.3 Planning Portal data
At the time of writing, one planning application had been submitted in the vicinity of the site, but it did not include any site investigation data.

7.2.4 Other BGS data
The BGS website offers for purchase a 'Civils Bundle', which comprises eight layers:

- Bulking volume
- Corrosivity (ferrous)
- Discontinuities
- Engineered fill
- Excavatability
- Foundation conditions
- Strength
- Sulphate/sulphide

These are provided on 1:50,000 scale maps, so are useful for guidance but need to be checked. There are also hazard layers which cover such things as radon, solubility, landslips, mining, methane and carbon dioxide.

7.2.5 Site visit/walkover survey
An early site visit can be invaluable, although access may not be straightforward.

Consider the ground surface and levels first. It might be observed that there is man-made material visible in the topsoil, unnatural landforms, areas of bare ground or particularly lush vegetation, ponding or reeds. Ideally, record all observations on a plan of the site or a screenshot from Google Earth. The first three observations would suggest that the ground has been disturbed, and that there may be contamination. The last three would suggest a high water table. On large sites, it is often possible at this stage to start to zone the site into better areas for buildings, car parking, attenuation ponds etc. Photographs are invaluable to refer to, once back in the office.

Often, more useful clues can be found from the surroundings. Walk the perimeter of the site and record the materials e.g. the presence of a high Victorian brick wall in good condition suggests that the ground conditions are likely to be reasonable for similar loadings, but a wall in poor condition should be of concern. Similarly, inspect any surrounding buildings for any signs of movement. Valuable information may be available from local residents e.g. flooded basements, springs appearing in their gardens, historical uses of the land etc.

Many companies have pro forma sheets for completion during a site visit. Alternatively, list what aspects will influence the particular project, and develop a project-specific record sheet, which can be particularly beneficial on complex projects. It is also useful to record non-geotechnical aspects e.g. tree size and species (there are Apps to help with this) that may be subject to tree protection orders, manhole covers, overhead lines, potential party walls etc., any of which might constrain the project.

7.3 Site analysis
Huge cost savings can be made by using the site most appropriately. Often, a preliminary site layout has been prepared by the architect or client, but the engineer should be prepared to challenge this.

7.3.1 Factors that can be used to zone sites for development
When developing brownfield sites, it is common to find that the depth to a good bearing stratum varies across the site. This can be due to:

- extractive working of clay or sands and gravels
- placement of landfill
- presence of mineshafts
- bomb damage

Logically, car parking would be placed in the poorer areas, and structures would be in areas with shallowest good bearing material. Similarly, the location of attenuation ponds or soakaways will be determined by the topography of the site, to avoid permanent pumping. However, if the preferred area is likely to flood, obviously it will not function for providing water storage, and allocation higher on the site will be necessary, or an alternative solution sought.

The positioning of structures is a 3-dimensional problem. If a basement is required, but the desk study indicates that a high water table is likely, consideration should be given to raising the building, so that perhaps there is only a half basement and a grand elevated entrance. However, there may be planning constraints on the building height.

As an example, two sites, owned by a local authority, were investigated for potential Park and Ride sites. The local authority records indicated that they were all historical landfill sites, therefore potential liabilities if they included hazardous waste. Desk studies were undertaken for all three sites, and the records confirmed the use as landfill sites. Inspection of historical aerial photographs showed which areas had been worked as clay extraction pits for brick-making. A walkover survey confirmed that some areas appeared to be undisturbed while others were clearly filled. Following a ground investigation, both sites were subsequently sold by the authority for retail parks, with buildings on the undisturbed areas and car parking over the landfills, after suitable sealing and treatment. A profitable outcome for the local authority.

7.4 Ground model and typical properties

It may be possible to draw cross-sections if there is sufficient information — alternatively, the ground profile may have to be based on one borehole. A site topographical survey may be procured at this stage, or it may be necessary to base the ground profile on estimates from the walkover survey. Having decided the profile, typical soil profiles can be attributed to them, concentrating on those that will be relevant to the design of the proposed structures.

The ground model also needs to consider if there is a risk of voids — either natural or man-made. If they are as a result of coal workings, and if the thickness of rock exceeds 20 × the thickness of the potentially extracted seam, further investigation would be required to confirm the actual thickness and depth to rock. It would be prudent to make an allowance for probing and grouting, until a minimum rock cover of at least 10 × the extracted thickness has been proven. CIRIA C758[45] is a useful guide.

Other causes of voids e.g. sinkholes are likely to be of less predictable dimensions. If the search of the BGS website indicates that soluble strata are present, an extra provision should be made for investigation, and flexibility in layout would be advantageous to avoid them.

Figure 7.6 shows a ground model for the St Phillips Marsh sites. Many assumptions are made at this stage which will need to be checked, so they should be added to the risk register. It is far too easy for these assumptions to become treated as established facts, so the risk register will keep track of the checks required.

Most of the typical properties in the section have either been decoded from the description or from the testing included on the borehole logs.

Figure 7.6: Cross-section of St Phillips Marsh site

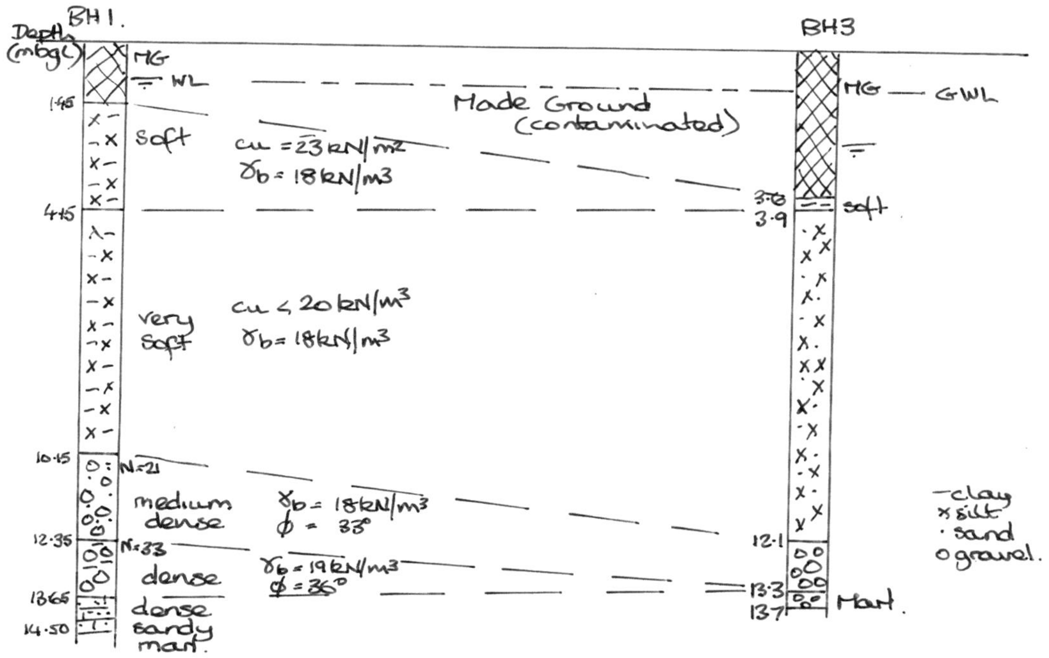

7.5 Substructure scheme design

7.5.1 Foundations — permissible bearing capacity for shallow foundations

Having collated all the available information, the interaction between the development of the site and ground conditions can be considered. Usually the most important decision is: Will it require shallow pad/strip foundations or deep piled foundations? For shallow foundations the next question should be: I have calculated a permissible bearing capacity, but is the estimated settlement and possible differential settlement tolerable for my building?

Shallow foundations are often preferable on the basis of cost anc ease of construction, so if a founding stratum, that can support your building loads on reasonable sized foundation is within reach of normal excavating plant (say 5m), then this is normally the best option. For this depth of foundation, the simplest option is to fill the trench with mass concrete, back up to close to the ground surface (which may not be preferred on environmental or even cost grounds). There are a few situations where piling is the best solution, even when a competent bearing stratum is available at shallow depth. Examples of these would be if the ground is severely contaminated, so piling would minimise excavation required and waste disposal, or for speed of construction, as precast driven piles can be used immediately, with no curing delays.

Having decided which of the four material types are present at the ground surface on your site, Table 7.10 can be used to determine a permissible bearing capacity and foundation size, an estimate of settlement, floor slab options, soakaways, CBR for access/car park design, concrete mix etc.

Table 7.10: Guidance on likely foundation options

	Fine-grained soil (cohesive and plastic)	Coarse grained soil (granular)	Rock	Made-ground or peat
Shallow foundation bearing capacity	Table 7.11	Table 7.12	Table 7.13	If <5m thick, try to found on material beneath, or alternatively pile
Minimum foundation depth below final ground level	1.0m ± may be deeper near trees or shallower if lower plasticity	0.6m	0.45m	Not applicable
Settlement estimate	Permissible bearing capacity assumes that 25mm of settlement will be acceptable (or see Section 7.5.2)	Figure 7.7	Depends on Young's modulus of rock and discontinuities in rock. Generally very low	Not applicable
Rate of settlement	Rate of consolidation settlement depends on grain size; is likely to take months and possibly years	Rapid — during construction	Rapid — not during construction	Not applicable
Floor slab	Groundbearing, except for soft or very soft deposits and any that might be affected by volume change e.g. trees, dewatering or gas venting	Groundbearing, except when on thick loose deposits, or suspend for gas venting	Groundbearing, except where void is required to vent gases, or on black mudstones due to pyrite	Suspended
Soakaway viability	Unlikely to work	Good option	Flow is likely to be in joints in the rock, which may cause flooding elsewhere. Use with caution	No — would add to leachate and spread contamination
CBR values for road design (%)	1–7, lowest for high plasticity clay, increases with sand content	20–60	>60	Too variable to estimate

A common mantra of geotechnical engineers is 'plan for the worst, and hope for something average'. This thinking should be put into practice by taking the lower bound values in the absence of any more definitive information. Caution at this stage may result in a conservatively designed foundation, but clients are generally happy if costs go down during construction, but not when they go up. This has been applied to the advice given for this concept design stage.

BS 8004 has been substantially rewritten between the 1986 and 2015 versions, largely to fit with Eurocode 7 (BS EN 1997)[46]. Typical presumed bearing values were given by BS 8004:1986[47] for the commonly encountered strata in the UK.

One-minute method for any ground conditions
Step 1 — Decide if the bearing stratum is cohesive, granular or rock
Step 2 — Select the appropriate strength or density from Tables 7.11–7.13, and note the presumed bearing capacity

Table 7.11: One-minute method to determine bearing capacity on cohesive soils

Cohesive soil strength description	Presumed allowable bearing capacity (kN/m^2)
Very stiff	300–600
Stiff	150–300
Firm	75–150
Soft	Not advisable without more data

Note: Assumes that load is within middle third of foundation and static.

Table 7.12: One-minute method to determine bearing capacity in granular soils for three foundation widths at a minimum depth of 0.75m

Granular soil density description	Presumed allowable bearing capacity (kN/m^2)		
	1m wide	2m wide	4m wide
Very dense	800	600	500
Dense	500–800	400–600	300–500
Medium dense	150–500	100–400	100–300
Loose and $N = 5$–10	50–150	50–100	50–100
Loose $N < 5$	Not advisable without more data		

Notes:
These figures should be halved if groundwater might be present within a depth equal to the width of the foundation below the foundation.
Derived/adapted from *Foundation design and construction*, M.J. Tomlinson and R. Boorman, 7th ed, 2001. Reprinted by permission of Pearson Business[48].

20-minute method on cohesive soils
If any strength data is available from prior testing, take an average value for a depth of twice the foundation width below the foundation. This is the depth that is assumed to be stressed by the applied load. If standard penetration test N values are available in cohesive soils, they can be converted approximately to undrained shear strength, simply by multiplying by 4.2 (Stroud[44]). The allowable bearing capacity q_{all} can be estimated with the following equation:

$$q_{all} = c_u \times N_c/3$$

Where:
N_c = 5.2 for a pad foundation and 6.1 for a strip foundation

Table 7.13: Presumed bearing capacity on intact unweathered rock selected (based on BS 8004:1986)

Rock			Presumed allowable bearing capacity (kN/m^2)
Strong igneous rocks			10,000 ($>$concrete strength)
Strong limestones and sandstones			4000
Schists and slates			3000
Strong shales, strong mudstones and strong siltstones			2000
Mercia Mudstone	Zone 1	Mudstone	750–1000
	Zone 2	Angular blocks of unweathered marl with almost no matrix	500–750
	Zone 3	Matrix with frequent lithorelics up to 25mm	250–500
Chalk	Grade I	Blocky hard, moderately weak joints, more than 200mm apart and closed	1000–1500
	Grade II	Blocky, medium hard joints, more than 200mm apart and closed	750–1000
	Grade III	Rubbly, unweathered joints, 60–200mm apart and closed	500–750
	Grade IV	Rubbly, partly weathered joints, 10–60mm open to 20mm and infilled soft chalk	250–500
	Grade V	Structureless, remolded, containing intact lumps	125–250

Note: BS 2004:1986 also includes guidance on presumed bearing values of various specific rock types, including chalk (Table 2), Mercia Mudstone (Table 3) and jointed rock (Figure 1 and Table 4). This is not included in the more recent editions.

Ten-minute method on granular soils

If any standard penetration test N values are available, take the average value for a depth of twice the width of the foundation below the foundation. Factor this value by 10 to obtain a presumed bearing capacity (the factor should be halved if the soil is below the water table). It would be cautious to limit this value, so that it does not exceed the lower bound of any softer bands above. This simplification assumes that the resultant of a column load is within the middle third of the foundation, so eccentricity is limited.

At the early stages of projects, this is often the best available estimate of the presumed bearing value. Further investigation is essential to confirm the ground conditions, strength and compressibility.

7.5.2 Settlement for shallow foundations
7.5.2.1 Cohesive soils
Five-minute method

Typically in the UK, presumed bearing value is based on the load which will result in 25mm of settlement, but this may not be acceptable to the structural engineer. Often, it is differential settlement which is more damaging to the structure, but it is unlikely that there will be sufficient information at concept design stage to establish the degree of variation across the site. The detailed bespoke ground investigation should be designed to address this.

Ten-minute method

Stroud established a relationship of m_v modulus of volume compressibility to N value, which for most clays is a factor of 0.43 to obtain m_v in m^2/MN. If SPT N values are available in cohesive deposits, it is possible to estimate consolidation settlement. A simple approximation for this is:

Settlement $= m_v \times 0.55q_n \times 1.5B$

Where

B = breadth of foundation
q_n = net foundation pressure

Modulus of compressibility values can also be obtained from typical values published by Tomlinson (Table 7.14 in this guidance).

Table 7.14: Modulus of compressibility for various soils

Type	Qualitative description	m_v (m²/MN)
Heavily over-consolidated boulder clays and stiff weathered rock, hard London clay, Gault clay and Oxford clay	Very low compressibility	<0.05
Boulder clays and very stiff London clay, Oxford clay and Mercia Mudstone	Low compressibility	0.05–0.10
Weathered London clay, Oxford clay, Mercia Mudstone and normally-consolidated clays at depth	Medium compressibility	0.10–0.30
Normally-consolidated alluvial clays	High compressibility	0.30–1.5
Very organic alluvial clays and peats	Very high compressibility	>1.5

Note: Derived/adapted from *Foundation design and construction*, M.J. Tomlinson and R. Boorman, 7th ed, 2001. Reprinted by permission of Pearson Business.

7.5.2.2 Granular soils

Burland, Broms and de Mello (Figure 7.7)[49] gathered measurements of settlements for pad and raft foundations, and plotted this against the density of the granular soils and applied load. This can be used to estimate foundation size, to support the known load with acceptable settlement if the density of the granular soil is known.

Figure 7.7: Observed correlation of settlement/load to breadth of foundation for changing densities of granular soils

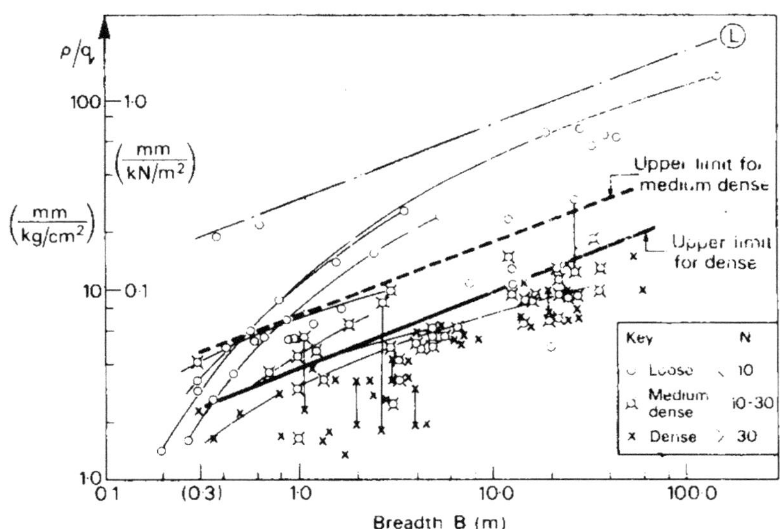

Generally, in the UK soil strength increases with depth, but it is always wise to check that there are no weaker layers at depth that could result in settlement or collapse.

There are three main categories to consider:

- **Weaker layers below the bearing stratum**
  There are commonly peat layers and lenses within alluvial deposits which can result in very large and often laterally variable settlement. Another common problem is alluvial clay which often has a desiccated crust, which may have sufficient strength to support residential housing, but the underlying very soft clay may result in bearing failure or excessive settlement
- **Man-made voids which might result in collapse**
  The most common of these is coal mine workings, but checks should also be made of other extractive processes e.g. the pumping of salt brine, workings of precious metals and flints. These workings should have been identified in the desk study stage from the BGS website or other sources.
- **Natural solution features**
  The presence of soluble rocks e.g. chalk, limestones, gypsum etc. should instigate a check of local stability. The BGS website is again a useful source of information, and local residents are generally aware of such problems

7.5.3 Shallow foundation construction
Actual foundation dimensions are usually determined by the excavator bucket sizes, which are generally 300/450/ 600/750/900mm. For mass concrete foundations, the thickness cf the concrete should be sufficient that a 45° line drawn from the applied load should intercept the side of the foundation and not the base so, approximately, the thickness is at least half the width.

7.5.4 Deep foundations (piles)
Pile design, based on limited information, is more of a challenge than shallow foundation design. Good information may be gained from local piling contractors, engineers of buildings in the vicinity of the site and Building Control.

The three main founding materials are rock and cohesive and granular soils. Assuming that at concept design stage the information available is limited to SPT N values and strength, either from tests or based on descriptions, the available design methods are simple. These assume that the piles have a minimum centre-to-centre spacing of three pile diameters.

The difficulty with pile design in the UK is that all of the land has been affected, to some extent, by glaciation, therefore the engineer has to deal with a mixture of granular and cohesive deposits, and deeply eroded river courses. At concept design stage, it can be necessary to calculate the shaft resistance and base resistance for both granular and cohesive soils (Table 7.15), and use the worst for each e.g. in the Bracklesham Beds in Southampton, which are interbedded sands and clays. Most areas of the UK are much simpler, being on stiff clays or rock. In areas on stiff clays, be careful to ignore any overlying strata, which will not be contributing to the skin friction.

Table 7.15: Pile resistance

	Cohesive soil	Granular soil	Rock
Shaft resistance	$Q_s = \alpha \times \bar{c}_u \times A_s$	$Q_s = K_s \times p_s \times \tan \delta \times A_s$	Assume zero
Base resistance	$Q_b = 9 \times c_{ub} \times A_b$	$Q_b = A_b \times p_d \times (N_q - 1)$	$q_{ub} = 2N_\phi \times q_{uc} \times A_b$

Allowable load $Q_a = (Q_b + Q_s)/2.5$

A factor of safety of 2.5 is used, as it is assumed that cautious values have been adopted.

Where α is normally taken as 0.45

A_s = area of shaft
A_b = area of base of pile
c_u = average undrained shear strength in cohesive soil
c_{ub} = undrained shear strength at toe of pile
q_{uc} = uniaxial compressive strength of rock
N_f = $\tan^2 (45 + \phi)/2$

p_s = effective overburden on side if pile
p_d = effective overburden at pile toe = $\gamma \times$ depth-water pressure at toe
N_q = Berezantsev's bearing capacity factor (Figure 7.8)
δ = angle of wall friction (Table 7.16)
K_s = earth pressure coefficient (Table 7.16)

Figure 7.8: Berezantsev's bearing capacity factor N_q

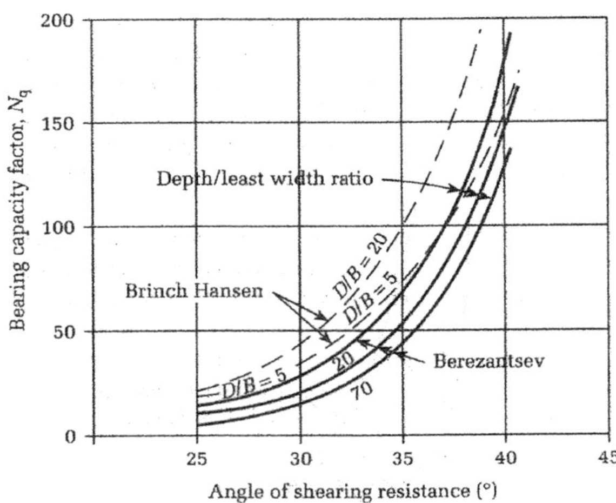

Note: Derived/adapted from *Foundation design and construction*, M.J. Tomlinson and R. Boorman, 7th ed, 2001. Reprinted by permission of Pearson Business.

Table 7.16: Angle of wall friction and earth pressure coefficients

Pile material	δ	K_s	
		Low density	**High density**
Steel	20°	0.5	1.0
Concrete	$\frac{3}{4}\phi$	1.0	2.0
Timber	$\frac{2}{3}\phi$	1.5	4.0

Note: Derived/adapted from Broms[50].

Displacement is required to generate skin friction and, as settlement on rock is likely to be minimal, skin friction in the overlying material should be ignored at this stage. Capacity should be limited to an acceptable stress in the pile concrete.

Negative skin friction (NSF)

Always consider whether there is a potential source of NSF e.g. a proposal to raise the level of the site on top of soft compressible clay, or peat layers which will decay over time, so the material above will settle, pulling the piles downwards.

NSF can be calculated by either the α or β methods.

The α method is the same as calculating the skin friction to support the pile, but assuming the highest likely strength to give the worst NSF.

Alternatively, the β method based on the effective stress on the sides of the pile as follows:

$$\text{NSF} = \beta \cdot \bar{\sigma}'_v \cdot A_s$$

Where β is often taken as 0.25.

These can differ significantly, so it is advisable to calculate both and take the worst, at this stage in design.

Be careful when choosing safe design values for this e.g. taking the upper bound of the strength range will give you the highest potential skin friction, and do not reduce by applying a factor of safety.

Tension piles
Tension piles are surprisingly common — under stability cores, holding down basements that try to float, or winged structures that try to fly. The skin friction element of the pile design can be used to calculate the potential capacity of tension piles. Be sure to check that none of this material is going to be removed in construction, in case the load on which you are relying isn't actually there at all.

Example design for foundations at the St Phillips Marsh site:

Starting from the ground surface and working downwards.

Made-ground *is unlikely to provide a suitable bearing stratum, and is likely to require a suspended floor slab and ground remediation.*

Soft clay *has a reasonable thickness in Borehole 1 of 2.7m and is encountered at 1.45m depth, so shallow strip foundations could be formed at this depth. Based on the simple calculation in Section 7.5.1, an allowable line load on a 1m wide strip foundation would be 46kN/m run, or alternatively a column load on a 1m square pad of 40kN. Risks associated with this choice, would be:*

- *Excessive settlement in the soft clay beneath, which would be expected to exceed 25mm. Undisturbed samples and oedometer tests would be required to quantify*
- *Excavation below the water table. The made-ground is described as a loose, granular material, so permeability can be expected to be at least as high as sand. Dewatering will require disposal of potentially contaminated water. Excavations are likely to be unstable and require full shoring*
- *Excavation of made-ground will result in material either to be reused on-site or taken for disposal which is likely to be costly*
- *The thickness of the made-ground and soft clay varies across the site, so while this might be an option at Borehole 1, it is not at Borehole 3. This will require closely spaced investigation, to define the suitable area*
- *Foundation size limited, so does not stress weaker layer beneath*

Very soft clay *is present between depths of about 4–10m. It is too soft for shallow foundations and would not provide sufficient support for piles.*

Medium dense sand and gravel *was 3.5m thick in Borehole 1 but only 1.2m thick in Borehole 3. In Borehole 3 there would be a risk of piles punching through the gravels into the marl beneath, which may be weathered Mercia Mudstone.*

Piles of 0.6m diameter with a capacity of 750kN is a possible foundation solution in the area of borehole, but detailed investigation would be required to confirm the extent of thick gravels (Figure 7.9).

Mercia Mudstone *is consistently found between 13–14m depth on this site. The only testing we have is an aborted SPT with 83 blows for only 0.2m penetration. The descriptions do not comply with BS EN ISO 14689 or with the weathering grades for Mercia Mudstone.*

Treating the Mercia Mudstone as an extremely weak rock gives an allowable pile capacity of 593kN. Pile diameter or length can be increased to provide the capacity required. This is probably an overly safe design, but there are less risks with this foundation option, therefore it is the recommended solution for this site (Figure 7.10).

7.5.5 Retaining wall design
The key design decisions are:

- Is the wall able to move?
- What water pressures may occur?

Figure 7.9: Medium dense sand and gravel pile design

$$Q_b = A_b \times P_d (N_q - 1)$$
$$\phi = 38°$$
$$\gamma_b = 18 \text{kN/m}^3$$

Take depth 12m, pile ϕ = 0.6 m.

$$A_b = 0.28 \text{ m}^2$$
$$P_d = (12 - 1.45) \times 18 - (12 - 0.9) \times 10.$$
$$= 189.9 - 111.$$
$$= 78.9 \text{ kN/m}^2.$$
$$N_q = 35 \qquad D/B = 12/0.6 = 20.$$

$$Q_b = 0.28 \times 78.9 (35 - 1)$$
$$= 751 \text{ kN.}$$

Figure 7.10: Mercia Mudstone pile design

Using c_u = 4.2 N and N=83
$$= 348 \text{ kN/m}^2 \text{ minimum}$$

Extremely weak UCS 0.6 – 1.00 MPa
$$c_u \quad 300 - 500 \text{ kPa}$$

Taking c_u = 350 kPa, although likely to be much higher.

Pile 3m into Mercia mudstone + 0.6m ϕ.

$$Q_a = \alpha \times \bar{c_u} \times A_s + 9 \times c_{ub} \times A_b.$$
$$= \frac{0.45 \times 350 \times 3 \times \pi \times 0.6 + 9 \times 350 \times \pi \times 0.3^2}{3}$$
$$= \frac{890}{3} + 890$$
$$= 593 \text{ kN.}$$

It is common to assume K_a conditions apply, but if the basement is a rigid box, then it is unable to move sufficiently for earth pressures to fall to K_a conditions, and it must be designed for at rest (K_0) earth pressures. Also consider 'short-term' and 'permanent case' during construction, as retaining walls are commonly cantilevered during construction, but propped by the floor slab in the final construction, so the earth pressures will change.

$K_0 = 1 - \sin\phi$ where ϕ can be established from earlier relations

$K_a = 1 - \sin\phi/(1 + \sin\phi)$

CIRIA C760[51] gives guidance on water pressure on basements.

How the current water levels might change must be considered, particularly with changes in water management and climate change. An example of how this can adversely affect design is the reduction of extraction of groundwater from the chalk underlying London for drinking water, which has resulted in a gradual rise in groundwater levels, flooding many basements.

For preliminary designs:

- Gravity retaining walls typically have a minimum base width of 66% of the wall height
- Embedded cantilever retaining walls have a ratio of $\frac{1}{3}$ cantilever with $\frac{2}{3}$ embedded
- Useful guidance is provided in:
    - BS 8002[52]
    - BS 8081[53]
    - CIRIA Report C760 (supersedes C580)

7.5.6 Dewatering

Dewatering is generally an expensive undertaking in granular soils or contaminated land. Permission may be sought from the local authority to discharge to the drainage system, or from the Environment Agency to discharge to a watercourse, but these will not automatically be permitted. For contaminated sites, any water will need to be tankered from the site for costly disposal. Guidance on dewatering methods is provided in CIRIA C750[54].

For the example site at St Phillips Marsh, it would be strongly recommended that a basement is not considered, due to the difficulty of dewatering and dealing with the contamination. On many sites this advice is ignored, due to the value of the space outweighing the costs.

7.5.7 Infiltration drainage

Disposal of surface water arising from hard surfaces as part of a scheme has to be considered under 'sustainable urban drainage systems' (SuDS).

Guidance is provided for soakaways in BRE Digest 365[55] for the undertaking of tests to establish if the ground is suitable and for its design. Beware of installing a soakaway on a site which could adversely impact on neighbouring sites, especially on slopes. CIRIA C753[56] takes a much wider view of surface water runoff and its management in terms of water quality, biodiversity and alternative solutions e.g. retention ponds etc.

7.5.8 Pavement design

Guidance on pavement design is provided in the *Design manual for roads and bridges*[57], using the CBR values that have been estimated. This will provide an overly conservative design for a car park.

If plasticity index data is available, the CBR value in clay can be refined using Table 7.17.

Table 7.17: Correlation of CBR value to plasticity index and water level

Plasticity index (%)	Description	CBR	
		Water >600mm below formation	Water within 600mm of formation
70	Clay	2	1
60	Clay	2	1.5
50	Clay	2.5	2
40	Clay	3	2
30	Silty clay	5	3
20	Sandy clay	6	4
10	Sandy clay	7	5

7.5.9 Gas protection measures for radon, methane

BS 8485[58] gives very detailed guidance on the protection measures.

UKRadon[59] provides useful guidance for establishing the levels of radon in a particular area, and what measures should be incorporated into building design.

The site that we have considered at St Phillips Marsh would need to include some of these measures due to the nature of the made-ground which could be expected to release methane.

7.5.10 Shrinkage/swelling

The NHBC Standards[60] give straightforward guidance for establishing the effects of trees on foundations, and how to avoid them. At concept design stage, if the trees have been identified during the site visit and the previous investigations have identified a soil type, it is possible to establish a sensible foundation depth. If the plasticity of the clay is unknown, it is sensible to assume that it is high.

BS 5837[61] gives helpful advice on building close to trees, and necessary precautions to protect the tree, building and subsurface utilities.

7.5.11 Contamination

It is unlikely that at concept design stage there will be sufficient information available to assess the contamination risk, beyond giving a likelihood for the requirement of a remediation contingency.

A code of practice for the investigation of potentially contaminated sites is given in BS 10175[62]. Once test results are available, they can be processed using the Environment Agency's *Contaminated land exposure assessment* (CLEA) tool[63], which comprises a handbook and software to help assess the risks of contaminated land exposure to human health.

Jon Carr
University of Sheffield and Jon Carr Structural Design

8 Developing a concept

8.1 The six key decisions you need to make simultaneously

Developing a structural design concept is similar in many ways to solving a set of simultaneous equations, except that there is often more than one solution which meets the brief.

We will look at the following six key items to be considered, and decisions to be made, when developing a conceptual design:

1. Influence of ground conditions.
2. Material selection.
3. Structural system — loadbearing, framed or hybrid.
4. Grids and structural layouts.
5. Spans of floor and roof structures.
6. On- or off-site/prefabricated construction.

There are also many other valid considerations e.g. the integration of structure and building services.

While each item is presented separately, these should all be considered at the same time, ideally. This is something that becomes easier with experience, and almost happens subconsciously after several years of practise. Concept design often tends to be an iterative process.

It is also important to note that making one decision, or selecting one parameter e.g. which material to use, will usually have an impact on the choices available in relation to other parameters. For example, if it was decided, possibly for reasons of sustainability, that a timber floor system was to be used in a particular building, this would have implications on the spacing of the supporting walls, beams and columns. Similarly, if loadbearing timber or masonry walls were adopted for the building, this would place limitations on the overall height of the building.

While not listed here as a 'key decision', it is also essential to think about stability and robustness when developing your concept design (Chapter 9).

8.1.1 Decision 1: Influence of ground conditions

It is easy to fall into the trap of carrying out the concept design of the superstructure of a building with little, if any, consideration of the ground conditions. The foundation design is often carried out after that of the superstructure, and is often a *'fait accompli'* i.e. the foundations will be whatever they need to be in order to resist the applied loads. This could however lead to significant issues later in the project, potentially resulting in foundation solutions being adopted which are both impractical and uneconomical. It is good practice to get an understanding of the likely ground conditions as early as possible, by carrying out a desk study (Chapter 7). Indeed, the ground conditions may have significant implications for the superstructure design.

A good example is Walsall College in the West Midlands, which is built over old limestone mine workings, which occur at a depth of approximately 40m local to the proposed building. Figure 8.1 is a plan showing the mine workings (note that the areas shaded in grey are the voids).

Figure 8.1: Mine abandonment plan local to Walsall College

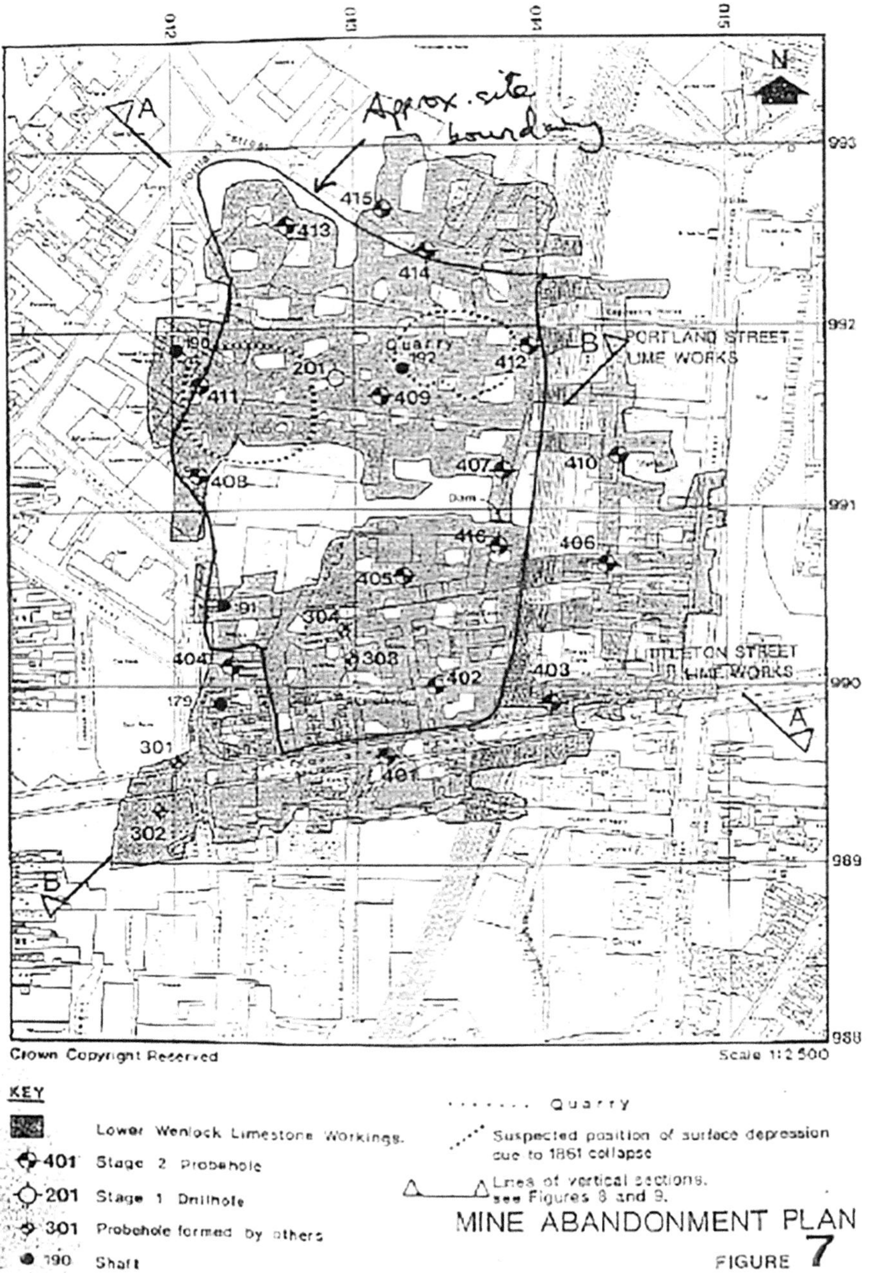

The local authority stipulated that the total dead and imposed load of the building should not exceed 40kN/m^2 at ground level. Given that the proposed building was five storeys high in some areas, and that the imposed floor load was typically either 4.0kN/m^2 or 5.0kN/m^2, this limited the floor dead load to approximately 3.0kN/m^2 at each level. The design team quickly concluded that a steel frame was the only viable solution, which not only complied with this stringent loading requirement, but also provided the flexible space requested in the client's brief. The completed project, which cost £40m, opened in 2009 (Figure 8.2).

Figure 8.2: Walsall College

8.1.2 Decision 2: Material selection
There are four commonly used structural materials to choose from when designing buildings in the UK:

- Steel
- Concrete
- Timber
- Masonry

In other countries, some of these materials may not be as readily available e.g. the steelwork industry in many countries is not as well developed, while other materials, such as lightweight steel framing, and even bamboo and earth may be used, particularly for domestic construction.

In many cases, a hybrid material solution will be appropriate e.g. steel frame buildings often use concrete shear walls or cores, while loadbearing masonry buildings commonly use timber floor construction.

Material selection depends on a wide range of considerations, typically:

- Building height
- Internal layout — open plan or cellular
- Structural form/spans — usually related to internal layout considerations
- Ground conditions and any limitations on the location of foundations e.g. adjacent buildings, underground services or tunnels
- New build, refurbishment or extension e.g. the additional loads are likely to be limited for a refurbishment or extension, whereas the floor-to-floor heights will also be dictated by the need to tie into existing floor levels
- Magnitude and nature of applied loads
- Sustainability/embodied carbon
- Cost, availability and lead-in time
- Fire resistance
- Acoustic properties
- Building physics e.g. lighting, ventilation etc.
- Distribution of building services
- Robustness/disproportionate collapse
- Aesthetics
- Design life, durability and maintenance requirements
- Preconceived ideas/previous experience

Table 8.1 shows indicative values for each of the four most common building materials.

Similarly, each material has its advantages and disadvantages, some of which are listed in Table 8.2.

Table 8.1: Key structural properties of common structural materials

Material	Young's modulus (N/mm^2)	Characteristic compressive strength (N/mm^2)	Characteristic bending strength (N/mm^2)	Characteristic tensile strength (N/mm^2)	Force density (kN/m^3)
Steel	$210,000N/mm^2$	Typically 275 or $355N/mm^2$ (but less if flexural buckling occurs)	Typically 275 or $355N/mm^2$ (but less if lateral torsional buckling occurs)	Typically 275 or $355N/mm^2$	$78kN/m^3$
Reinforced concrete	$15,000N/mm^2$ (long-term)–$30,000N/mm^2$ (short-term)	Typically 12–$90N/mm^2$ (but usually use 30–60 N/mm^2 in UK)	Dependent on amount of steel reinforcement provided	Anecdotally is equal to approximately 10% of its compressive strength, but generally assumed to be zero in design	$24kN/m^3$ (for dry mass concrete). Add $1kN/m^3$ for reinforcement plus $1kN/m^3$ for wet concrete, as appropriate
Timber (softwoods)	Varies from 8,000 N/mm^2 (mean value for Class C16 timber) to $13,700N/mm^2$ (mean value for Class GL32 glulam)	Varies from $17N/mm^2$ (for Class C16 timber, parallel to grain) to $29N/mm^2$ (for Class GL32 glulam, parallel to grain)	Varies from $16N/mm^2$ (for Class C16 timber, parallel to grain) to $32N/mm^2$ (for Class GL32 glulam, parallel to grain)	Varies from $10N/mm^2$ (for Class C16 timber, parallel to grain) to $22.5N/mm^2$ (for Class GL32 glulam, parallel to grain)	Varies from $3.7kN/m^3$ (Class C16 timber) to $4.75kN/m^3$ for Class GL32 glulam
Masonry	Not applicable since masonry is primarily designed as compression only, and the strength characteristics are based on the masonry/mortar bond	1-$10N/mm^2$ based on Table 2 of BS 5628-1 (the equivalent ULS values can be found in BS EN 1996-1-1)	0.1–$0.2N/mm^2$ based on Table NA.6 from NA to BS EN 1996-1-1	Assumed to be zero in design	$6kN/m^3$ (aerated blockwork) to $22kN/m^3$ (brick/stone)

Note: Derived/adapted from BS EN 1992-1-1, BS EN 1993-1-1, BS EN 1995-1-1, BS EN 1996-1-1[64–67] and the corresponding National Annexes[68–71], BS 5268-2[72], BS 5268-1[73] and IStructE manuals[74–77].

Table 8.2: Advantages and disadvantages of common structural materials

Steel	
Advantages	**Potential disadvantages**
High rate of recycling	Low inherent fire resistance — applied fire protection is usually required
Consistent material properties/isotropic	Corrosion if not suitably protected/detailed (depending on exposure/environment)
Good strength-to-weight ratio	Members/elements in compression prone to overall and local buckling effects
Speed of erection	Long lead-in time (typically 12–20 weeks)
Long-span floors achievable	Typically requires a hybrid system e.g. incorporating concrete floors, foundations and potentially shear walls/cores
Connections can transfer very high forces between elements	Stability during construction e.g. often need to provide temporary bracing
Dimensionally stable (not affected by creep or moisture effects, notwithstanding corrosion)	Vibration governs very long spans ($\geq$10m)
Reinforced concrete	
Advantages	**Potential disadvantages**
Very good inherent fire resistance – applied fire protection not normally required	Variable material properties/anisotropic
Flexibility of form	Corrosion if inadequate detailing/construction (depending on exposure/environment)
High thermal mass can be used to reduce, or even obviate, the need for mechanical ventilation	Members in compression prone to overall buckling effects
In situ RC connections can transfer high loads between elements	Not perceived as environmentally friendly compared with other materials, although the use of recycled aggregates and cement replacement materials (e.g. PFA and GGBS) partly addresses this
Dimensionally stable (not affected by moisture effects, notwithstanding corrosion)	Strength-to-weight ratio not as good as steel or timber
Medium- to long-span floors achievable	RC frames heavier than steel or timber frames
Relatively short lead-in time for *in situ* concrete (typically 4–8 weeks)	Speed of erection/need for propping for *in situ* concrete (precast is quicker, although lead-in times are longer)
Strong visual identity	Low tensile and bending resistance if unreinforced
Good acoustic properties due to high mass	Long-term creep effects needs to be considered

Table 8.2: *Continued*

Timber	
Advantages	**Potential disadvantages**
Natural and renewable	Variable material properties/anisotropic
Lightweight	Fire resistance during construction
Good strength-to-weight ratio when loaded along its grain	Movement/creep effects due to shrinkage and moisture (depending on type of timber product specified and moisture content)
Flexible (can be easily cut to length in timber yard or on-site)	Challenging to satisfy robustness/disproportionate collapse requirements for tall buildings i.e. Class 2B upwards
Short- to medium-span floors achievable	Can only transfer modest loads between members
Short lead-in time for sawn timber	Sawn timber products slow to construct on-site
Engineered and prefabricated timber products e.g. CLT, SIPs, are quick to construct on-site	Long lead-in time for engineered and prefabricated timber products e.g. CLT, SIPs
	Vibration governs long spans ($\geq$6m)

Masonry	
Advantages	**Potential disadvantages**
Very good in compression	Variable material properties/anisotropic
Good inherent fire resistance	Thermal expansion and contraction effects e.g. cracking if inadequate allowance for thermal movement
Low maintenance	Poor in flexure/bending
Appearance is generally popular with the public	Not perceived as environmentally friendly compared with other materials, due to use of natural materials (clay, aggregate etc.) and high energy required during production
	Labour intensive to construct
	Availability/cost of suitably skilled labourers
	Challenging to satisfy robustness/disproportionate collapse requirements for Class 2A buildings, and especially challenging for Class 2B upwards
	Can only transfer modest forces between members

One way of quickly shortlisting potential structural materials is to consider building height (Table 8.3).

Steel and concrete are usually suitable for any height of building, although they tend not to be used for low-rise housing with a cellular layout.

Timber is being used for increasingly taller buildings e.g. the ten-storey Dalston Works building in Hackney, London (at the time of construction, the world's largest CLT building)[78] and the 18-storey TallWood House building in Vancouver (at the time of publication, the world's tallest mass timber building)[79], with a concept stage proposal to build as tall as 80-storeys. The vast majority of timber framed buildings in the UK are currently four storeys or less.

Loadbearing masonry buildings up to three storeys are still commonly used in the UK residential sector. While it is possible to build higher in loadbearing masonry, this is now relatively rare in the UK.

For timber and masonry buildings higher than four storeys, it can be very challenging to satisfy the robustness/disproportionate collapse requirements outlined in Building Regulations Approved Document A[80]. For further details, refer to Chapter 9.

Table 8.3: Building height limitations for common structural materials

	One or two storey buildings	Buildings up to four storeys	Buildings with 5–18 storeys	Buildings over 18 storeys
Steel	Y e.g. portals	Y	Y	Y
Concrete	Uncommon	Y	Y	Y
Timber	Y	Y	Y	N
Masonry	Y	Potentially	N	N

8.1.3 Decision 3: Structural system

We essentially have three fundamental options when it comes to the structural system — the system which transfers all the loads down to foundation level:

- Framed
- Loadbearing
- Hybrid of the two

As discussed, the height of the building may well determine which structural system is to be used, with loadbearing systems not currently suitable for buildings over ten storeys (and rarely used for many buildings taller than four storeys, although the introduction of solid wood solutions, such as CLT, has provided structural engineers with more options). For buildings up to ten storeys, the proposed use of the spaces and nature of the architectural floor plans (open plan or cellular) should inform the structural system.

Framed structures

An open-plan floor layout is one with no internal walls to be used as permanent loadbearing elements. It relies on a framed-type structure (beam and column or flat slab), to transfer gravity loads down to the building's foundations (Figures 8.3; a naturally ventilated shallow plan building and 8.4, a mechanically ventilated deep plan building).

Figure 8.3: Indicative column layout/grid arrangement for a shallow-plan/open-plan steel framed office building

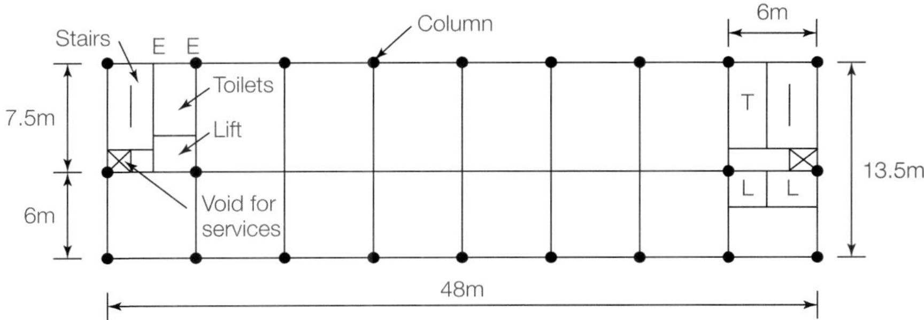

Figure 8.4: Indicative column layout/grid arrangement for a deep-plan/open-plan steel framed office building with a central atrium

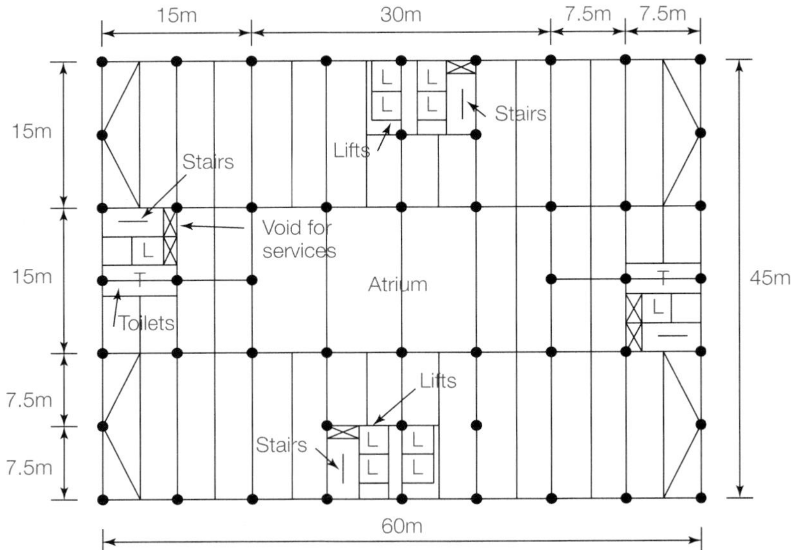

The frames shown in Figs 8.3 and 8.4 could potentially be made from steel, concrete or timber (depending on the height of the building). This form of construction is typically suited to commercial buildings, such as offices, retail and mixed-use buildings, but could also be used for hotels, student accommodation and other types of residential building. There are various ways in which framed structures can resist lateral loads, as discussed in Section 8.4 and Chapter 9.

Loadbearing structures

A cellular layout is one with a sufficient number of suitably located internal loadbearing walls to support the gravity loads from the floors and roof (Figure 8.5). These walls could be in masonry, reinforced concrete, timber or even

Figure 8.5: Typical floor layout for Stadthaus, London, showing CLT wall panel locations

Fifth floor plan (apartments for private owners)

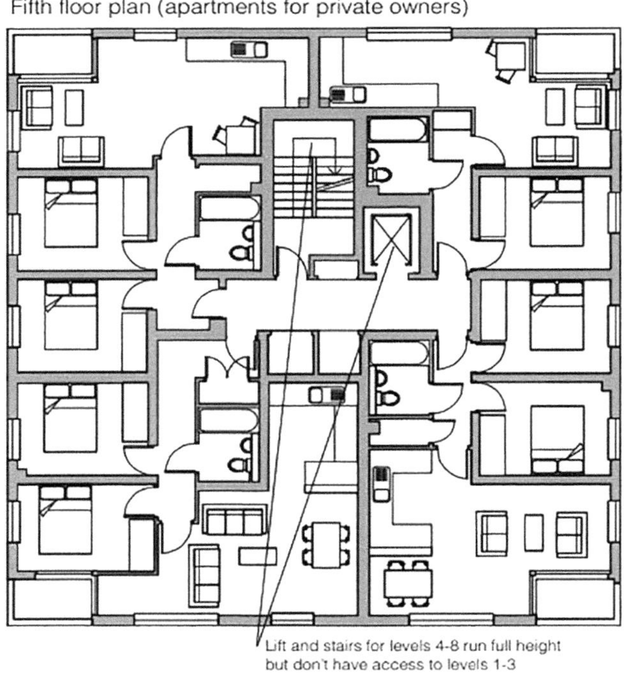

light gauge metal studs (Figures 8.6–8.9) and should ideally be ccntinuous for the full height of the building, to avoid the need for transfer structures/beams. Such walls will often be used to transfer lateral loads to the foundations, although other options are available (Section 8.4 and Chapter 9). This form of construction is probably best suited to residential buildings (apartment blocks), and potentially to hotels, student accommodation and prisons, for example.

Hybrid structures

In some cases e.g. apartment blocks, hotels and student accommodation, where a more open-plan area is required at ground floor than at upper floor levels for example, a hybrid solution could be adopted. The TallWood House student residence building in Vancouver is a good example of this — the bottom storey is a concrete frame with columns at relatively large centres, on top of which is a timber framed structure, with columns at closer centres (Figures 8.10–8.11).

Figure 8.6: Loadbearing masonry walls

Figure 8.7: 'Tunnel form' RC walls

Figure 8.8: CLT walls

Figure 8.9: Light gauge steel stud walls

Figures 8.10 and 8.11: Structural frame of TallWood House, Vancouver

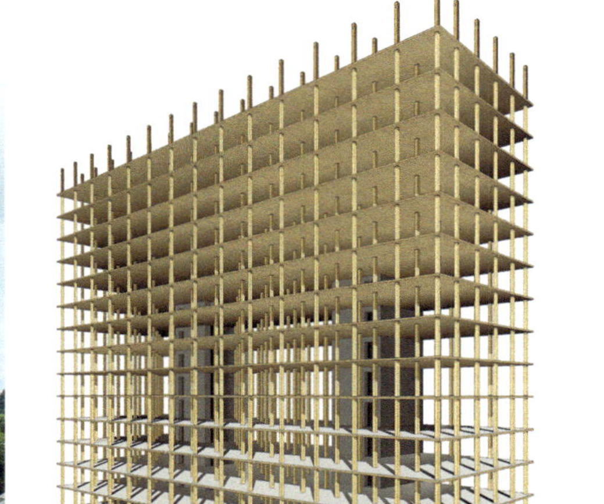

8.1.4 Decision 4: Grids and structural layouts

When we talk about grids, we are usually referring to where the floor and roof supports occur, and how far apart they are, and whether these supports take the form of columns or loadbearing walls. The gridlines shown on engineer's and architect's drawings should coincide with column and/or loadbearing wall locations (although changes made during the design development period can result in members being moved off-grid).

Grids should ideally be regular/uniform and perpendicular, to simplify fabrication/construction and maximise economies of scale etc. Gridlines should also be consistent over the full height of the building, in order to avoid the need for transfer structures.

There is often a trade-off between flexibility and economy when determining the structural grid. While placing supports as far apart as possible might seem like a good idea, as it maximises the future flexibility of a building, this comes at a cost. A practical and affordable compromise needs to be reached.

Grid dimensions generally depend on various factors, including:

- Proposed building use/gravity loads
- Proposed room layouts
- Future flexibility requirements
- Presence of car parking under building
- Spanning capability of proposed floor and roof structures
- Spanning capability of proposed cladding solution

Further guidance regarding optimum grid dimensions is given on:

- Offices
- Residential
- Car parks
- Transfer structures
- Optimum module sizes and construction materials

Offices

The British Council for Offices (BCO)'s *Guide to specification*[81] provides extensive guidance on the preparation of scheme designs. For naturally ventilated offices, a building width of 12–15m is typically used, which can be achieved by two spans of 6–7.5m — with a column placed adjacent to a central corridor, similar to the arrangement shown in Fig. 8.3. Natural lighting can also play an important role in determining the width of floor plate (Figure 8.12). This, and many other helpful rules of thumb relating to sustainable building design, can be found in *101 Rules of thumb for low energy architecture*[82] and *101 Rules of thumb for sustainable buildings and cities*[83].

Figure 8.12: Depth of daylight penetration

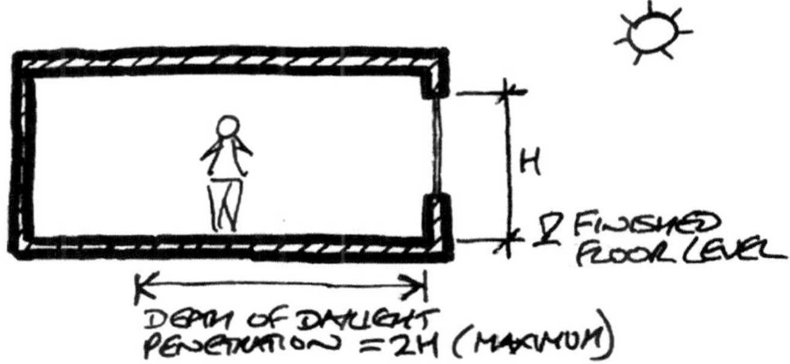

In larger buildings, a long-span solution provides a considerable enhancement in flexibility of layout. For air-conditioned offices, a clear span of 15–18m is often used. Fig. 8.4 gives an indicative column layout/grid arrangement for a mechanically ventilated steel framed office building with a central atrium.

Residential (including bedroom floors to hotels and student accommodation)

Residential buildings can vary significantly in their size and nature, which complicates giving advice on their associated structural grids/solutions. Large apartment blocks can be designed in a similar way to office blocks, as described, with correspondingly similar column centres/structural grids. Alternatively, the requirement for walls at relatively close centres (to separate rooms/apartments) can be utilised structurally, with loadbearing walls constructed from a range of materials (Figs. 8.6–8.9).

Fig. 8.5 shows a typical floor layout to a multi-storey residential building, in which CLT wall panels have been used. A similar floor layout could easily apply to structures utilising loadbearing masonry, reinforced concrete, or light gauge steel stud walls.

Car parks

The structural grid/column locations to multi-storey car parks, and any buildings containing a car park within their footprint e.g. in the basement will, as a minimum, need to comply with the layout shown in Figure 8.13. However, clear-span construction would be preferable, with columns at 7.5m i.e. 3 × 2.5m centres in one direction and 16.0m i.e. 2 × 5.0m + 6.0m in the other, which should accommodate larger vehicles e.g. SUVs.

Figure 8.13: Indicative column layout to multi-storey car parks

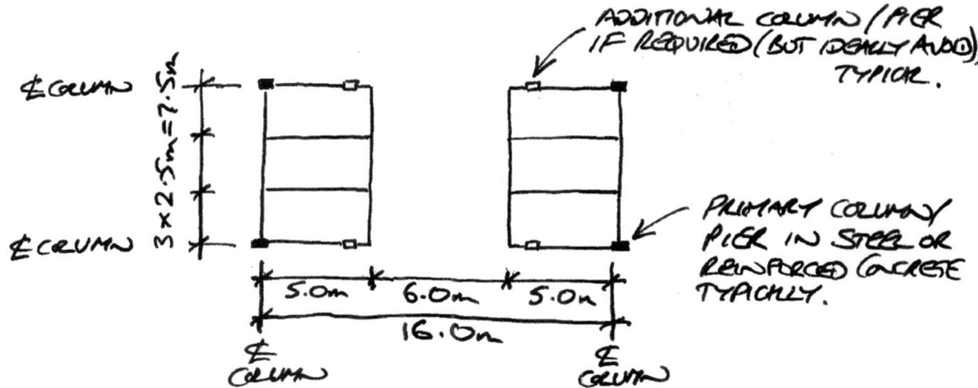

Transfer structures

A common solution is to adopt a transfer structure at ground floor level where an office, retail, residential or mixed-use building is constructed over a basement car park. As an alternative to using a 7.5m × 16.0m column grid at all levels (which would result in relatively deep beams spanning the 16.0m dimension, so relatively high floor-to-floor heights), a 7.5m × 8.0m column grid could be adopted at upper levels, with the additional columns picked up by transfer beams or trusses above the car park (Figure 8.14).

Another good example of a transfer structure is the use of a concrete podium slab at first floor level of the TallWood House student residence building (Figs. 8.10–8.11), in order to create an open plan space at ground floor level.

Figure 8.14: Transfer structure over a basement car park

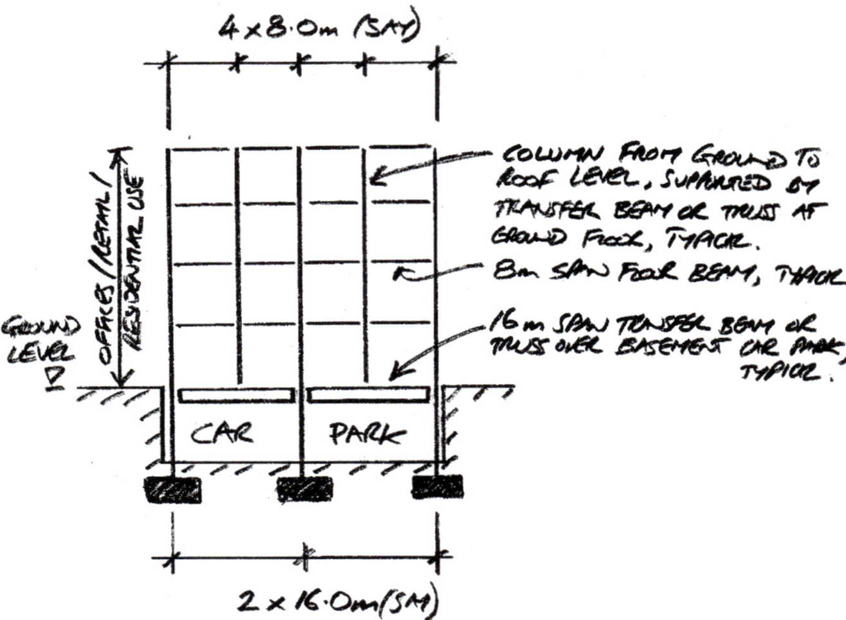

More elaborate transfer structures include raking columns at ground floor level (Figure 8.15), and systems which use the whole height and width of the building as a transfer structure e.g. when spanning over a road or underground railway lines (Figures 8.16–8.19).

Figure 8.15: Exposed concrete transfer columns in Spinningfields, Manchester

Figure 8.16: Deep truss at roof level spanning over road, Bristol

Figure 8.17: Exchange House, London — four large steel arches to span over train lines adjacent to Liverpool Street station

Figure 8.18: Cannon Place, London — a series of trusses span over a combination of train lines, archaeological remains and a scheduled ancient monument

Optimum module sizes for construction materials
The materials adopted may come in optimum module sizes, which should ideally inform the structural grid
e.g. hollow-core precast floor units typically come in 1.2m wide modules, while timber panels typically come
in multiples of 0.6m, 1.2m and 2.4m. Buildings with masonry cladding (loadbearing or non loadbearing)
should ideally have grids and floor-to-floor heights based on brickwork dimensions i.e. 75mm high × 225mm
long, allowing for 10mm mortar joints, and blockwork dimensions i.e. 225mm high × 450mm long, allowing
for 10mm mortar, as appropriate. Proprietary cladding and roofing systems may also come in standard
size panels.

8.1.5 Decision 5: Spans of floor and roof structures
The spans of floor and roof structures are closely related to the structural grid/supports described in Section
8.1.3. The first question to ask is: Could the floor span the distance between support (whether columns or
loadbearing walls) without using any intermediate beams? — which obviously depends on the type of floor
being considered. While the comments that follow do not explicitly refer to roof structures, they are also valid
for roofs.

Roof structures are generally subject to lower variable loads than floor structures, which provides the option
of using lightweight roof structures for many buildings. It is relatively common for concrete frame buildings to
have a lightweight steel roof structure, and for some (if not all) internal columns to be omitted at roof level
(Figure 8.19).

Figure 8.19: Lightweight roof structure to multi-storey building

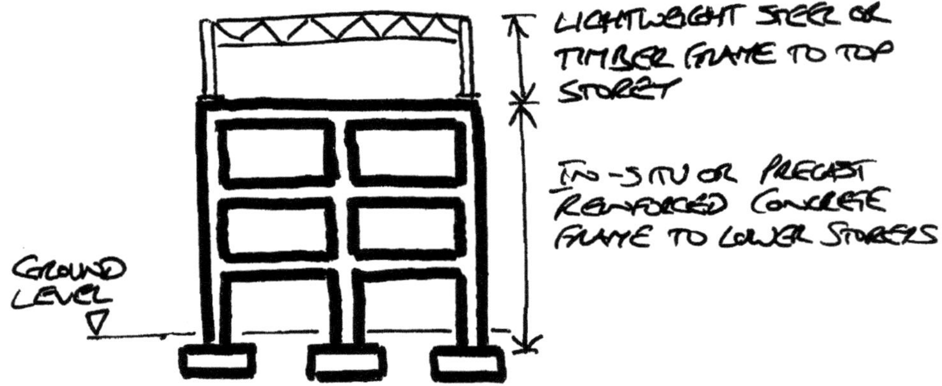

Floors
Table 8.4 provides guidance on suitable floor systems, depending on the imposed loads and the span. Where
N/A is shown this does not necessarily mean the solution isn't possible, rather that it is not a standard solution.
Note that this table is not exhaustive, and only gives an indication of how far typical floor systems used in the
UK can span.

In instances where Table 8.4 is not applicable, the span-to-depth ratios given in Table 8.5 can be used to determine
an initial size for the floor system. Further details can be obtained from the literature[3–6,74–77,84–89], from where the
values in Tables 8.4–8.6 have been derived/adapted.

It is interesting to note from Tables 8.4 and 8.5 the significant increase in allowable span between single and multiple
span floor systems e.g. by comparing the allowable spans for one-way spanning, *in situ* RC slabs. It is much more
structurally efficient to use slabs/floor systems with two or more spans wherever possible.

Table 8.4: Maximum spans for common types of floor structure (indicative values)

Floor system	Allowance for super-imposed permanent loads due to finishes, services etc. (unfactored)	Maximum clear span, depending on value of unfactored variable floor load					Relevant sources
		1.5kN/m^2	2.5kN/m^2	3.5kN/m^2	5.0kN/m^2	7.5kN/m^2	
220 × 75 Class C24 timber joists at 400mm c/c	0.25kN/m^2	5.55m	N/A	N/A	N/A	N/A	84
220 × 75 Class C24 timber joists at 400mm c/c	1.25kN/m^2	5.0m	N/A	N/A	N/A	N/A	84
100 deep CLT floor panels	1.0kN/m^2	3.8m	3.4m	3.1m	3.0m	2.7m	4
240 deep CLT floor panels	1.0kN/m^2	6.9m (governed by vibration, so a function of dead load and span)				6.4m	4
140mm deep composite metal deck slab (60 mins fire)	Need to consider allowance as part of variable floor load	4.3m	4.3m	4.3m	4.3m	4.15m	86–89
295mm deep composite metal deck slab (60 mins fire)	Need to consider allowance as part of variable floor load	6.0m	6.0m	6.0m	6.0m	5.95m	86–89
150mm deep beam and block floor	1.8kN/m^2	6.4m	6.0m	5.5m	5.1m	N/A	86–89
150mm deep non-composite hollow-core precast units (60 mins fire)	1.5kN/m^2	7.5m	7.5m	7.1m	6.7m	5.85m	86–89
200mm deep non-composite hollow-core precast units (120 mins fire)	1.5kN/m^2	10.0m	9.5m	8.55m	8.05m	7.1m	86–89
300mm deep non-composite hollow-core precast units (120 mins fire)	1.5kN/m^2	14.6m	13.7m	12.5m	11.9m	10.7m	86–89
200mm one-way spanning C30/37 *in situ* RC slab (single span)	1.5kN/m^2	≥6.0m	6.0m	5.7m	5.4m	5.0m	6
200mm one-way spanning *in situ* C30/37 slab (multiple span)	1.5kN/m^2	≥7.0m	7.0m	6.7m	6.4m	6.0m	6

Table 8.4: *Continued*

Floor system	Allowance for super-imposed permanent loads due to finishes, services etc. (unfactored)	Maximum clear span, depending on value of unfactored variable floor load					Relevant sources
		1.5kN/m^2	2.5kN/m^2	3.5kN/m^2	5.0kN/m^2	7.5kN/m^2	
300mm two-way spanning C30/37 *in situ* RC flat slab (multiple span)	1.5kN/m^2	9.2m+	9.2m	8.5m	8.2m	7.0m	6

Note: Values are all based on unpropped construction except for *in situ* RC slabs.

Table 8.5: Guidance on preliminary sizing of common types of floor structure (indicative values)

Floor type	Ideal span-to-depth ratio	Efficient span range	Relevant sources
Softwood timber office floor (softwood joists at 400mm c/c)	L/10–L/15	≤8m	3
One-way spanning RC slab (simply-supported)[a,b]	L/23–L27	4–10m	85
One-way spanning RC slab (multiple span)[a,b]	L/27–L/32	4–10m	85
One-way spanning RC slab (cantilever)[a,b]	L/6	≤4m	85
In situ RC flat slab (multiple span)[a,b]	L/23–L/28	4–10m	85
In situ PT flat slab (multiple span)[a,b]	L/30–L/40	6–13m	85

Notes:
[a] Span-to-depth ratios for RC/PT slabs are based on overall depth.
[b] Span-to-depth ratios for RC/PT slabs are based on imposed loads of 2.5–10.0kN/m^2.
Actual span to depth ratios may significantly differ from the ideal values quoted here, depending on the value of the applied loads, for example.

Beams

The type of floor to be used will inform whether or not any floor beams are required. Floor (and roof) beams will be required in almost all framed structures, the only exception to this being *in situ* RC flat slab structures. Even loadbearing structures may require some beams locally e.g. where more open-plan areas are required, such as at ground floor level.

Due to the significant number of potential variations, including load, beam span and centres, support and propping conditions, it is not practical to produce a version of Table 8.4 for beams. Table 8.6 provides indicative span to depth ratios for a range of timber, steel and concrete beams which are commonly used in the UK. More detailed information can again be obtained from the literature.

Table 8.6: Guidance on preliminary sizing of common types of beam (indicative values)

Beam type	Ideal span to depth ratio	Efficient span range	Max. span	Comments and relevant sources
Sawn timber beams[a,b]	L/10– L/15	≤5m	5m	3,4,77
Glulam beams[a,b]	L/10–L/15	≤8m	30m	Consider transport and erection for long-span beams[3,4,77]
Flitch beams (steel/timber)[a,b]	L/10–L/20	≤6m	Potentially 10m, but 6m in practice	Expensive, so ideally used for repairs and alterations. Max. span governed by max. length of timber available[4]
Simply supported rectangular RC beams[a,b,c]	L/12	4–10m	12m	3,6,74,85
Simply supported RC T-beams and L-beams[a,b,c]	L/10	5–12m	14m	3,6,74,85
Continuous rectangular RC beams[b,c]	L/15	4–10m	≥12m	3,6,74,85
Continuous RC T-beams and L-beams[b,c]	L/12	5–14m	≥15m	3,6,74,85
Cantilever rectangular RC beams[b,c]	L/6	≤4m	4m	Likely to be governed by deflection[3,6,74,85]
Cantilever RC T-beams and L-beams[b,c]	L/5	≤4m	4m	Likely to be governed by deflection. No benefit in using flanged cantilever beams to floors, since flange is in tension[6,74,85]
Secondary steel floor beams/trusses (subject to UDLs)[a,b]	L/15–L/25	4–20m	≥20m	Consider transport and erection for long-span beams/trusses[3]
Primary steel beams/trusses (subject to heavy point loads)[a,b]	L/10–L/15	4–12m	≥20m	Consider transport and erection for long-span beams/trusses. May need to use plate girders, since standard UB sections may not be adequate[3]
Steel transfer floor beams/trusses[a,b]	L/10	16–30m	≥30m	As above (also note that splicing may be required)[3]
Lightweight steel roof beams[a,b]	L/18–L/30	6–60m	≥60m	As above (also note that splicing may be required)[3]

Notes:

[a] All beams and trusses are considered to be simply supported floor beams unless noted otherwise.

[b] Timber and steel beams are assumed to be unpropped, while *in situ* RC beams are propped.

[c] Span to depth ratios for RC beams are based on overall depth.

Actual span to depth ratios may significantly differ from the ideal values quoted here, depending on the value of the applied loads, for example.

8.1.6 Decision 6: On- or off-site construction

Are there opportunities (particularly on confined city centre sites, but also elsewhere) to prefabricate structural elements off-site? Prefabrication is certainly not a new idea, and was used extensively for post-war construction in the UK, albeit with mixed success. To avoid the baggage associated with the term 'prefab', many of the latest systems are generically referred to as 'modern methods of construction' (MMC) or 'off-site manufacture'.

The extent to which pre-fabrication/MMC can be adopted varies from a simple node/connection detail to an entire building (Figure 8.20), from precast floors and walls (Figures 8.21–8.22), to structurally insulated panel systems (SIPs) (Figure 8.23) to complete modular units, as used for accommodation, healthcare and education, in which services, doors, windows and finishes are all installed in the factory (Figure 8.24).

Figure 8.20: 'Matrix of prefabrication'

Figure 8.21: Precast concrete floor units

Figure 8.22: Precast concrete walls

Figure 8.23: Structurally insulated panel system (SIPs)

Figure 8.24: Use of modular units/volumetric construction for a school

The key benefits of pre-fabrication/MMC/off-site manufacture include:

Social
- Improved Health and Safety conditions/less accidents
- Improved working conditions

Environmental
- Reduced road traffic movements (congestion and pollution benefits)
- Reduced waste
- Reduced energy use on-site
- Reduced energy use in operation

Economic
- Faster construction with up to 60% reduction in on-site construction programme e.g. due to reduced wet trades, and less days lost to inclement weather
- Higher quality construction/finish (e.g. of exposed concrete surfaces)
- More consistent structural properties/performance
- Reduced snagging and defects
- Reduced issues associated with drying shrinkage and creep of concrete and timber
- Higher tolerances

For more information, refer to Buildoffsite[90].

Figure 8.25 shows an indicative example of the programme benefits of pre-fabrication/MMC/off-site manufacture.

Figure 8.25: Indicative programme benefits of pre-fabrication/MMC/off-site manufacture[91]

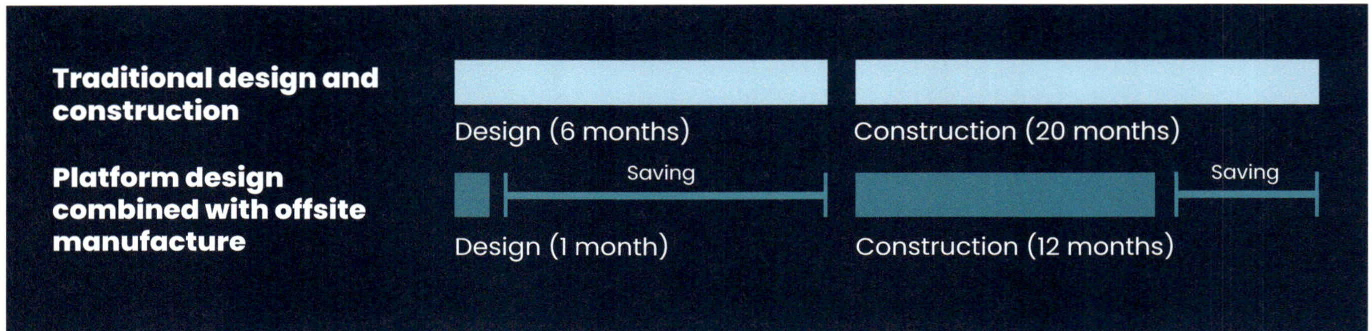

8.2 Structural layouts

Once you have evaluated the six key decisions discussed in Section 8.1, you can start producing structural layouts, in order to communicate your ideas to the rest of the project team. At concept design stage, it is likely that you will produce a number of options for discussion, and potentially initial costing. Unless there is a very clear preference for a particular solution from the project team, and/or you are very experienced, there is a danger that just considering one option will result in you missing a better one.

The form of the structural layouts can vary at this stage, and options could be drawn up by hand or in CAD. If buildings are very repetitive in nature, it may be sufficient to show just a few typical bays as opposed to the structure of the whole building. Some examples of concept stage structural layouts are presented in Figures 8.26–8.31, as well as in Figs. 8.3–8.5, presented earlier in this chapter.

When we talk about layouts, we don't just mean floor plans, roof plans and foundation layouts, as these do not usually tell the full story of how the structure works — it is a good idea to include at least one full-height cross-section through the building (and more if the cross-sections vary along the length of the building). It would also be helpful to include any non-standard details you are proposing, so that the main contractor and associated subcontractors can make appropriate allowances e.g. for cost and programme.

Information which should ideally be shown on structural layout drawings at concept design stage includes:

- Typical member sizes e.g. beam, column and loadbearing wall sizes
- Floor slab type, and span direction
- Roof construction, and span direction
- Foundation/substructure type(s) and sizes
- Average weights of steel frame members at each floor and roof level
- Reinforcement weights/m^3 of concrete for any RC elements
- Key dimensions (column-to-column dimensions, floor-to-floor heights etc.)
- Locations and sizes of lateral stability systems e.g. shear walls/cores, vertical bracing, plan bracing

- Locations of any movement joints
- Any non-standard details e.g. Figure 8.32
- Strategy for integration of structure and services
- Key assumptions and data e.g. permanent and variable loads assumed for the design
- Information relating to construction sequence, buildability and (residual) design risks which might not be obvious
- Pros and cons of the option(s) shown

Figure 8.26: Don Valley Stadium — isometric sketch of typical bay

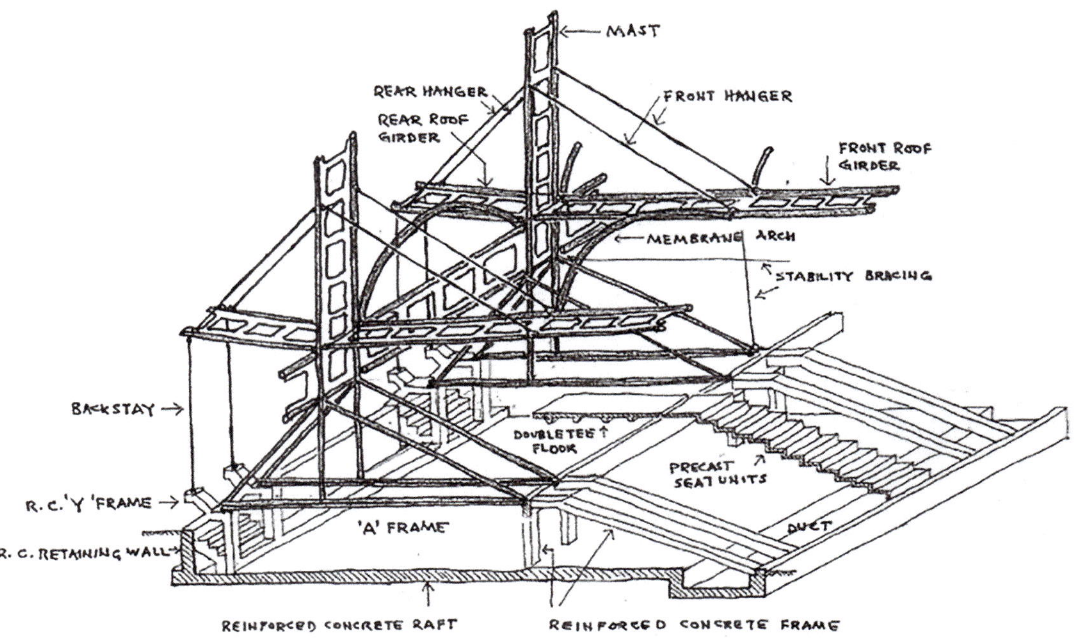

Figure 8.27: Research laboratory — typical floor layout for steel framed option

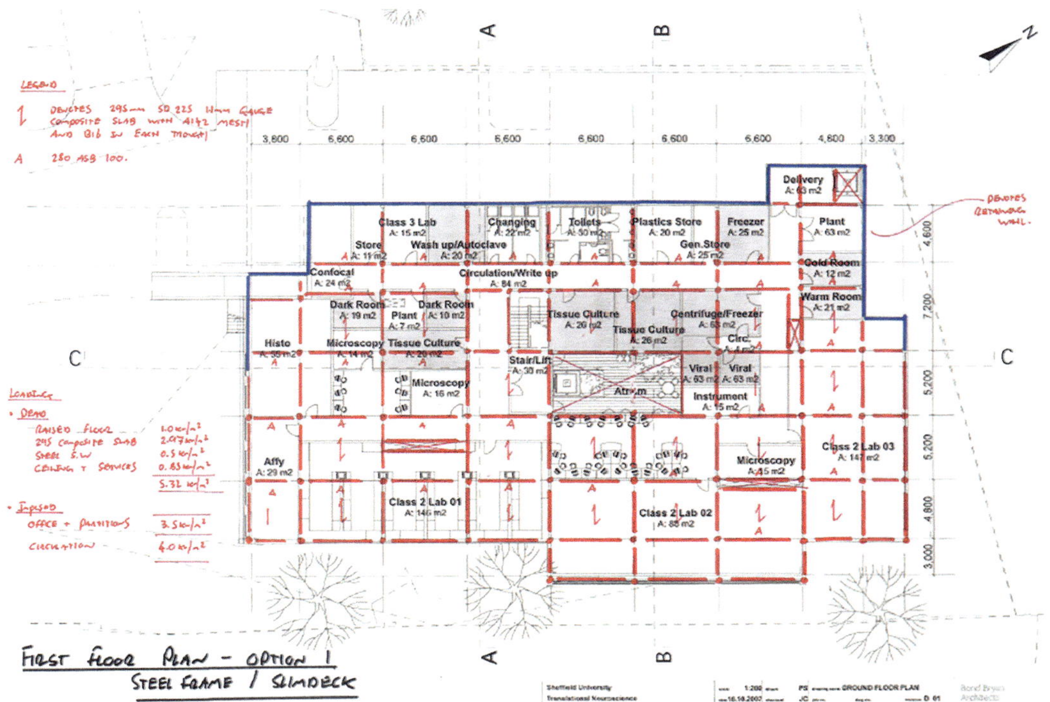

Figure 8.28: Research laboratory — typical floor layout for RC flat slab option

Figure 8.29: Education building — part-floor layout and full-height section for steel frame option

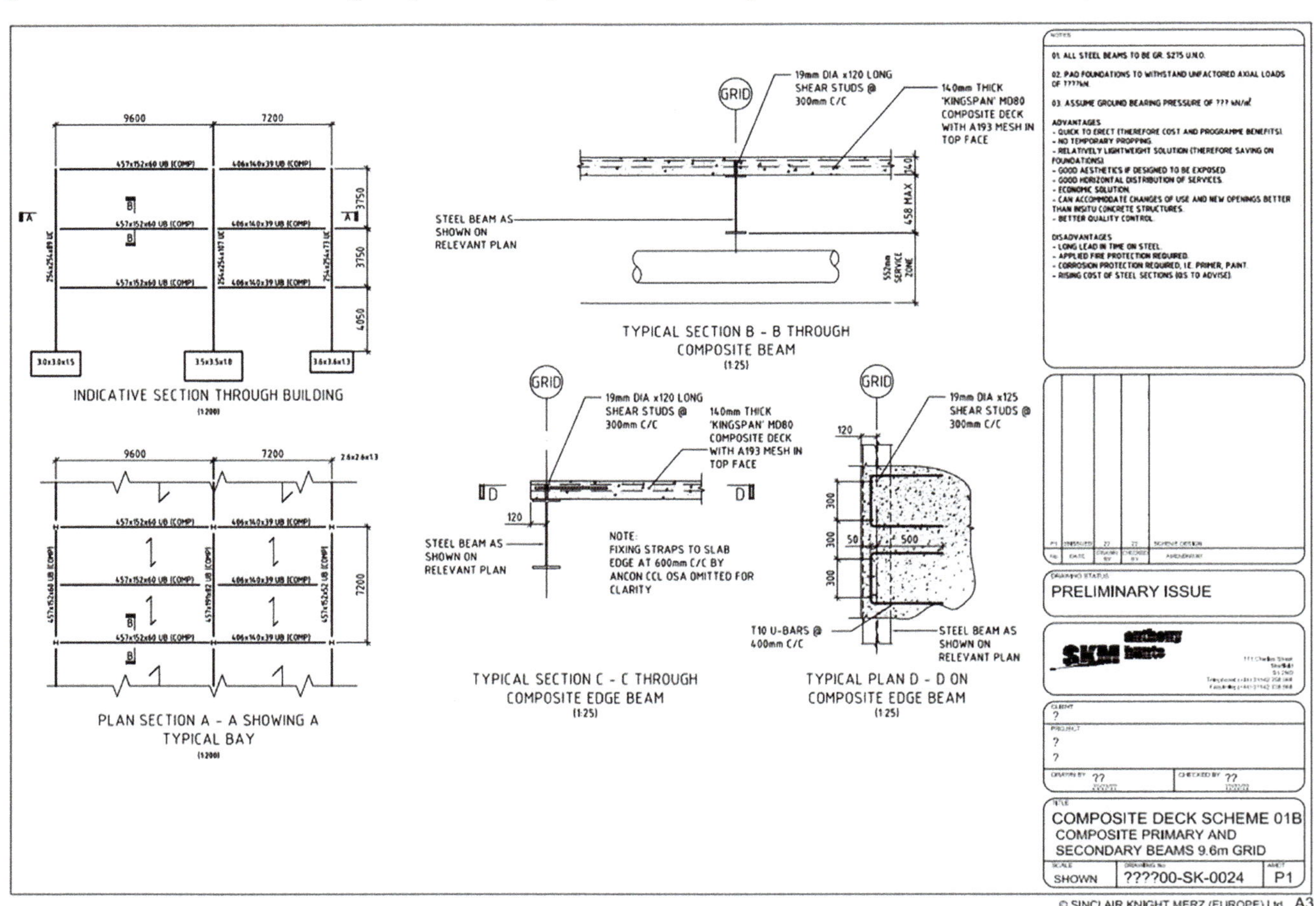

Figure 8.30: Education building — part-floor layout and full-height section for RC frame option

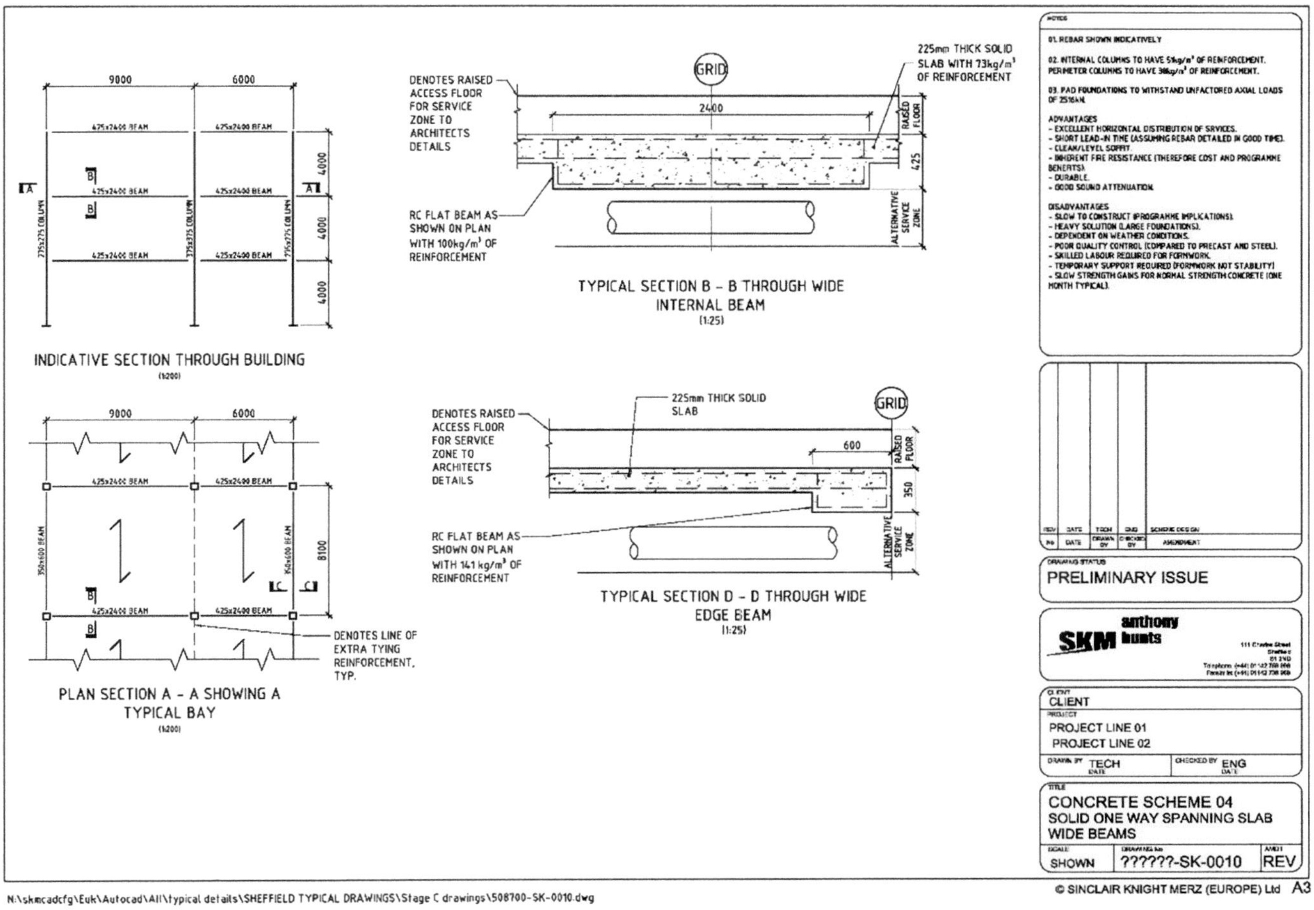

Figure 8.31: Isometric view of new mezzanine floor structure to an existing church

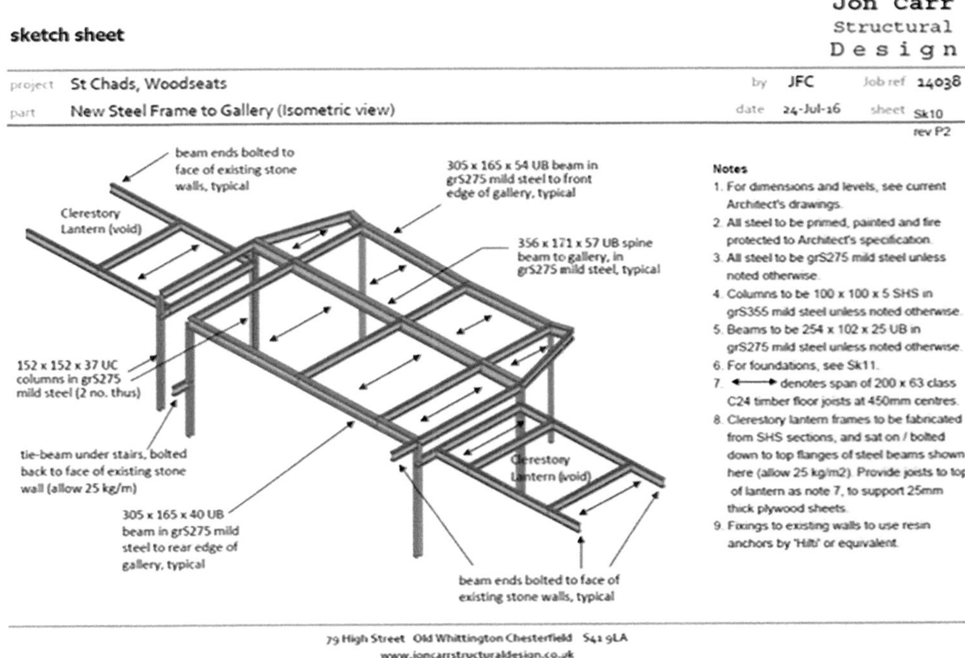

Figure 8.32: Notched steel beam detail to accommodate ducts/services

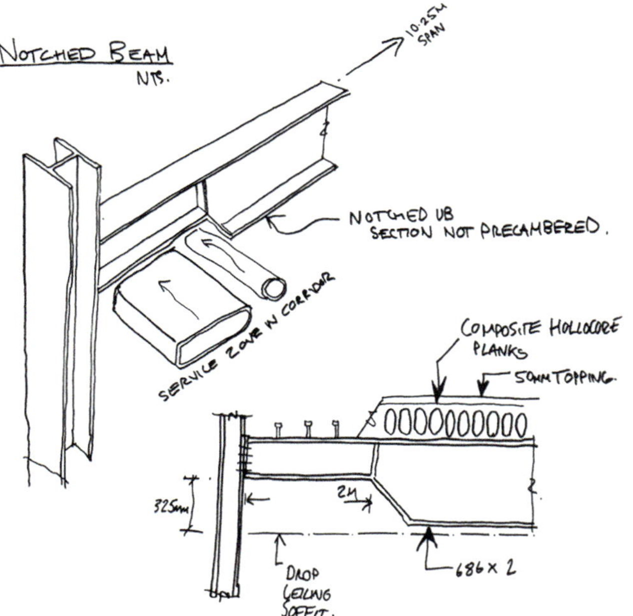

8.3 Roofs — where structural engineers get to have some fun!

The work carried out by structural engineers is often hidden in the final condition. So when we get the opportunity to show off our creativity and ingenuity, let's make the most of it. Good examples of where we can have some fun include roof structures, atria, entrance canopies, pedestrian bridges/walkways and staircases.

We will focus here on roof structures. Whether a roof to a multi-storey building or a long-span single storey structure, the relatively low variable loads on roofs, and with no requirement for the roof to be flat/level (as is required for a floor), opens up a diverse range of structural solutions to explore. In terms of structural form and materials, options include the following (with some examples in Figures 8.33–8.37):

- Conventional beams (straight, curved or tapered) in steel, timber and reinforced concrete
- Flitch beams (straight, curved or tapered) using a hybrid of steel and timber
- Trusses (straight, curved or tapered) in steel, timber or a hybrid of the two
- Portal frames in steel, timber and reinforced concrete
- Cable stayed roofs in steel, timber or a hybrid of the two
- Arched roofs in steel, timber and reinforced concrete
- Concrete shell roofs (singly- or doubly-curved)
- Timber grid-shell roofs (singly- or doubly-curved)
- Folded-plate structures (typically in either reinforced concrete or timber)
- Membrane/fabric structures (with double/anti-clastic curvature)
- Cable-nets
- Air-supported structures
- Tensegrity structures

Engineers are therefore encouraged to research and collect images and details of inspirational and innovative solutions, which they can potentially use on their own projects at some point in the future.

Figure 8.33: Selection of roof structure options

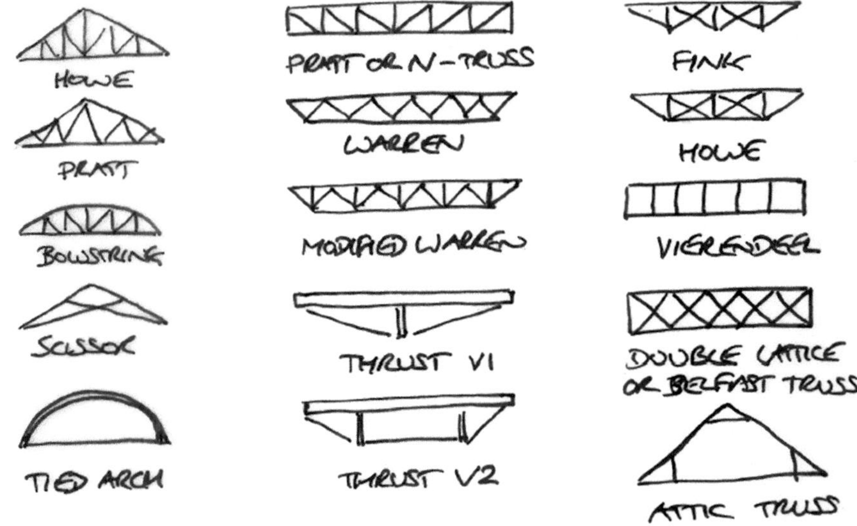

Figure 8.34: Sketch showing roof structure options to Bradford Academy

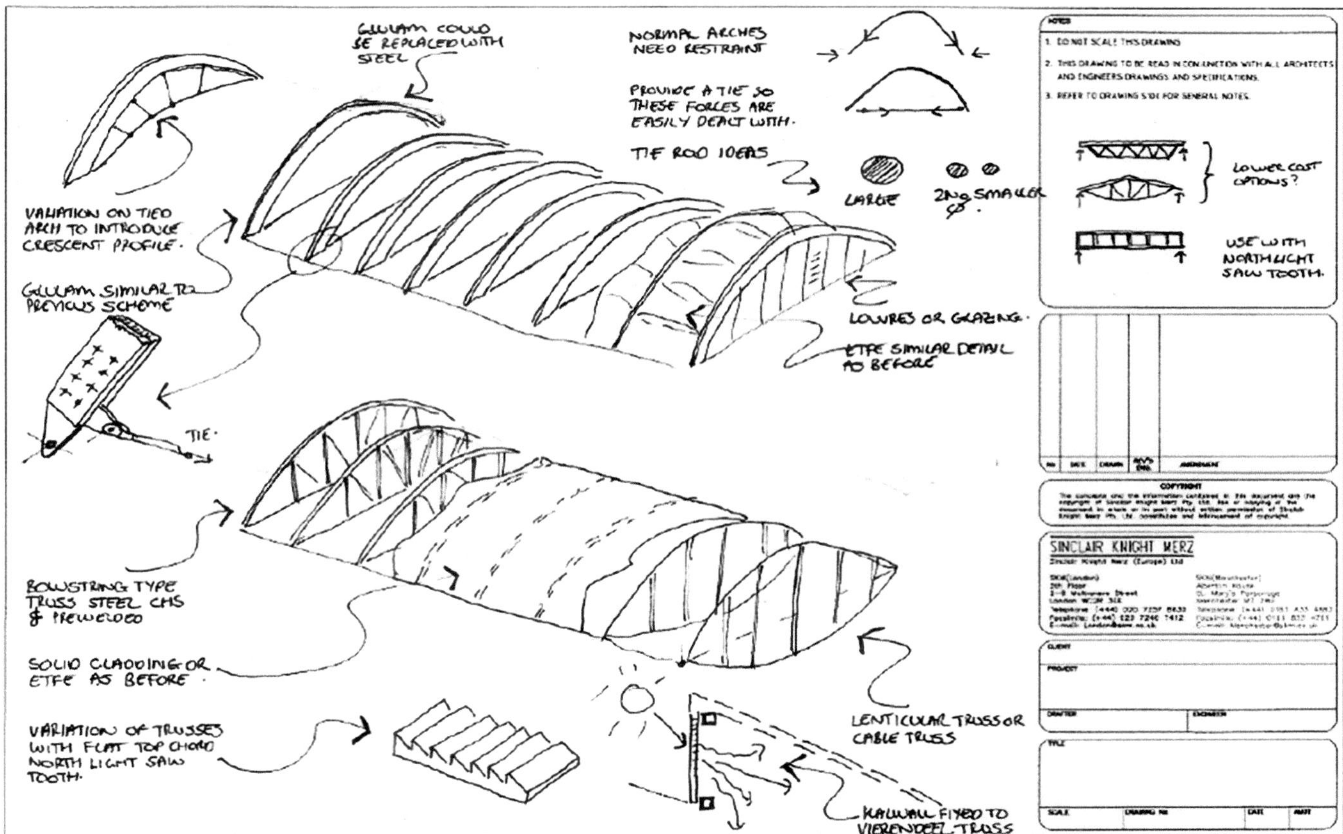

Figure 8.35: Adopted roof structure to Bradford Academy (tied steel arch with ETFE pillows)

Figure 8.36: Exposed concrete roof structure to York College

Figure 8.37: Steel 'trees' supporting steel roof structure to Teesside University

8.4 Movement joints, lateral stability and robustness

Due to their complexity, movement joints, lateral stability and robustness are covered in detail in Chapter 9, but they are fundamental structural issues which need detailed consideration at concept design stage. Making changes later is likely to involve a significant amount of re-design, re-drawing and re-modelling for the design team, and potentially subcontractors, which somebody will have to pay for.

8.5 Foundations

Foundations are covered in detail in Chapter 7. It is important to think about potential foundation solutions at concept design stage, since subsequent changes could prove costly and/or delay the project. In some cases, the foundation/substructure design could have a significant influence on the design of the superstructure being supported. Good examples of this are shown in Figures 8.17–8.18, where the presence of significant constraints below ground (train lines and archaeological remains) informed the structural frame solutions.

8.6 Practical examples

In Chapter 11 we look at three relatively common types of building in the UK — a ten-storey office block, a 30m span single-storey building, and a three-storey residential building. For each building type, two potential structural concepts are presented.

While these solutions cover a range of structural forms and materials, consider that many other solutions are possible. Furthermore the 'best solution' for any given project (if one exists, since in reality there will often be a number of potentially valid solutions) will depend on a diverse range of considerations, many of which will be project-specific.

James Norman
University of Bristol

9 Stability, robustness and movement joints

The IStructE *Practical guide to structural robustness and disproportionate collapse in buildings*[92] states that:

"robustness is not a commodity readily defined".

However, detailed reading of the guide shows that robustness can loosely be split into three categories:

- Lateral stability (as opposed to local stability and buckling)
- Robust design and details
- Disproportionate collapse

While it is not possible to go through each of these in detail at conceptual design stage, many of the decisions made will impact them. As a result, a clear and well-considered approach to each early on will save engineers a lot of time later in the design process.

This chapter provides guidance for low- and medium-rise buildings. The stability design of high-rise buildings is a specialist area beyond the scope of this book.

9.1 Lateral stability

While it could be argued that the provision of lateral stability is not so much a case of robust design, but of design more generally, as in the design of beams and columns, we would argue that the function of lateral stability *is* to ensure a robust design. In simple terms, lateral stability prevents the collapse of buildings subjected to lateral loads. If inadequate lateral stability is provided, the effect is not just the failure of the bracing or shear wall, as it would be for a beam or column, but the failure of the entire building. Lateral stability, while a minor part of the cost and material use of the building, is a major part of the structural design of the entire building.

As with all other elements of design, the aim at the concept stage of the project is not to complete a full design of the building's lateral stability, but to do enough to convince yourself, as an engineer, that the solution is feasible and should work. Depending on the type of lateral stability provided, this might be as simple as looking at a plan, or it may require quite significant calculation, even at the early stages. This chapter will help highlight:

- different stability mechanisms
- what is an appropriate level of design at this stage for the proposed solution
- how to approximately calculate the associated loads

Ultimately, the aim is to be able to conceptualise in three dimensions how the lateral stability works, and communicate this to the rest of the design team.

9.1.1 Horizontal loads

Horizontal loads can be separated typically into:

- internally applied loads
- externally applied loads

Internal loads are caused by a variety of factors e.g. out-of-plumb of the building and lack of fit. These forces are typically dealt with as lateral loads, known as 'equivalent horizontal forces' (EHF) in the Eurocodes, or 'notional horizontal loads' (NHL) in British Standards, rather than through the load eccentricity of element loads, although other approaches are possible.

Externally applied loads include earthquakes, explosions, large horizontal impacts and wind loads. In the UK, only the last of these typically needs to be considered at concept design stage, however accidental loads in the form of impacts and explosions may need to be considered when thinking about the effects of disproportionate collapse. For buildings in seismic regions, the consideration of earthquakes is as important, if not more important, than wind loading, depending on the region. The consideration of loads due to earthquakes is beyond the scope of this book but *Seismic design of buildings to Eurocode 8*[93] and the IStructE *Manual for the seismic design of steel and concrete buildings to Eurocode 8*[94] provide a comprehensive description of the design process.

Typically, external loads result in a reaction at foundation level, equal and opposite to the applied external load. However, internal loads, if a function of gravity loads only, must be resolved within the building (sometimes using the foundations), and do not exert horizontal loads on the ground around.

Estimating wind loads

There are three levels of wind load estimate (note these are UK-specific but a similar approach may be suitable elsewhere):

- **The five second guess**
  This is useful when you need a really quick answer. It won't be correct, but it will be a good guess. Most laterally applied loads are in the order of 1kN/m^2. To give some conservatism, you could use 1.3kN/m^2, and if your building is on high ground or near the coast (especially if exposed to south-westerly winds), you could use 1.5kN/m^2.
- **The ten minute calculation**
  If you have a little longer, it isn't very hard to get a conservative wind load. You need to take the following steps:

  Step 1 using Figure NA.1 from the National Annex to BS EN 1991-1-4[95], locate your building and choose $v_{b,map}$
  Step 2 using NA.2.5, calculate c_{alt} using equation NA.2a (for buildings taller than 10m the other equation gives a less conservative value). Unless your building is temporary or going to be standing for a very long time you can ignore c_{season} and c_{prob}. This equates to c_{season} and c_{prob} both being equal to 1.0. Likewise, we will ignore wind direction as it almost always reduces the wind speed. We can now multiply $v_{b,map}$ by c_{alt} to get our final, conservative wind speed
  Step 3 calculate the pressure using $q_b = 0.625.v_b^2$ (Equation 4.10 of BS EN 1991-1-4[96])
  Step 4 choose the worst distance to town and sea and use these values to look up the $c_e(z)$ and $c_{e,T}$ from Figure NA.7 and NA.8 in the National Annex (if you can't work out the distance to the edge of town, you can assume it is zero as this will be conservative)
  Step 5 multiply these values and the basic wind pressure to get the site wind pressure
  Step 6 using Table NA.4, calculate the building c_p. If you don't have time to calculate h/d or it is not clear use 1.3. This gives the lateral load on a roughly rectangular building in kN/m^2

At this stage, it's worth noting that if your building is an open canopy, or wall, dome, cylinder or giant asymmetric blob, there are other pressure coefficients you should use. Some are given in BS EN 1991-1-4, but often reference to Cook[97] will be required. For most structures, the horizontal load will be less, as the shape is more aerodynamic than a rectangular cuboid. The notable exception is a wall which has particular properties under wind load. For uplift, structures such as open canopies exhibit larger values than bluff bodies such as rectangular buildings, but this won't affect the lateral load.

- **The one hour calculation**
  If you have time, you can do a full wind calculation. It doesn't take long and you will need to do it at some point. Most offices either have access to an automated calculation package, or will have their own spreadsheet that can do this, and produce pressure maps for the 12 directions of wind load. Once the building location is agreed to a set site, the site values will not change even if the building moves or changes orientation.

Estimating equivalent horizontal force

BS EN 1991-1-4 provides detailed guidance for EHF, depending on the number of floors and columns. However, as a simple guide 1/200 of the vertical load should be applied as an EHF. For many buildings, EHF is a small proportion of the total applied horizontal load, and can be ignored at the concept design stage. However, this is not always the case. In the following circumstances, EHF should be considered at the concept design stage:

- If there is no wind load e.g. for mezzanines or new structures infilling an existing courtyard
- If the building is tall and thin e.g. the Flatiron Building in New York only attracts a small amount of wind load on its narrow face but the EHF remains the same regardless of direction
- If the building is heavy. Typically, *in situ* concrete frames are heavy, so the EHF should be considered

Table 9.1 provides some simple percentages of EHF to wind load, for typical floor build-ups. The breadth of the building (the width of the face with wind load applied) is not a consideration, as the wind load and EHF are both a function of this, so if you double one you double the other.

Table 9.1: Approximate percentage of EHF load

Floor-to-floor height (m)	Floor depth (m)	Timber	Steel	Concrete	Typical weight (kN/m^2)	
		1	4	8	Floor weight	
		2	5	10	Combined floor and wall	
3	10	2	5	10	% of EHF to horizontal load, assuming 1.3kN/m^2 base pressure and C_{pe} of 1.3. (values in **bold** are at or above the 30% cut-off)	
3	20	4	10	20		
3	30	6	15	**30**		
3	40	8	20	**40**		
4	10	2	4	8		
4	20	3	8	15		
4	30	5	12	23		
4	40	6	15	**30**		
5	10	2	3	6		
5	20	3	6	12		
5	30	4	9	18		
5	40	5	12	24		

Typical wind loads are roughly $1kN/m^2$, so if you take a value of $1.3kN/m^2$ as suggested assuming the five second guess, this allows for up to 30% EHF so under most cases the EHF can be ignored at this early stage (Table 9.1).

The EHF is a function of the vertical load. When considering load combinations, we need to decide if the wind load or vertical live load (variable action) is the leading action. When we take wind as the leading variable action, the EHF reduces as the magnitude of the vertical load reduces, due to us not taking the fully-factored vertical load, but a reduced factored vertical load (typically the factor of safety is 1.5 for the leading action, and 1.05 for the vertical variable load if it is not the leading variable action).

9.1.2 Stability design

At the concept design stage, it is essential to consider the location of the lateral stability elements, and communicate these clearly to the rest of the design team. This may not be a simple conversation, as often there is a desire for open-plan floor plates, and walls around cores often hide service risers which may clash with the stability.

The first thing is to ask the architect for a set of sketch plans. You then need to locate bays where walls/bracing could go. The obvious location is around stair and lift cores. These are ideal as they tend to run the full height of the building. However, you need to be careful with stair cores as the head room at first floor is often greater than at upper floors, therefore the depth of the stair core can change. Also, look for areas of perimeter wall where there are no windows. If these don't exist, you will need to negotiate some, or find another way of providing stability. Often, the building services engineers are also looking for riser locations, so it can make sense to work together, but be careful as vertical risers often contain large penetrations through the slab, making it harder to get the force into your stability system.

Ultimately you need a set of plans, or even better a 3D sketch that shows the locations of stability right through the building. This is often enough at the concept design stage.

Selecting the stability system

You may have already decided what type of structure you are going to have (Chapter 8). Once you have a rough idea of where your lateral stability is going to be located, you need to propose a system. There are typically three ways of stabilising a frame:

- Shear walls — concrete, masonry or timber
- Bracing — typically steel
- Moment frames — steel, concrete or timber

These approaches are summarised in Table 9.2. Shear walls are stiffer than the other systems, and are therefore more suited to taller structures, but they also use more material and provide little opportunity for windows or openings. Bracing is lighter, and with careful detailing can sit in front of windows and include penetrations, especially smaller ducts. Moment frames are very flexible and should ideally be avoided. If there are no possible locations for walls or bracing, moment frames can be utilised and can produce a very attractive finish. However, caution must be applied when using these systems. Compression/tension systems are more structurally efficient than moment systems, so a wall/braced system is preferred to a moment frame system.

Setting out the stability system

You are hopefully now in a position to set out your lateral stability.

Here are some simple rules for stability design — the first rule is essential, the others are all 'best practice', rather than 'absolute musts'. We have all worked on projects where there is only space for a few paltry walls, in absolutely the wrong places, and you have to make it work.

For a low cost, efficient solution the tips in Figure 9.1 should be followed:

Figure 9.1: Different options for stability systems

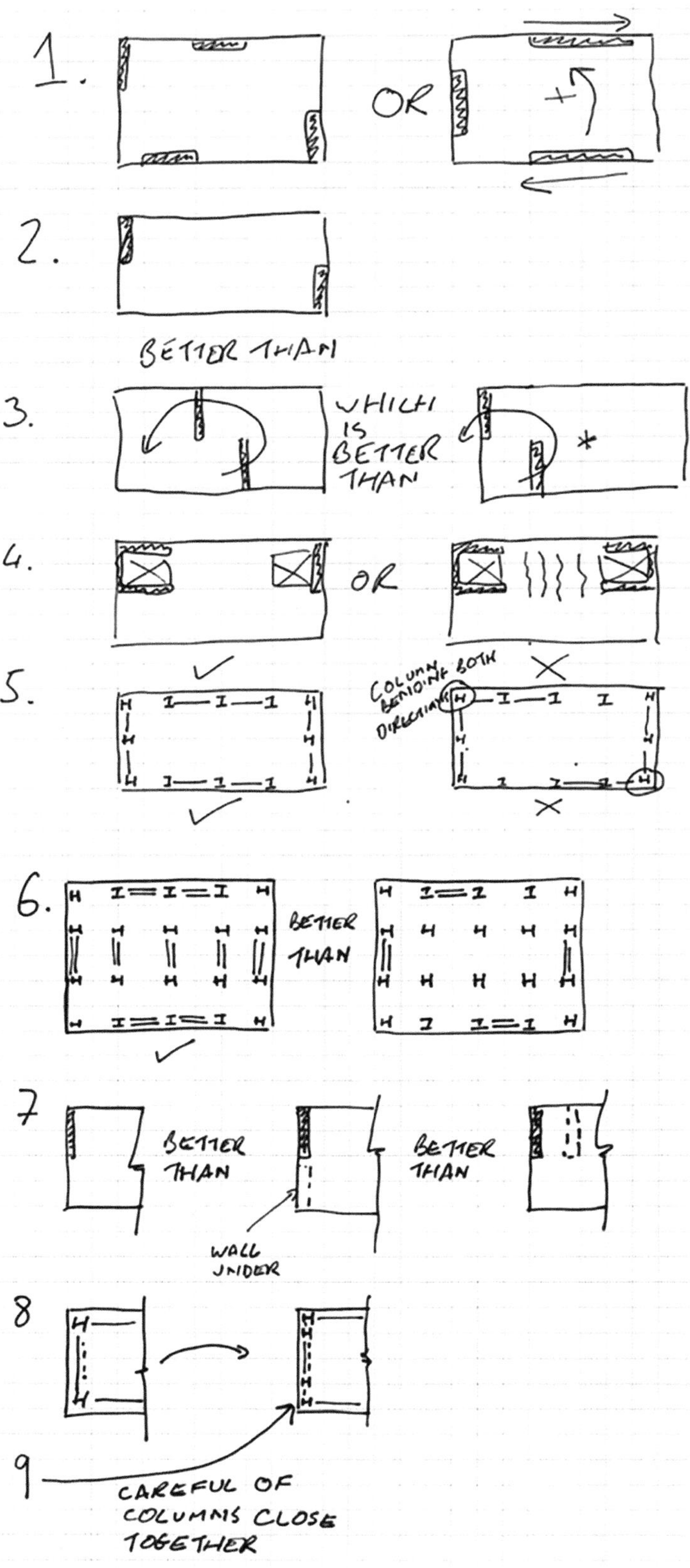

Essential

1. For rectangular frame buildings, we need a minimum of two bays in each direction, or one bay with two bays designed to remove the associated twist (the term 'bays' is used here generically to mean either braced bays, shear walls or moment frames), although careful consideration of robustness is required.

Best practice

2. Bays should be located as far away as possible, to minimise the eccentricity of the load.
3. Bays should not be clumped in the centre, as this will not provide good torsional resistance.
4. Ideally avoid trapping slabs between stiff walls, as this can lead to cracking (see section on 'Shrinkage').
5. For moment frames, every column should be used for one frame direction only — avoid using the same column in both directions.
6. For moment frames, try to use as many consecutive frames as possible, either parallel or perpendicular.
7. Avoid changing the position of stability bays from floor to floor (see section on 'Changing lateral stability locations' and Figure 9.4).
8. Remember for braced frames, you need a column at either end of the bracing.
9. Try not to have too many columns near your bays, as these reduce the axial load on your bay, which helps prevent uplift in the system.

Table 9.2: Types of stability systems for different frame types

	Frame type				
	Steel	**Concrete**	**Timber**		**Masonry**
Shear wall	Concrete cores	Concrete cores and walls	Solid timber walls	Racking panels	Solid masonry walls
	OK by inspection	OK by inspection	OK by inspection — provide double the expected number for concrete	OK by inspection — provide lots of walls — suited to loadbearing wall construction	OK by inspection — provide lots of walls — suited to loadbearing wall construction
(typically)	4–20 floors	2–20 floors	2–6 floors	1–3 floors	1–2 floors[a]
Bracing	Steel bracing	N/A	Steel rods (tension)	Timber struts (compression)	N/A
	Size worst		Size worst	Size worst for connection and strut buckling	
(typically)	1–20 floors		1–4 floors	1–2 floors	
Moment frame	Steel moment frame	Concrete moment frame	Timber moment frame		N/A
	Detailed calculation for worst frame to check deflection and provisional moment connection sizing	Detailed calculation to check floor depth at column head	Detailed calculations at the earliest stage as connections will govern member sizing and careful consideration of sway stiffness to ensure limited deflection		
(typically)	1–4 floors	1–2 floors	1–2 floors		

Note: [a] Masonry is limited not by stability, but the robustness requirements for all building types except single residential dwellings, which may go up to four floors.

Diaphragms

All lateral stability systems rely on the loads being transferred from the perimeter walls and floor plates to the vertical components (Table 9.3). For floors, this is typically achieved through diaphragm action. A diaphragm is a stiff plate which is deep enough to transfer load without significant flexure occurring, so for the sake of analysis, can be assumed to be infinitely stiff. For roofs, this can also be done through diaphragm action, but this isn't always the case, so at concept design stage you may want to consider adding plan wind bracing as an alternative method.

Shrinkage

In item 4 of our best practice list, we mentioned avoiding trapping slabs between stiff walls. It is worth just pausing a little longer on this point. All materials expand and contract e.g.:

- Steel, under variations in temperature
- Timber, from changes in atmospheric moisture
- Concrete, as it cools following its initial chemical reaction

These movements, if we are not careful, can be trapped by stiff stability elements, either leading to cracking, or worse, increased stresses in members that we haven't designed for. Ideally, when setting out stability, we should avoid trapping the slab and allowing stress to build up (Figure 9.2). This may not be possible e.g. if we have lift cores at each end of the building, and require them to provide adequate strength in the other direction. If this is the case, we should consider how this stress can be relieved. For concrete structures, we can leave a section of slab and pour it 28 days later, once much of the shrinkage has occurred (although this will require careful consideration as further movement due to shrinkage, thermal effects etc will take place, especially around the construction sequence). In steel and timber structures, we can use slotted holes to enable a small degree of movement at a designated position, with the finishes detailed to accommodate this movement.

Figure 9.2: Avoiding locking in stress in the slab

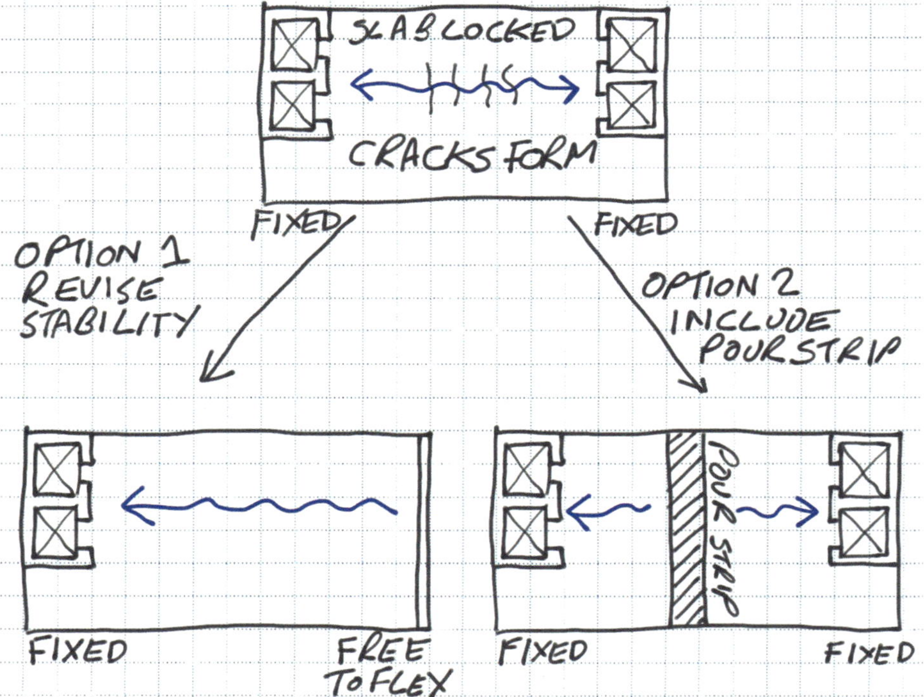

Table 9.3: Rules of thumb for spacing of different stability systems

Material	Type	Typical location	Notes	Max. spacing between lateral stability bays
Concrete	*In situ*	Roofs and slabs	Acts as diaphragm. Ideally floor plate aspect ratio is between 1:1 and 5:1	40m
	Precast biscuits with *in situ* topping	Slabs		
	Metal deck with *in situ* topping	Slabs		
	Precast planks	Slabs	Acts as diaphragm, as long as structural topping is included. Ensure topping does not have services ducts in its depth as this will limit diaphragm action	
Steel/ aluminium	Deep metal deck	Roofs	Does not typically act as diaphragm, so provide plan wind truss	40m
	Cold-rolled purlins and deck	Roofs	Does not typically act as diaphragm, so provide plan wind truss. It may be possible to use purlins as wind truss but they will need significant increase in weight/size	
Timber	CLT	Roof and slabs	Acts as diaphragm. Ideally floor plate aspect ratio is between 1:1 and 4:1. Ensure halfing joints specified to ensure shear transfer	20m
	Boarding on joists	Slabs	Acts as diaphragm. Ideally floor plate aspect ratio is between 1:1 and 2:1. It is important this is correctly detailed with board joints sitting on joist lines and nailed both sides to ensure shear transfer	10m
		Roofs	Can act as diaphragm as above. For traditional inclined truss roof more typical to provide timber in-plane bracing for wind	10m

Irregular plan buildings

For non-rectangular buildings, you will need to think a little more carefully about stability. It is not essential that you have two walls acting in two perpendicular directions (although this is ideal). It is essential that you think about the horizontal load as acting in all directions, not just two perpendicular directions. This is also true for rectangular buildings, but the assumption is that the load from any direction can be turned into a vector in the perpendicular directions of the walls, therefore each set resists loads in that direction. However, where walls are required in the perpendicular direction to resist twist of the frame, it may well be that the worst load case is not in either orthogonal direction, but somewhere in-between e.g. the CitiCorp Center, New York[98].

Figure 9.3 shows three types of building. For circular buildings with two orthogonal sets of walls, the design approach is the same as for rectangular buildings, but with a revised pressure profile. Where non-orthogonal walls are provided, the wind needs to be considered from all 12 directions to ensure the worst load case is considered for each wall. Where walls are orthogonal but only one wall is provided in any direction, then as with rectangular buildings we need to consider the additional load from the twist as well as the significant reduction in redundancy.

For the triangular building, it can be seen that wind in the direction shown puts a load into all three walls, each one providing a vector of resistance to the wind. In this case, the largest load goes into the wall in the direction of the wind, with the other walls also taking a share and providing resistance to twisting. By just considering two directions, the maximum load on each of the three walls won't be captured.

For the rectangular building, which has an equivalent layout to the right-hand circular building, the load on the three walls for the two orthogonal directions can be seen. However, in this case, by considering a wind load at 45° to the building (the right-hand case), we can see that the load on the bottom wall goes up, and the load in the top wall reduces. This demonstrates the importance of considering multiple wind directions even if walls are orthogonal.

Figure 9.3: Stability solutions for non-orthogonal buildings

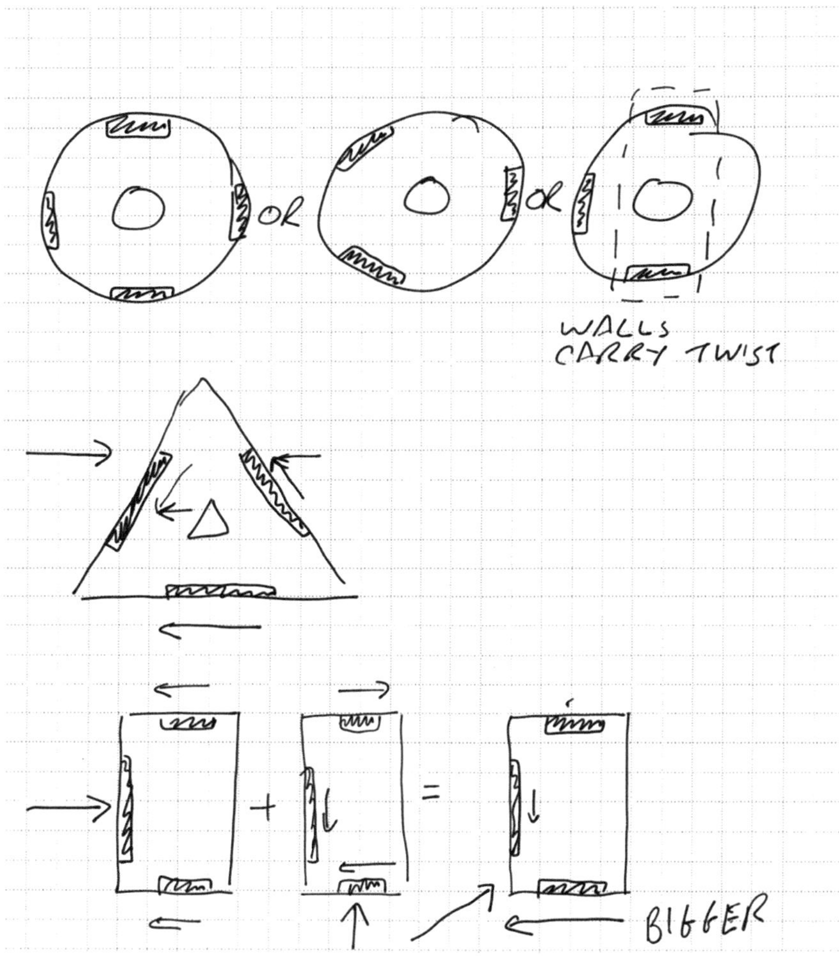

Changing lateral stability locations

It is not ideal to change stability locations on-plan floor-to-floor. However, sometimes there are no other options. It may be helpful to consider stability frames as cantilevers, with their base acting as a fixed support. The fixed support carries both shear and a moment. When we move a frame, the shear can be transferred (sometimes with difficulty) from one location to another. The moment however, normally in the form of a push/pull in the columns, needs to be carried all the way down to the foundations. For this reason, when a stability bay moves it is essential that the columns are maintained to ensure the push/pull is transferred to ground. For the shear, if the bay moves in the direction of the shear, it is reasonably straightforward as the centre of stiffness of the structure remains the same (middle illustration in Figure 9.4). If the bay moves perpendicular to the bay, the centre of stiffness shifts, creating an eccentricity in load and a potential twist. The slab will need to carry this force, which can sometimes prove problematic (right-hand illustration in Fig. 9.4).

Figure 9.4: Effect of moving stability locations floor-to-floor

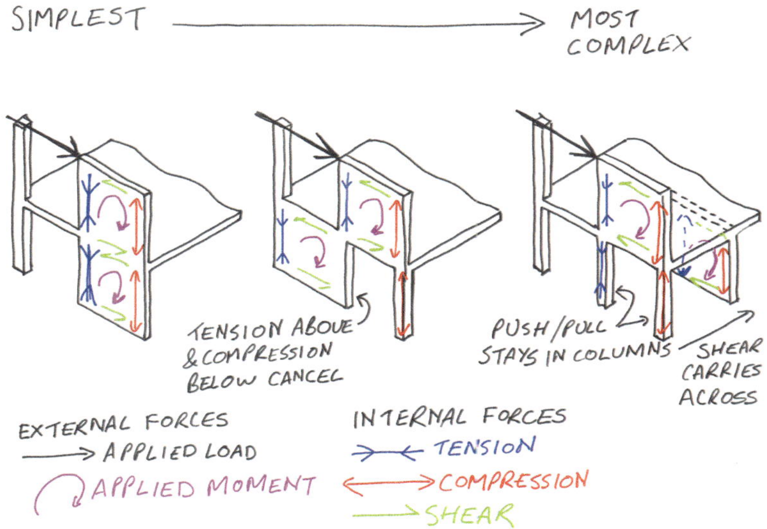

Structural movement joints

For large-plan buildings, where there is a significant change in building height or variable ground conditions, it may be necessary to introduce structural movement joints (Figure 9.5). These split the building into smaller parts, and prevent excessive build-up of stress due to thermal movement and shrinkage. Ideally, each section of the building should be treated as an individual building, and lateral stability should be provided in both directions for each section. This will typically lead to double columns at the interface where there is a movement joint.

Sometimes it is not possible to completely separate the sections — there are two typical reasons for this:

- It is not possible to provide stability for all portions in all directions. In this case, the shear may need to be carried in the short direction (we are trying to separate the floor plates in the long direction). This can be done through the use of dowelled connections
- Double columns are not architecturally acceptable – in this case a sliding joint will be required. This can be achieved in the same way or through a halving joint which will enable movement in both directions but provides vertical support

Figure 9.5: Strategies for stabilising structures with movement joints

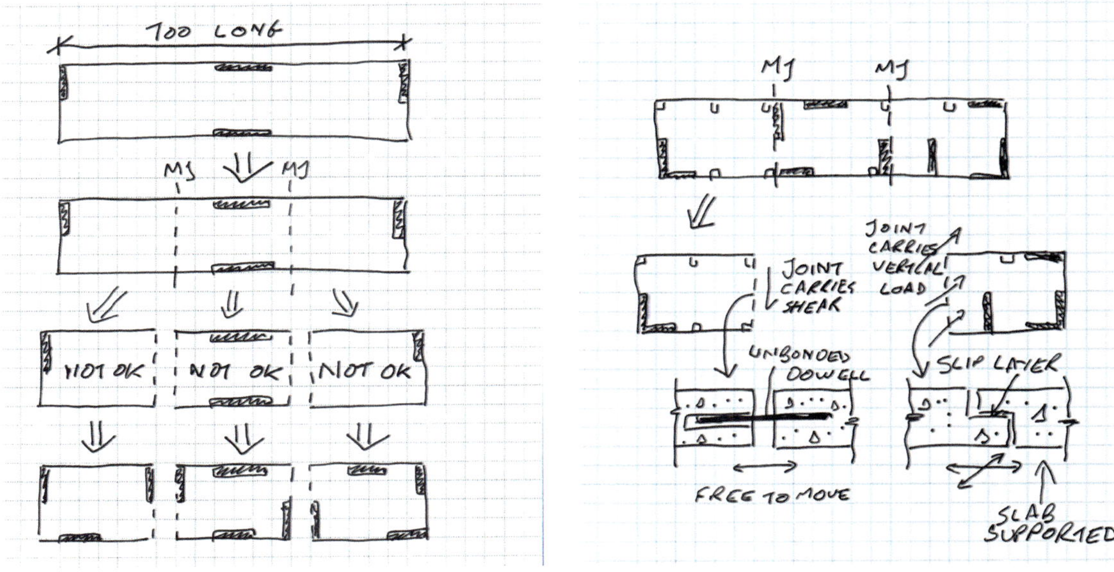

It is important, before finalising stability, to agree movement joints with the architect (they will need to be reflected through the facade, so the location is key), as they will affect the location of your stability walls.

For masonry facades (brick, block or stone), we often require movement joints for movement, due to thermal and moisture effects. While these should match the structural movement joints, they are required at much closer centres (typically 6m for blockwork and 12m for brickwork), and are typically set out by the architect with input from the engineer. If we are designing a steel or concrete frame with a masonry facade, we do not need movement joints in the frame at every point where the masonry movement joints occur in the facade.

Drawing the solution

At this stage, you should have a stability scheme and be ready to communicate it. The best way to show a stability solution is to draw a 3D frame and show the location of walls, bracing and moment frames (Figure 9.6). If there are movement joints, consider exploding the diagram, so that you can convince yourself of the adequacy of each element. While an experienced engineer may be able to understand and appreciate the stability by simply marking it onto plans, the complexity of stability means that a 3D drawing is more likely to highlight challenges e.g. changes in location.

Figure 9.6: Stability scheme

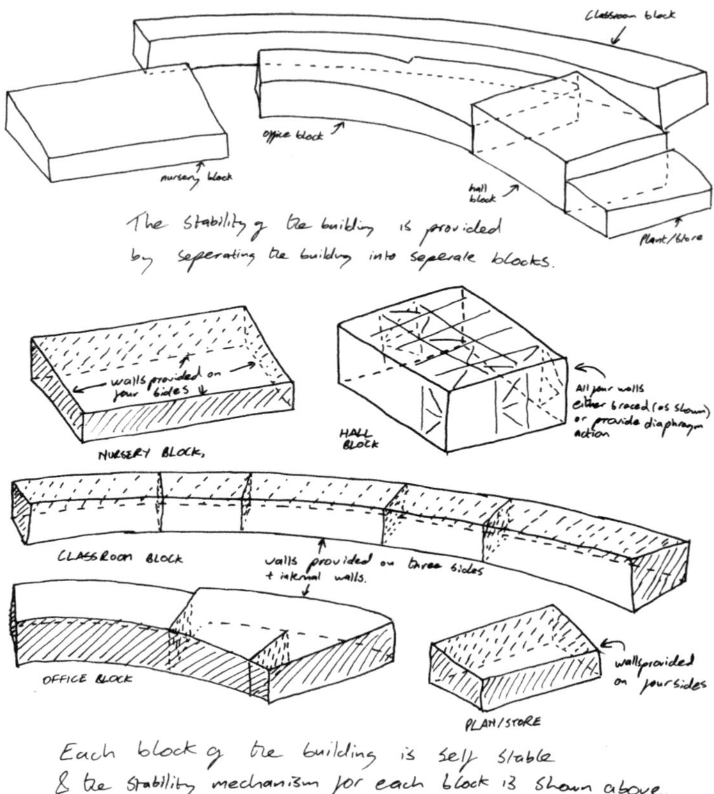

Checking global design for overturning and uplift

Having drawn the stability system, we now want to make sure we have considered overturning and uplift at the base. Uplift occurs where the tension force from the moment exceeds the dead load of the building acting on that stability element.

While a detailed analysis will not be required for all walls/braced bays, walls with a high aspect ratio (tall and not very long/wide apart) would benefit from further consideration at this stage. There are two simple questions to ask:

1. Do we anticipate uplift?
2. If we do, what mechanism will hold it down?

Ideally, we want to avoid uplift altogether, and at this stage this should ideally be our strategy. A quick analysis of the moments, shear and vertical load (Figure 9.7) will reveal whether uplift is a concern. If we think we have uplift, we should push the ends of the wall or columns further apart, if the architecture will allow. This has two effects:

- Increases the lever arm, reducing the uplift force
- Leads to greater loads on the columns/stability elements, helping to resist the uplift

If we can't do this, we need to ask what will hold down our building? If we have pads and footings, we will need mass concrete. Every cube of concrete provides 24kN of resistance, so if you need to overcome 1000kN of uplift, that will be more than 40m^3 of concrete — or seven trucks full!

If we have piled foundations, we will need to add tension piles. Unlike compression piles, tension piles cannot utilise end-bearing, so work on skin-friction only. We can anchor them into rock, but this becomes expensive. The capacity of tension piles can vary significantly from zero to roughly the same as compression piles. It is not possible to estimate the tension pile capacity until you have carried out the ground conditions, and the more accurate your information, the better your tension pile estimate will be.

Figure 9.7: Calculating shear, moments and stresses on shear walls

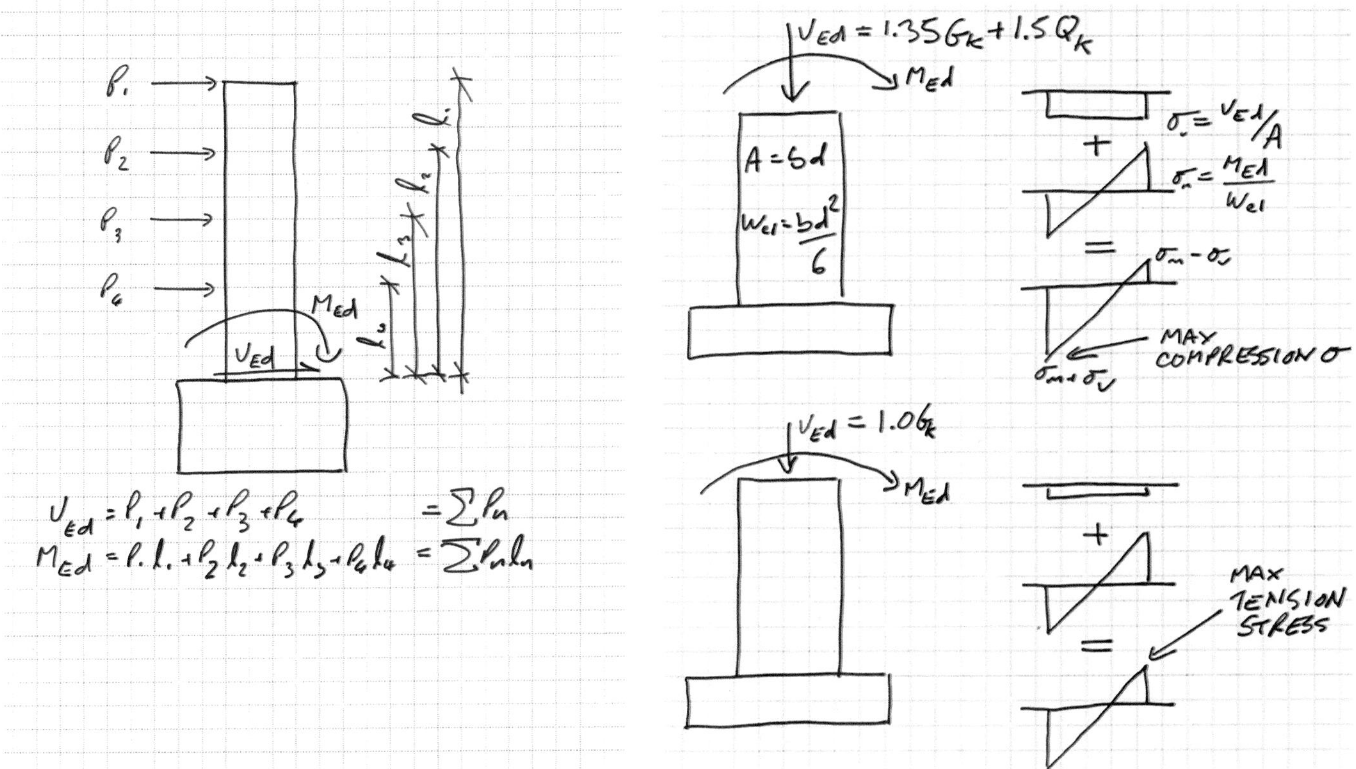

9.2 Element design (and when you can ignore it)

Having chosen your stability system and checked it for global overturning, it may be necessary to size some elements (Table 9.4). This should only be done near the end of the process, once the design team is confident about the location of stability bays. At concept design, you may only need to do a simple height-to-width check, or you may have to do a full detailed design.

If in doubt, or if you exceed the rules of thumb in Table 9.4, it is always better to do more design work now, and avoid a design that you are not confident works, or requires structural gymnastics to prove it stands up later.

Table 9.4: A quick guide to stability element design

Element	Level of design at scheme design	Rule of thumb/points to note
Concrete shear wall	Draw, ensure walls are adequate and check rule of thumb	Walls to have height-to-length of min. 4:1 (or equivalent stiffness if lift/stair cores). Typical thickness is 200mm but check cover for fire
Masonry shear wall	Draw, ensure walls are adequate and check rule of thumb	Walls to have height-to-length of min. 2:1. If large openings occur consider a steel picture frame
Timber shear wall	Draw, ensure walls are adequate and check rule of thumb	Walls to have height-to-length of min. 2:1
Steel bracing	Draw, ensure braced bays are adequate and size most highly loaded	Ensure bracing can fit within wall build-up if not expressed
Timber bracing	Draw, ensure adequate number of braced bays are provided and size most highly loaded in both tension (connection check) and compression (buckling check)	Connection design may govern bracing size and/or size of vertical elements bracing is connected to
Steel moment frame	Draw, ensure adequate number of moment frames are provided, size most highly loaded and check deflection/ensure you clearly draw and label moment connections	Moment connections will have an impact on cost and ensuring they are all clearly labelled on drawings will avoid problems later. Note that connections may be deeper than main sections so the architect should be advised
Timber moment frame	Draw, ensure adequate number of moment frames are provided and size most highly loaded, paying particular attention to the connection size which will most likely govern the size of both beams and columns. Check deflection	Ensure all moment connections are drawn on GAs

9.3 Robust design and the pitfalls to be avoided

In addition to lateral stability, the other areas of robust design that need to be considered are:

- Ductility
- Redundancy
- Insensitivity
- Uncertainty
- Fire

Ductility

Ductile designs are always preferred to brittle ones, as they provide warning signs of failure and often enable redistribution of load, if the assumed structural model does not match the built reality. Most major building materials (steel, concrete and, arguably, timber) can be designed to ensure they act in a ductile manner. However, some more unusual materials e.g. glass, ceramics and fibre reinforced polymers may act in a brittle way. In these cases, a much higher level of care is required, and simply increasing sizes of sections if we are not sure about the magnitude of stresses will not be enough and, in some cases, may prove disastrous as the increased stiffness will attract more load, which may lead to sudden brittle failure.

Redundancy

The concept of redundancy is that if a member were to fail, there is another load path which can be used. It is the gold standard of robust design, but is not always possible. A typical structure with no redundancy is a cantilever which, when the support becomes a hinge, turns into a mechanism. There is no alternative load path or catenary action that can be achieved. Another example is a truss, as, if one element were to fail, it will typically not have an alternative load path, and the whole system will fail. This does not mean that we cannot use a cantilever or truss, but they are less robust, and therefore require careful consideration.

The same is true of stability systems. Figure 9.8 shows a number of different layouts. We can consider the redundancy, by removing each wall in turn and seeing the impact. When we have more than two walls or braced bays acting in the same direction, the loss of one wall will typically lead to a loss of stiffness, but the structure remains stable. When we only have two walls or braced bays acting in each direction, the loss of one wall will lead to a large reduction in stiffness, but the system remains stable, as the orthogonal walls will still act to resist the twist. Sometimes, when we design a stability system we find that one wall or braced bay carries almost all the load, with the other walls or braced bays just removing the eccentric load. In this case, we are very reliant on this wall and removal will lead to both a huge loss of stiffness and possibly an unstable structure, depending on the adequacy of the remaining walls. We may only be able to provide three walls due to other constraints, often architectural ones. In this case, the loss of a wall would lead to a complete loss of stiffness, and the system will become unstable.

We must go through these thought processes, as we consider the redundancy of our stability system. If we can only provide three walls, it is important to recognise the lack of redundancy, and ideally we should try and provide more than two walls of similar stiffness, especially in the dominant wind load direction (most buildings are rectangular on-plan, therefore the wind load is likely to be larger when acting on the long face). The aim is to make the structure efficient (so not adding walls unnecessarily) but also robust with, ideally, a good level of redundancy.

Figure 9.8: Redundancy in stability systems

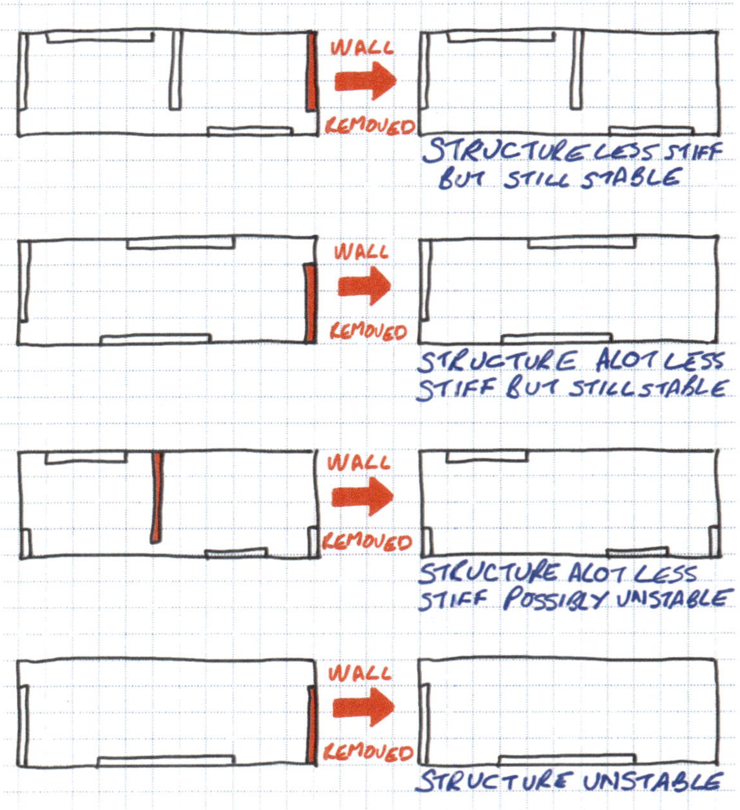

Insensitivity

A robust structure is insensitive. This often (but not always) means that small changes in construction and/or use, do not alter the structural performance.

Most modern structural forms are insensitive to minor changes, whether ground movement, shrinkage or thermal movement. However, some are very sensitive to small movements. A typical example is a masonry arch, especially flat arches, where ground movement can lead to small horizontal movements, which can move the thrust line in the arch leading to cracking and, in some circumstances, collapse. Other arching structures, such as grid shells and large steel arches, can be similarly sensitive to small movements, and these need to be carefully considered.

Precast connections, which have very small bearing areas, can be highly sensitive to small changes. These should be avoided, and reasonable bearing areas provided.

Uncertainty

Designing for uncertainty is an essential part of the design process. Many things are unknown as we design our building:

- What will the loads really be?
- What will the material strength really be?
- Will our building be truly built to plumb?

Actually, many of these unknowns are, in fact, well known and documented, and are already accounted for in the design, both by using characteristic values (rather than mean values), and by factors of safety. We certainly don't need to make a further allowance for them in the design.

Other uncertainties arise from site specific constraints e.g whether we might find 'gulls' in the ground (large voids that can occur in limestone rock). Sometimes, these are completely unpredictable, but often we can make an allowance for them in our design e.g. by designing strip foundations to span over them.

Foreseeing these unknowns is a skill of the designer, and comes from a combination of asking great questions and experience. The trick is not to design a building that can withstand every eventuality, but to devise a design which is adaptable and can be quickly and easily modified, as and when the uncertainties are discovered. In the 'gulls' situation for example, if we had used pad footings, we may have struggled to quickly react to discovering them, whereas strip footings just require the addition of a rebar cage, and would span over the gull, avoiding delays on site.

Fire (with input from Jonathan Hall of Buro Happold)

Designing for fire is a complex issue, made more confusing by the variation of scope for different members of the design team from project to project. Typically, the scope will be as follows but this will need clarifying on every project you work on. Most projects, particularly larger projects, will have a project fire engineer. For smaller projects in the UK, the architect may take ownership of meeting Approved Document B of the Building Regulations[98].

- **Fire engineer**
  Sets performance criteria for the design to meet Building Regulations requirements for fire safety:
  - Confirms structural fire resistance and required compartmentation within the building
  - Confirms the requirements for fire safety systems e.g. detection, suppression, smoke control
  - Confirms the travel distance and occupant capacity limits for the building
  - Highlights the criteria for internal and external linings
  - Advises on external fire spread and firefighter access, to and within the building
- **Architect**
  - Confirms safe egress in line with the criteria specified by the fire engineer
  - Confirms how fire rating and details for fire compartments are achieved
  - Ensures strategy for limiting spread of fire (including, but not limited to, specifying finishes on exposed timber)

- **Building services engineer**
  - Confirms smoke ventilation strategy, if deemed necessary by the fire engineer, especially if mechanical forcing is involved
  - Ensures mechanical systems are designed to prevent spread of fire and smoke
- **Structural engineer**
  - Ensures structure remains sound for duration of the fire resistance period
  - Supports architect in ensuring structural materials provide adequate fire compartmentation
  - If providing facade engineering (which is typically a separate appointment to structural engineering), ensures facade is designed to prevent spread of fire

At concept design, the engineer should be asking these two questions, as a minimum:

1. What is the fire resistance requirement for the structure, and for different spaces in this building?
2. How is the fire resistance achieved?

For exposed structure, this can be even more challenging. Table 9.5 gives some typical strategies.

Table 9.5: Fire-proofing strategy for exposed structure

Element material	Protection type/strategy	Fire resistance[a]	Comments
Concrete	Provide adequate cover for rebar	30–240 mins	
Steel	Intumescent paint	30–120 mins (up to 240 mins with specialised products)	Requires maintenance. Check visual appearance with architect
	Exposed steel fire designed using concrete infill	30–60 mins (up to 180 mins with concrete reinforcement if designed as a composite member to BS EN 1994-1-2[100])	Requires very careful consideration and specialist design advice
Timber — joists, beams and columns	Charring of exposed timber	0–30 mins	Requires design for fire using reduced charred section. Requires flame flame-retardant[b]
Timber— solid timber such as unidirectional laminated timber (ULT) and CLT	Charring of exposed timber	30–60 mins	Requires design for fire using reduced charred section. Requires flame-retardant coating[b]

Notes:
[a] Fire resistance period may be reduced by use of mechanical restraint systems such as sprinklers. This is typically specified by the project fire engineer.
[b] There is a debate about use of timber for structure and facade of medium- and high-rise buildings, so latest guidance should be consulted before advising further.

9.4 A brief guide to disproportionate collapse and when it needs to be considered

Disproportionate collapse is where the *degree* of failure is considered disproportionate to the *cause* of failure.

Building Regulations Approved Document A categorises buildings into four classes (1, 2A, 2B, and 3). Depending on the class, the requirements of disproportionate collapse vary. Reference should be made to the Building Regulations for the full definition of the classes. Some typical cases are summarised in Table 9.6 (which is an abbreviated version of Table 4.1 in the IStructE *Practical guide to structural robustness and disproportionate collapse in buildings*[92]).

Table 9.6: Definition of building class for typical building types

Building type	Building class			
	1	2A	2B	3
Agricultural	All			
Houses	1–4 storeys	5 storeys (single occupancy)	≥6 storeys	
Hotels, flats and other residential buildings		1–4 storeys	5–15 storeys	≥16 storeys
Offices		1–4 storeys	5–15 storeys	≥16 storeys
Retail		1–3 storeys and $<2000m^2$ floor area per storey	4–15 storeys and $<2000m^2$ floor area per storey	≥16 storeys or $>2000m^2$ floor area per storey
Building to which public are admitted		1–2 storeys and $<2000m^2$ floor area per storey	≥3 storeys and $<5000m^2$ floor area per storey	≥3 storeys and $>5000m^2$ floor area per storey
Educational		1 storey	2–15 storeys	≥16 storeys

At scheme design stage, it is important to consider disproportionate collapse to ensure a feasible approach is available. Full detailed design can then be carried out at a later stage. Table 9.7 provides typical approaches for different building classes and different construction methods. They are based on the recommendations of Section 9 of the IStructE guide.

Under Class 2B, there are often multiple agreed strategies — from element removal to providing horizontal and vertical ties. We are required to provide at least one of these in our design. At concept stage we should consider which strategy would be most appropriate, and record our approach and any critical details that should be considered. If we choose to use the 'key element' approach, it is important to do some initial checks, to ensure that the elements that we are treating as 'key' are suitably sized, and to make a record. This will be particularly important on the Stage 2 drawings.

Table 9.7: Approaches to disproportionate collapse

Construction type		Building class			
		1	2A	2B	3
Steel frame		No additional considerations	Provide horizontal ties	Provide horizontal and vertical ties	Carry out a risk assessment but as a minimum provide the same as 2B
				Or notional element removal	
				Or key element design	
Concrete frame	In situ	No additional considerations	Rebar continuity typically provides adequate ties	Rebar continuity typically provides adequate ties	Carry out a risk assessment but as a minimum provide the same as 2B
				Or notional element removal	
				Or key element design	
	Precast		Ensure connections provide adequate horizontal ties	Ensure connections provide adequate horizontal and vertical ties (vertical ties can be hard to form)	
Timber	Loadbearing stud and joist floor (platform timber frame)	No additional considerations	Typically, connections provide requirements for horizontal ties and no further consideration is required at scheme design	Provide rim beam, and sequentially check wall removal	Carry out a risk assessment but as a minimum provide the same as 2B
	Solid timber (CLT and unidirectional laminated timber (ULT))			Ensure solid wall panels can act as beams, and connections can carry hanging floor load	
	Frame (sawn timber/glulam/ laminated veneer lumber (LVL))		Provide horizontal ties	Horizontal and vertical ties	
				Or notional element removal	
				Or key element design	
Masonry		No additional considerations	Provide horizontal ties	Not recommended. Provide horizontal and vertical ties — in practice this is very difficult to achieve	Not recommended

Richard Harpin
University of Sheffield

Rachael De'Ath
University of Bristol and Arup

10 Concept design calculations

10.1 Introduction

The purposes of calculations at concept stage are:

- to provide sufficient confidence that the proposed structural scheme is feasible
- to provide an initial estimate of member sizes/structural zones to allow other aspects of the design to be coordinated with the structure
- to provide sufficient structural information to allow the quantity surveyor (QS)/cost consultant to develop a robust initial cost plan

The number of calculations and the level of detail required in the concept design calculations will depend on:

- the onerousness of the constraints
- the complexity of the proposed structural system
- the familiarity of the design team with the proposed system

For a building with no unduly onerous constraints, using a structural system which the engineers have experience of designing, and the QS has experience of pricing, few, if any, explicit calculations may be required at concept stage. Conversely, for a building with very tight restrictions on structural zones, and/or a complex/unfamiliar structural system, more detailed calculations will be required to provide the required level of confidence at concept stage. If this is the case, it should be reflected in the programme, and the fee allowed for the concept design stage.

Before starting the concept design calculations, it is worth spending time reviewing the structure and constraints to establish what calculations should be carried out at this stage.

If you start with a plan and a clear view of what you are going to do at this stage, it is much easier to avoid endless calculations later. During concept design there will be lots of change. A concise set of key calculations which can be readily updated for changing geometry and requirements as the design evolves, will be of value to you as the structural engineer and to the rest of the team. It will also help to prevent excess and unnecessary expenditure in design fees for this stage.

10.2 Pre-calculation checks — load paths and construction sequences

10.2.1 Load paths

Structural calculations can be viewed as a way of demonstrating that the proposed load paths have sufficient capacity to transfer the applied loads to the ground, without undue deflection/vibration/settlement etc.

Before starting any calculations, it should be verified that the proposed structural scheme contains at least one continuous load path for transferring each applied load to the ground. This is best done by providing annotated load path diagrams. As a minimum, these will be required to show how gravity loads (due to dead and live loads) and lateral loads (due to notional horizontal loads and/or wind, earth and water pressures) are transferred to the ground. Other loading situations may need to be considered e.g. uplift on lightweight roofs. It may be helpful to note any assumptions made about member restraints etc. on the load path diagrams.

Figure 10.1 shows the gravity (a) and lateral (b) load paths for a simple steel framed building. Ideally, load paths should be as simple and direct as possible, but complex load paths may sometimes be required to meet the architectural constraints. No matter how complex the load paths become for a new building, as the concept engineer you have the luxury of being able to choose them, but this is not the case when working with existing buildings.

Figure 10.1a: Gravity load path for steel framed structure

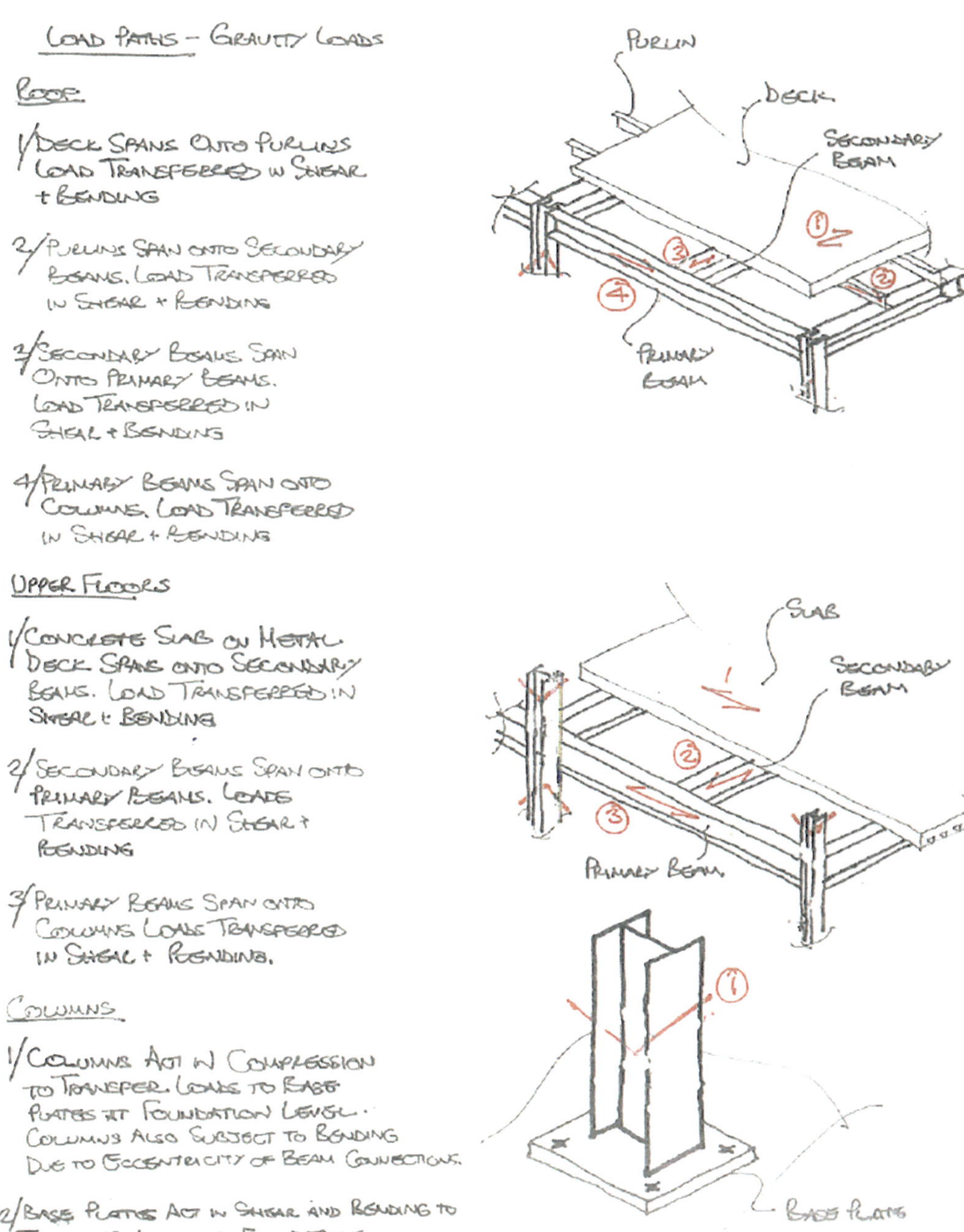

Figure 10.1b: Lateral load path for steel framed structure

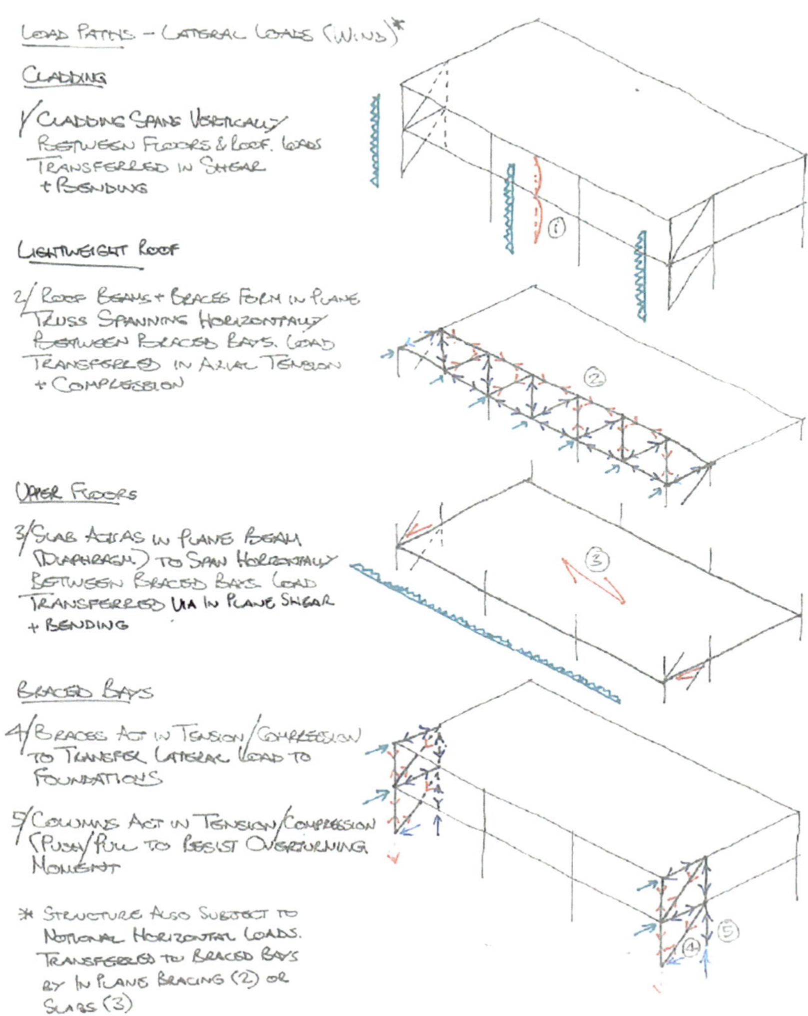

Before any alterations to the building are considered, the existing load paths must first be understood. This can sometimes require a considerable amount of detective work, particularly if the building has been altered over its lifetime. If load paths cannot be identified for all loads, or the capacity of the elements forming the load path to resist the existing loads cannot be justified, remedial works may be required. If the building is showing signs of distress these may be required as a matter of urgency.

Once the existing load paths have been understood, the impact of the proposed works on them needs to be considered. They may need to be strengthened, or new load paths provided to accommodate the works. It must be ensured that it is possible to maintain the required load paths at all times during the construction of the new works.

10.2.2 Construction sequences

Structural engineering is not an abstract discipline. It must be possible to translate the proposed design into reality. This is likely to be much easier if construction, or 'buildability', as it is sometimes referred to, is considered from the very start of the design process, rather than leaving it as something for the contractor to sort out, after the design for the permanent case has been completed.

It has always been good practice to take buildability into account in the design process but in the UK designers are now explicitly required to do so under Regulation 9 of the CDM Regulations[101]. This requires designers to eliminate, reduce and control risk at all stages of a project's life through design. Carpenter[102] states that to comply with this regulation, the permanent works designer must:

- understand how the structure can be constructed, and temporary works erected, used and dismantled safely
- determine if, by altering or supplementing the permanent works design in some way (so far as is reasonably practicable), temporary works risks arising from construction, use or dismantling of temporary works can be eliminated or reduced
- consider what useful information should be passed on to the contractor (via the pre-construction information)

The best way to demonstrate that the proposed design can be built safely is to produce a 'storyboard', showing a viable construction sequence (Figures 10.2–10.6). The contractor may ultimately choose to adopt a different sequence, but it will still be helpful for them to understand your assumptions regarding construction, and to quote Bill Baker, lead designer of the Burj Khalifa, *'A designer should never propose a structural system without at least one idea of how to build it, because if there is at least one way to build a structure, there very well may be several ways to build it. The unfortunate corollary is that if the designer has not or cannot think of a way to build it, it may be unbuildable'*[103].

Credible does not necessarily mean straightforward, or low cost, but it does mean safe. While the construction sequence should be kept as simple as practically possible, complex structures, or highly constrained sites, may well require complex construction sequences.

It is very likely that temporary load paths will need to be provided to transmit loads acting on the structure during construction to the ground. In many cases these are best resisted by temporary support structures (usually referred to as 'temporary works'). While the design of the temporary works is usually the contractor's responsibility, any assumptions made by the designer about the provision of temporary works should be clearly indicated in the construction sequence.

Alterations to existing structures can often require complex construction sequences and significant temporary works, to maintain load paths through construction and limit movement of the existing structure.

A good example is the 2016 conversion of the Commonwealth Institute in London into the Design Museum. Fig. 10.5 shows the significant temporary works required to limit movement of the 1960s roof to 5mm during the construction of a new basement.

Another good example is the refurbishment of the Arkwright Chemistry Building at Nottingham Trent University (Fig. 10.6). Demolition of the surrounding buildings meant that masonry walls, that had previously been internal, were now exposed to external wind loading, while the site investigation revealed that some of the walls were founded on fill, meaning there was a risk of unacceptable settlement. The decision was taken to underpin the existing foundations and rebuild the walls, which was complicated by the fact that English Heritage had deemed that the timber roof was historically significant and needed to be retained *in situ* during the works. Fig. 10.6 shows the construction sequence adopted, together with photos of the works in progress and after completion.

In some cases, it will not be feasible, or economical, to provide temporary works, so the permanent structure will need to be designed to resist the construction loads. This may require different load paths to those in the permanent structure e.g. steel edge beams supporting precast units will be subject to torsion during construction until the structural topping has set. Similarly, prefabricated elements, which will be supported at both ends in the permanent case, may be required to act as cantilevers while being lifted/launched into place. If the requirements for the construction case are more onerous than the permanent case, they will need to be taken into account in the concept design calculations.

10.3 Empirical design — does it look right?

Until the Industrial Revolution, structural design was largely empirical, with the layout and sizes of structural elements based on rules of thumb, which had been developed, refined and handed down by generations of master builders. While the development of structural analysis methods over the last 250 years has allowed for the construction of ever more ambitious structures, with greater economy and a higher level of confidence in their safety, empirical design still has its place in concept design.

Empirical design is essentially designing a structure so the size and layout of the structural elements 'looks right'. This sense of what looks right comes from experience and observation. While most people will have developed some feel for what looks right structurally, whether through childhood experiments with building blocks, or casual

Figure 10.2a: Concept stage construction sequence 'storyboard' for a proposed gridshell

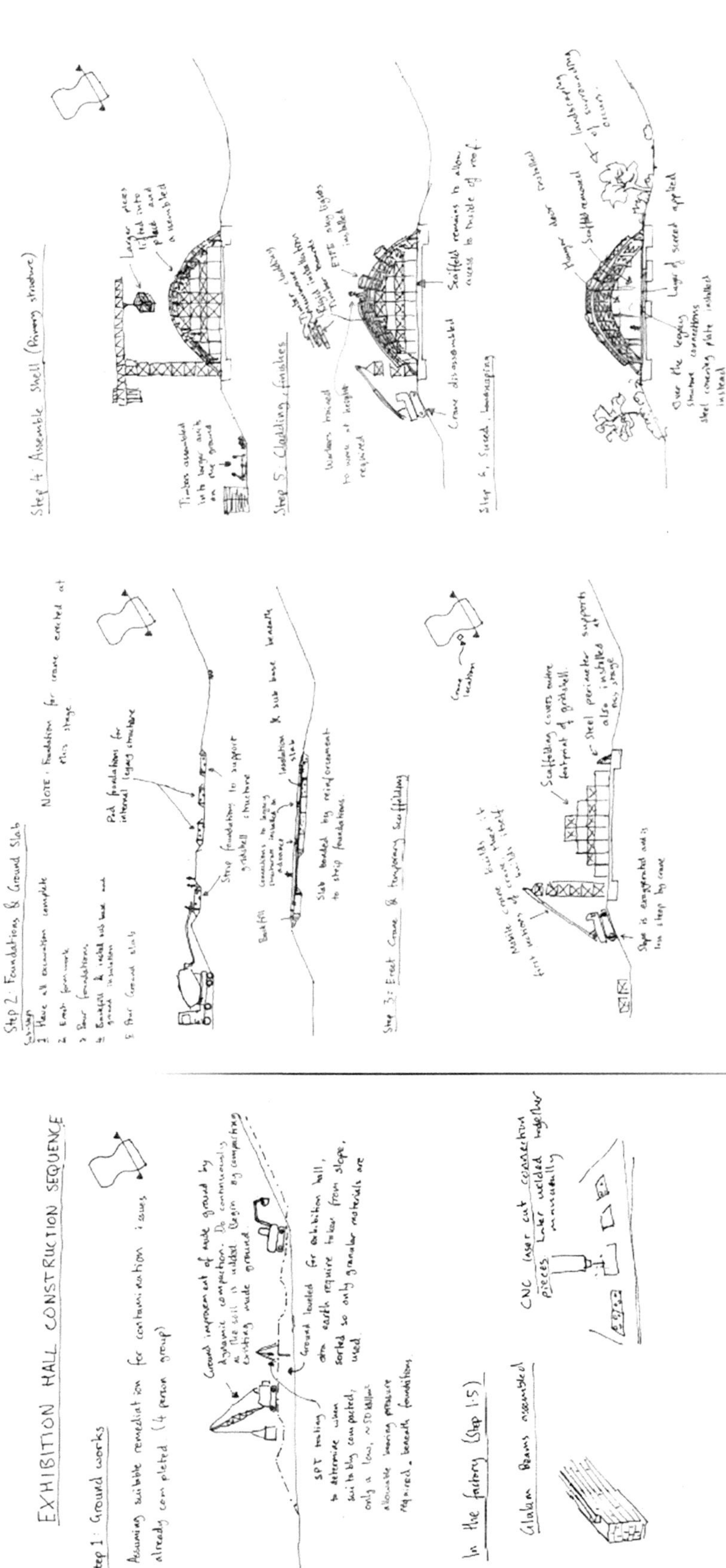

Figure 10.2b: Concept stage construction sequence 'storyboard' for King Abdullah Sport City, Saudi Arabia

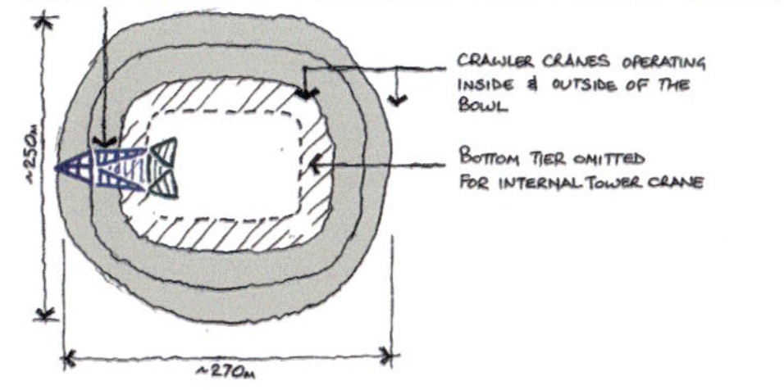
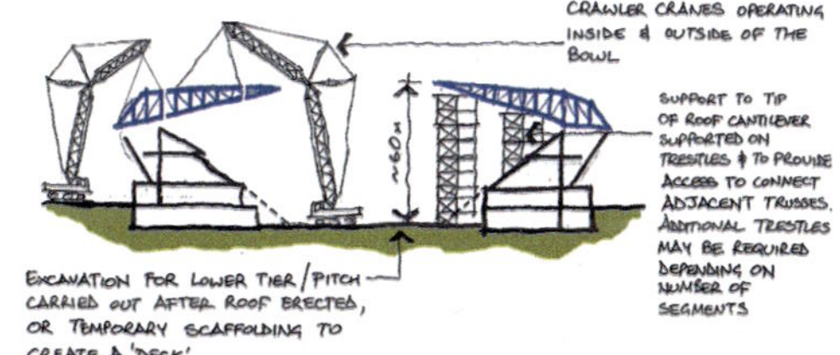
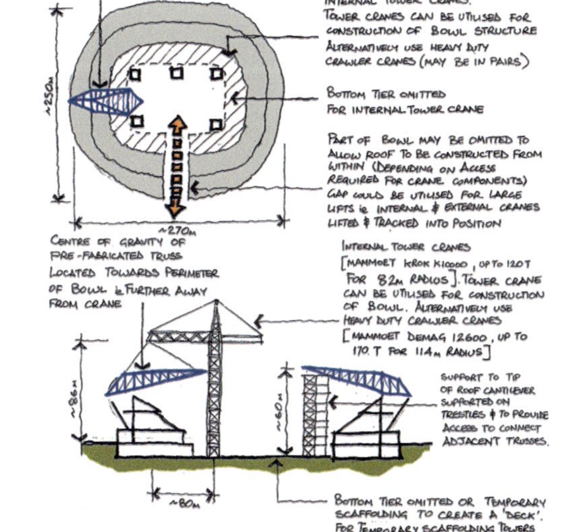
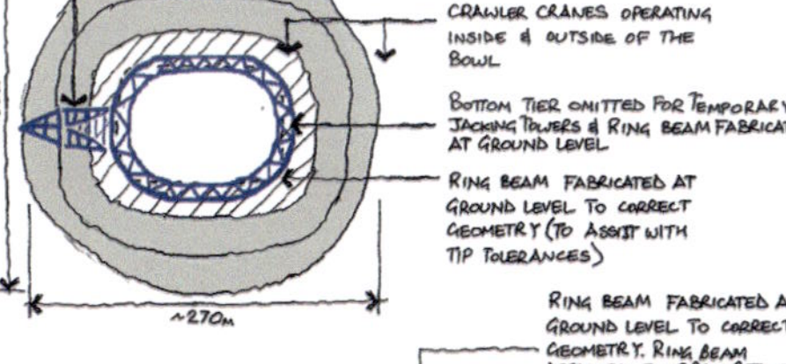
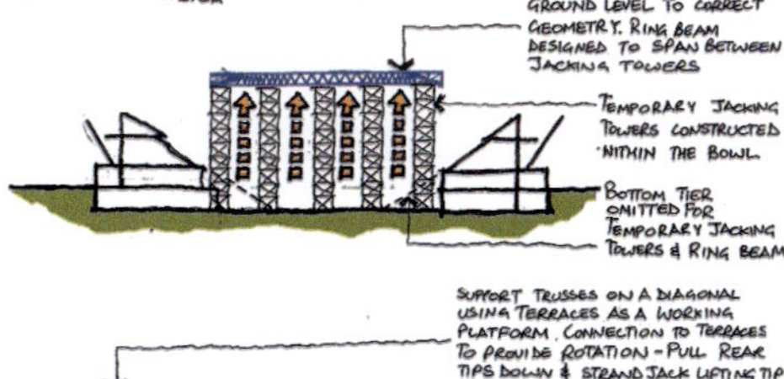

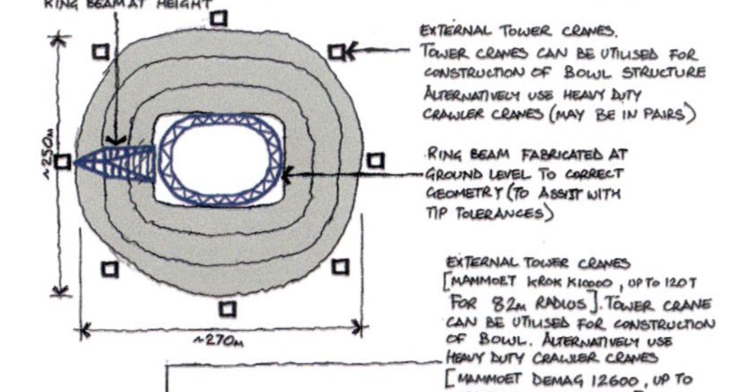
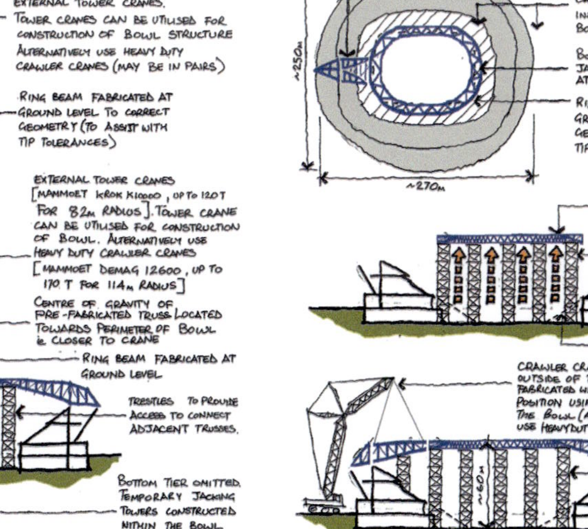
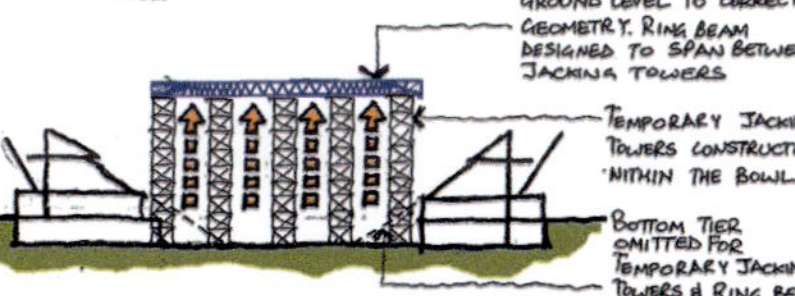
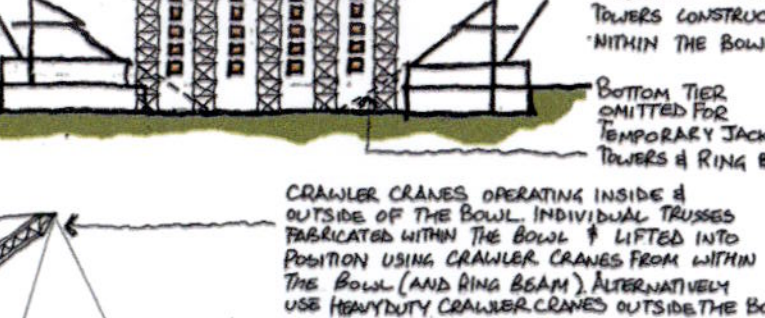
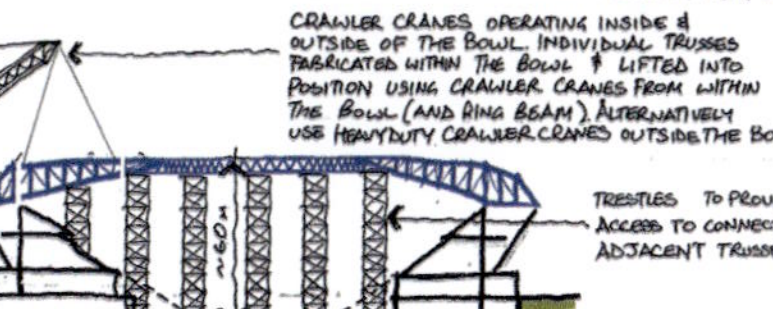
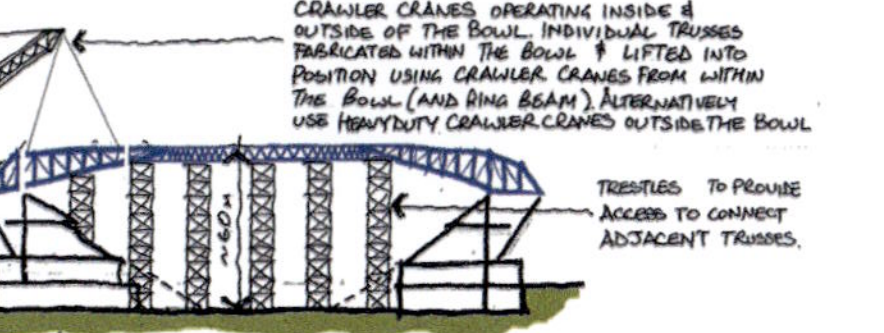

Figure 10.2c: Concept stage construction sequence 'storyboard' for Northampton Bridge

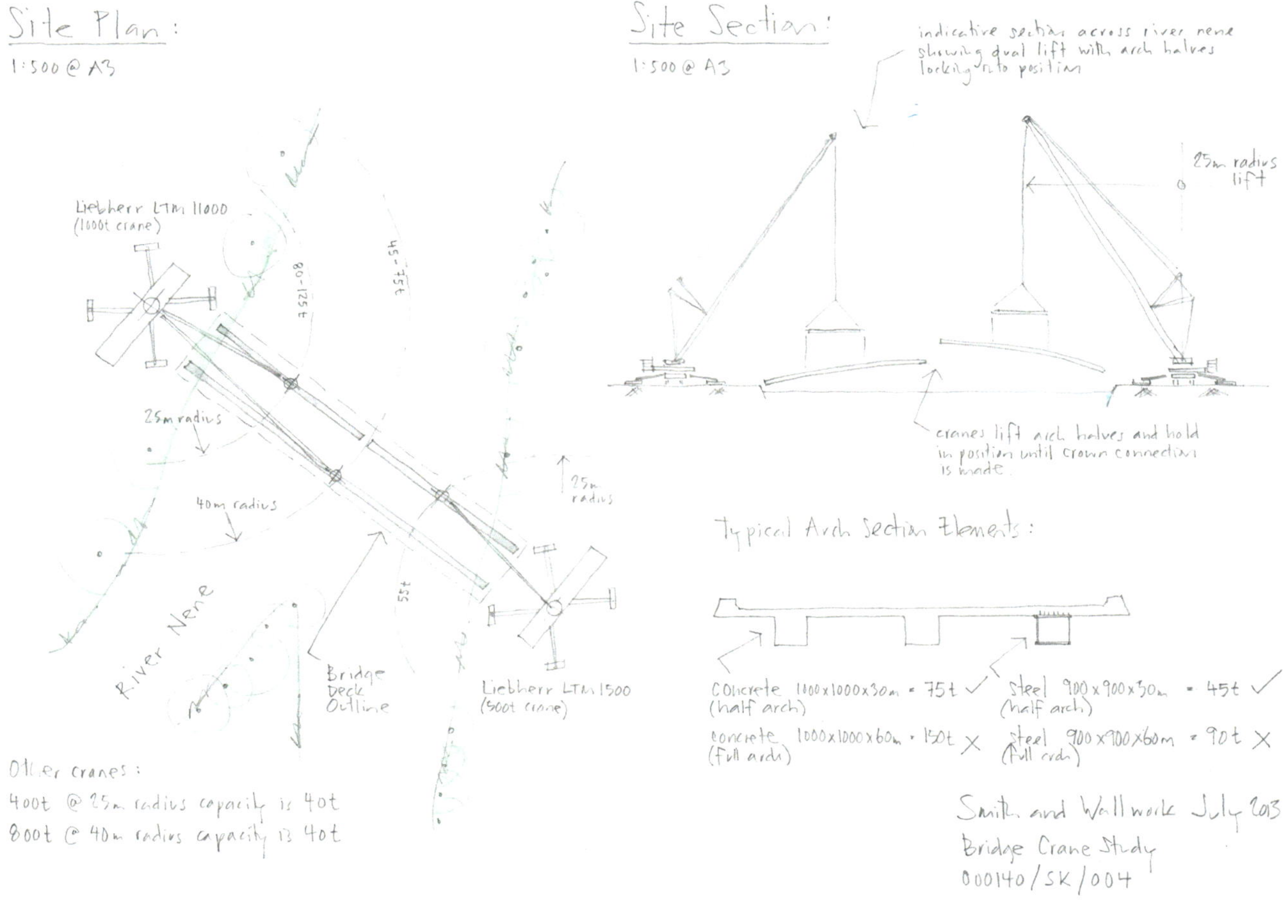

Figure 10.2d: Installation of Northampton Bridge

Figure 10.3: Structures with complex construction sequences — Aquatics Centre, London

Figure 10.4: Structures with complex construction sequences — Weald and Downland Gridshell, West Sussex

Figure 10.5: Temporary works during conversion of Design Museum, London

Figure 10.6a: Construction sequence 'storyboard' for remedial works to Arkwright Chemistry Building, Nottingham Trent University

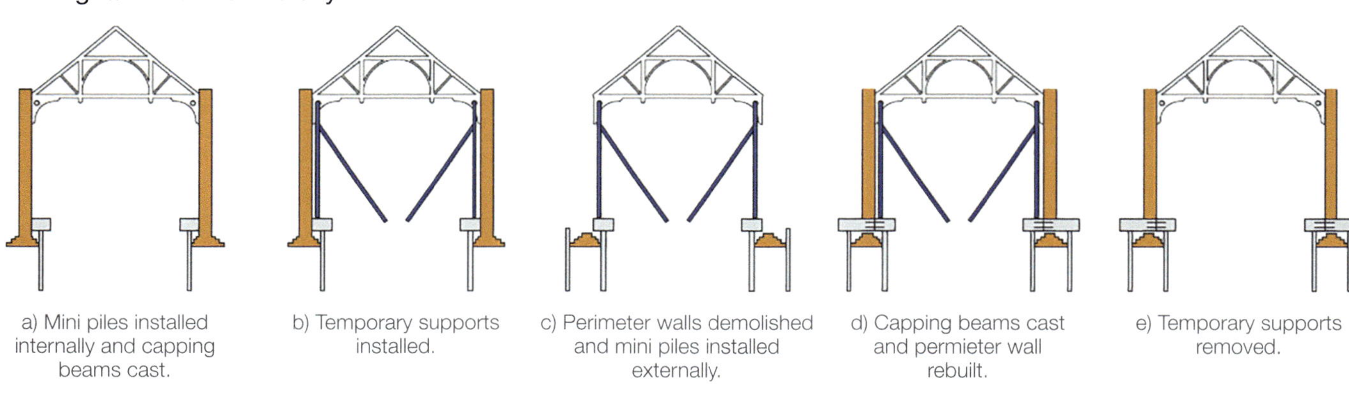

observation of the built environment, structural engineers wanting to successfully use empirical methods for the concept design of structures need to develop their intuition for what is structurally correct to a much deeper level. This can be done by drawing directly on the designer's personal experience, critical observation of the built environment, reference to published guidance or research into the literature.

One reason that concept design is usually carried out by experienced engineers is that they can draw on a much greater range of personal experience, or direct precedents, to inform their sense of what feels right for a particular project. Junior engineers can accelerate this process by critically reflecting on any project they work on. Some questions to ask include:

- What are the particular site/building constraints affecting the structural design?
- What are the loads applied to the structure and how does the layout of the structural elements provide load paths to transfer these to the ground? What alternative structural layouts/systems could you use to provide the required load paths? Why do you think the chosen solution was preferred to the alternatives?

Figure 10.6b: Remedial works in progress, Arkwright Chemistry Building, Nottingham Trent University

Figure 10.6c: Completed Arkwright Chemistry Building, Nottingham Trent University

- Why has a particular structural material been adopted? Why do you think this has been preferred to other materials?
- Are the sizes of the structural elements as you would expect, based on your experience and/or published guidance? If the element sizes are not as you expect, why do you think this is?
- Can you think of a viable construction sequence for the project? If your sequence is different to that proposed/adopted, why do you think this is?

Having asked yourself these questions, you should take advantage of any opportunity to discuss your thoughts with the engineers who carried out the concept design.

Another excellent way to develop your structural intuition is to attend design reviews, particularly concept reviews (Sections 10.6–10.7). To get the most out of these, you should prepare by critically reflecting on any information available in advance. While you may not have much to contribute to reviews initially, do not be afraid to ask questions. Sometimes it is the simplest, most fundamental, questions that need to be asked, and it is always possible that you may have picked up something that the design team has overlooked.

Critical observation of the built environment is essentially the same process as previously outlined, but applying it to buildings you see around you, rather than projects you are directly involved in. The advantage of this approach is it gives you a much wider range of projects to hone your concept design skills on. The disadvantage is that you won't have as much information available to help you understand the project constraints and the solutions adopted. With a little research, it may be possible to find more information in published case studies etc. Projects under construction are often easier to understand than completed projects, where the structure is often hidden behind finishes. Many experienced engineers cannot help but critically observe any new building or construction site they encounter.

Published guidance, such as span/depth tables, is the modern day equivalent of the master builders' rules of thumb, and represents the collected experience of the profession. Guidance can be empirical e.g. span-to-depth tables for steel beams, or based on calculation e.g. manufacturer's load-span charts for precast units. Examples of published guidance are given in Chapter 8 (Tables 8.4–8.6). Further sources of guidance include:

- **General**
  - Cobb, F. *Structural engineer's pocket book: Eurocodes (3rd edition)*. Boca Raton, FL: CRC Press, 2015
- **Reinforced concrete**
  - The Institution of Structural Engineers. *Manual for the design of concrete building structures to Eurocode 2*. London: IStructE Ltd, 2006
  - Brooker, O. *Concrete buildings scheme design manual: a handbook for the IStructE chartered membership examination, based on Eurocode 2*. Camberley: MPA – The Concrete Centre, 2009
  - Goodchild, C.H., Webster, R.M. and Elliott, K.S. *Economic concrete frame elements to Eurocode 2: a pre-scheme handbook for the rapid sizing and selection of reinforced concrete frame elements in multi-storey buildings designed to Eurocode 2*. Camberley: The Concrete Centre, 2009
  - 'Element size estimation. Technical Guidance Note 17 Level 1'. *The Structural Engineer*, 90(10), October, 2012, pp. 32–35
- **Structural steelwork**
  - The Institution of Structural Engineers. *Manual for the design of steelwork building structures to Eurocode 3*. London: IStructE Ltd, 2010
  - 'Element size estimation'. Technical Guidance Note 17 Level 1. *The Structural Engineer*, 90(10), October, 2012, pp. 32–35
- **Timber**
  - The Institution of Structural Engineers. *Manual for the design of timber building structures to Eurocode 5 (2nd edition)*. London: IStructE Ltd, 2020
  - Norman, J. *Structural timber elements: a pre-scheme design guide (2nd edition)*. High Wycombe: Exova BM TRADA, 2018
- **Masonry**
  - The Institution of Structural Engineers. *Manual for the design of plain masonry in building structures to Eurocode 6 (2nd edition)*. London: IStructE Ltd, 2018
  - HM Government (2013). *The Building Regulations 2010. Approved Document A. Structure*. Available at: https://www.gov.uk/government/publications/structure-approved-document-a [Accessed: March 2020]

Most published empirical guidance will come with explicit caveats regarding the limits of its applicability. However, in some cases the caveats may be implicit e.g. span/depth ratios for beams are not suitable for transfer beams subject to large point loads. Particular care must be paid if empirical guidance is being used to size structures at a larger scale than it is was developed for, to ensure that the design is not governed by effects that are not significant at a smaller scale e.g. dynamic behaviour of long-span structures.

For structural elements/systems which are outside the design team's direct experience, and are not covered by published guidance, precedents, which could provide a starting point for the concept design, may be found in the relevant literature. These could include:

- Peer-reviewed papers in academic journals
- Peer-reviewed papers published by learned societies e.g. *The Structural Engineer*
- Papers in company journals e.g. *The Arup Journal*
- Papers/articles in industry publications e.g. *Building*
- Papers/articles in trade publications e.g. *Concrete Quarterly*, *New Steel Construction*, *TRADA Yearbook*
- Books on structural engineering/architecture
- Industry/project websites

As with any literature review, the relevance and reliability of any material found will need to be critically evaluated, to determine how applicable it is to the proposed structure.

10.4 Concept design calculations

If empirical design methods are not considered to be applicable, or more certainty is required, explicit calculations will need to be undertaken. In many cases, it will only be necessary to provide calculations for critical elements, which are not covered by the empirical guidance. For instance, if we consider the structure in Figure 8.14, it would be reasonable to initially size the typical elements above ground floor using empirical guidance, but calculations will be required to size the transfer beams at ground floor.

The extent of calculations undertaken will also depend on the designer's experience and confidence in using empirical guidance. Inexperienced designers may choose to carry out more calculations for a range of standard elements e.g. most heavily loaded beam/column, to build up their confidence in empirical guidance. More experienced designers may be happy to rely on empirical guidance where it is based on calculations by others e.g. load-span tables, but will choose to supplement it with their own calculations where the guidance is based on rules of thumb e.g. span-to-depth ratios, or rely solely on the empirica guidance for simple structures e.g. domestic buildings, fully covered by Building Regulations Approved Document A[80].

As with any structural engineering calculations, there are three elements to concept design calculations:

1. Establishing the loads acting on the structure.
2. Determining the forces in the structural elements, due to the applied loads.
3. Confirming the structural elements can resist the forces acting on them, without undue deflections, vibrations etc.

The purpose of concept design calculations is not to fully analyse or design the structure, but to provide enough confidence that it will be possible to show the structure works during detailed design i.e. to demonstrate that the proposed load paths are feasible. At concept stage, the overall design of the project is generally still quite fluid and a number of iterations will normally be required to reach the optimum design. Trying to produce detailed calculations for each iteration will result in aborted work and wasted fees, and will not allow you to respond sufficiently quickly to the developing/changing design. You need to be able to produce some simple calculations, so that you can confidently provide structural advice. No engineer wants to be handed a project to take to detailed design stage, where everything in the scheme design was 25mm too small. This will make the rest of the project really hard work. At a later stage of design, if you can make small reductions in section size it will generally be viewed as a benefit — nobody will thank you for needing to make things slightly bigger.

10.4.1 Establishing the loads

Loads on structures fall into two categories — permanent and variable. As the name suggests, permanent loads/ actions are not expected to change during the life of the building. They can be split into 'dead' loads, due to the weight of the structural elements, and 'superimposed' loads, due to the weight of non-structural elements e.g. cladding, floor finishes, services in ceiling/floor voids. Similarly, variable loads/actions are those which are expected to change during the life of the building e.g. loading due to people, furniture, machinery, or environmental loads such as snow and wind. Variable loads are sometimes referred to as 'imposed loads' or 'live loads'. It is important to have a feel for the order of magnitude of loads from all three of these categories when carrying out scheme design, as this will help you to understand when changes are likely to have a more significant impact.

The dead loads will depend on the structural system and material adopted. Definitive guidance on the density of construction materials is given in BS EN 1991-1[104], and reproduced in condensed form in Cobb[3] and Brooker[85].

For concrete structures, it is necessary to estimate the depth of the structure using span-to-depth ratios, to determine an initial dead load to use in calculations. The unit weight of reinforced concrete should be taken as 25kN/m^3. For steel buildings an allowance is often made on plan area for the weight of the steelwork — typically 0.5kN/m^2 at floor levels and 0.3kN/m^2 for lightweight roofs. These allowances may need to be increased for structures with complicated framing layouts, which require more elements than in a typical floor/roof. Cobb recommends an allowance for plan area of 0.3kN/m^2 for timber floors (joists and boards/plywood) and 0.2kN/m^2 for timber roofs.

If sufficient detail is available, superimposed dead loads can be calculated from material densities. However, this is rarely the case at concept stage. Cobb and Brooker give guidance on loads due to cladding, finishes and services for typical buildings. Typical practice used to be to take an allowance of 2.5kN/m^2 on-plan for buildings with heavyweight (blockwork) internal walls/partitions — however, these loads depend on the storey-height of the given building and Brooker recommends an allowance of 3.0kN/m^2 on plan.

For buildings with lightweight internal walls/partitions e.g. stud walls, it is usual to take an allowance of 1.0kN/m^2 on-plan. However, this should be treated as a variable load/action, rather than a permanent one.

Definitive guidance on imposed loads for structures in the UK is given in the UK National Annex to BS EN 1991-1-1[105], summarised in Cobb. If the use of parts, or all, of the building is unclear it would be prudent to adopt a relatively generous loading allowance for concept design. The initial loading allowances should be reviewed, and reduced if possible, as more information on the use becomes available, to prevent wasteful design. Future flexibility, including possible change of use, should also be considered. In some cases, the client will provide explicit guidance on this e.g. for speculative offices, the developer may require an imposed loading allowance higher than 2.5kN/m^2 to be used, to allow for future flexibility, but consideration should be given to how much future flexibility is really required, to prevent wasteful design. An allowance of 7.5kN/m^2 should normally be made for services plant rooms, although this may need to be increased if heavier items (e.g. water tanks, large generators, UPS etc.) are required. For buildings with specialist uses every effort should be made to establish if there are any items e.g. heavy machinery or large sculptures, which may govern the design of the floor slabs.

In all cases, the imposed load allowances adopted should be clearly stated in the concept design documentation, with separate loading plans produced if necessary.

It can be difficult when discussing loading with the client or other members of a project team, for them to appreciate what 'loading allowances' actually means. In these situations, it can be helpful to relate the loads to something tangible, such as the images in Figure 10.7.

Imposed load reduction is not normally taken into account for the design of structural elements at concept stage, but may be taken into account when determining the loads applied to the foundations.

10.4.2 Determining the forces — approximate structural analysis

All structural analysis is an approximation of reality, but this is particularly true at concept stage. The limited time and resources available, and the probable need to consider a number of iterations of the design, requires the analysis model to be made as simple as possible, while still representing the behaviour of the structural element/s under

Figure 10.7: Variable (live) loads/actions

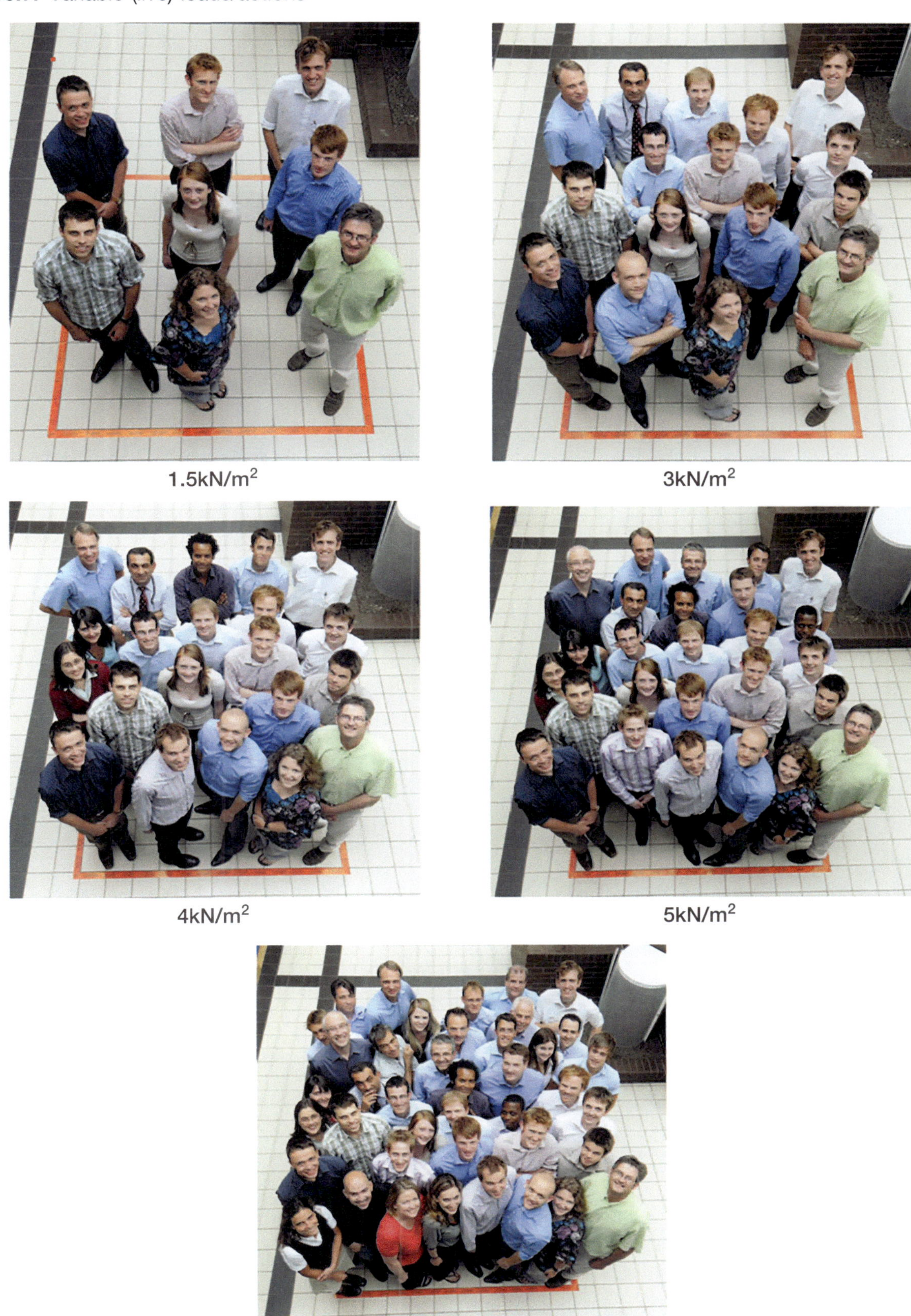

consideration. The resulting analysis may not give the closest approximation to reality possible, but will have the virtues of being quick to execute and easy to check, and in the words of John Maynard Keynes, quoted by Hugh Morrison in *Structural engineering art and approximation*[106]:

"it is better to be roughly right than precisely wrong".

The starting point of any approximate analysis is to reduce the structure to its simplest possible form that is still consistent with the assumed load paths. Even the tallest skyscraper can be approximated to a cantilever, to calculate the forces due to wind loading (Figure 10.8).

Vertical loads are generally considered through a load trace or run-down. This simple approach of attributing all of the loads to vertical elements of structure, provides a simple method of verifying the vertical load path, as well as providing approximate foundation loads. When carrying out a load trace, it is helpful to work with unfactored, serviceability loads, introducing load factors only at the end, if necessary.

Figure 10.8: IFC, Hong Kong (and simplified structural diagram)

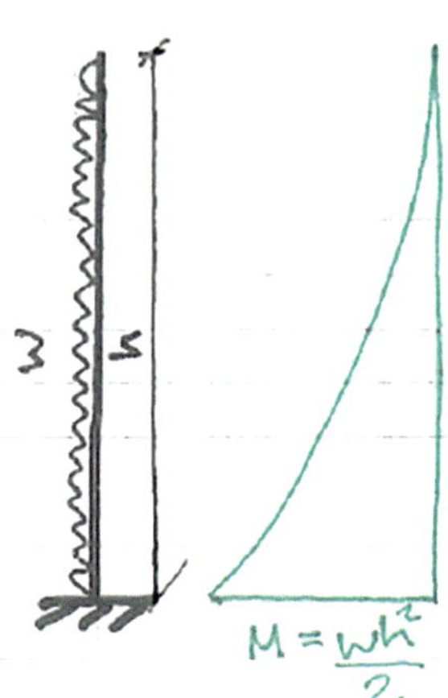

Care must be taken that the structure is not oversimplified. For example, a first approximation of the forces in a cable-stayed structure can easily be obtained, if it is reduced to a statically determinate structure, by assuming the deck/roof is simply supported between cables, the cables are inextensible, and rotation/deflection at the top of the masts is insignificant. In practice, none of these assumptions are likely to be entirely true, but for a structure with relatively short cables and stiff masts they should give a reasonably accurate approximation of the structural behaviour. For longer-span structures e.g. the cable-stayed bridge in Figure 10.9a, the extension of the cables and the rotation/deflection of the top of the mast is likely to be significant, resulting in much bigger sagging moments in the deck (Figure 10.9c), than predicted by the overly-simplified model (Figure 10.9b).

Initial design of shell structures can be carried out by considering a strip of the shell as an arch (see Section 10.4.4.4 for more on approximate analysis of arches). John Chilton's book on Heinz Isler[107] includes a copy of a single page concept design for a shell by Isler using this method. As with any compression structure, the possibility of buckling needs to be considered. Chilton states that Isler had a formula he used to satisfy himself at concept stage, that

Figure 10.9a: River Aire Centenary Footbridge, Leeds

Figure 10.9b: Overly-simplified structural diagram

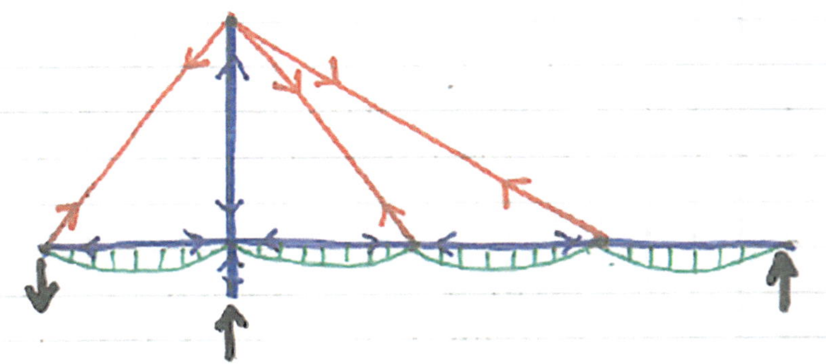

Figure 10.9c: More realistic structural diagram

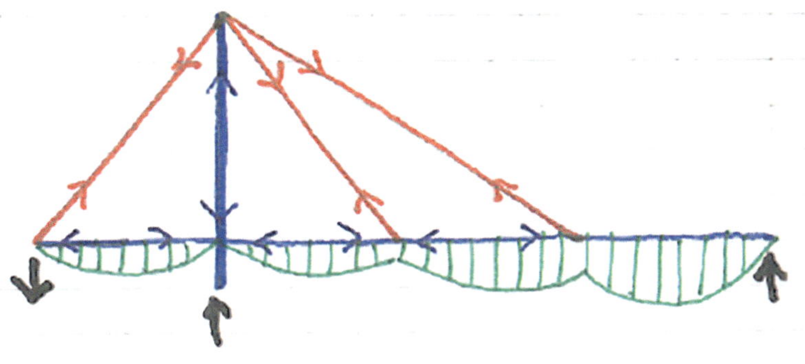

buckling was unlikely to be a problem, but does not provide any further details. In the absence of this, or similar information, the best way to check buckling of a shell structure is likely to be a nonlinear computer analysis, but providing the stresses are low (as is usually the case for a shell structure), this can reasonably be left until scheme design stage.

A similar method can be used for the initial analysis of a grid shell, with the forces in the arch strip being divided between the discrete members. An analogous method can be used for the initial analysis of tensile structures, by considering a strip of the structure as a hanging chain, or funicular, to determine the forces. Morrison contains a good example of the approximate analysis and design of a membrane structure, along with a wealth of other information on approximate structural analysis and design. Byfield[108] and Millais[109] also contain some very useful information and examples. Cobb, Brooker and the IStructE publications (including its Essential Knowledge Series) also provide useful methods for approximate analysis. Some of these methods are discussed in more detail in Section 10.4.4.

Simple models lend themselves to hand calculations, but if a number of variations on the structural system need to be investigated e.g. looking at different column grids to find the best solution, it might make sense to automate the analysis process, either by using a spreadsheet or parametric software, such as Grasshopper or Dynamo. If this is the case, the automated approach should first be verified against the hand calculations and simple checks carried out on the results from subsequent analyses e.g. if the span doubles, do the reactions double and the moments quadruple?

If the structural system is very complex, or it is difficult to evaluate the critical design criteria by hand e.g. floor vibrations in a research facility with very sensitive equipment, it may be necessary to carry out more detailed computer modelling at concept stage. Before starting any modelling, the reasons for doing it, and the expected outcomes, should be clearly defined. If the reason is because the proposed structural system is very complex, can it be simplified? The model should be kept as simple as possible while capturing the required behaviour. Does the whole structure need to be modelled or just the critical elements? Is a 3D model necessary or will a 2D model suffice? If possible, the model should be built up in stages, starting with a simple approximation, which can be verified against hand calculations. At each stage, any changes in the behaviour of the model should be explained before proceeding to the next stage.

The conversion of the Commonwealth Institute into the Design Museum (Fig. 10.5) provides a good example of this process. Simple hand calculations were vital in allowing the engineers, Arup, to develop an understanding of the existing structure, before producing a series of increasingly sophisticated analysis models to fully justify the proposed interventions (Figures 10.10a–b).

If the model is relatively simple it should be possible to justify the behaviour numerically, but as it becomes more complex, the emphasis will need to be on developing a qualitative understanding of the structural behaviour. Qualitative analysis is essentially checking if the analysis results looks right. Some questions to ask include:

- Is the model deflecting as expected?
- Are the shapes of the moment diagrams as expected?
- Are elements in tension/compression as expected?
- Is the relative magnitude of forces in the elements as expected?
- Is the direction and relative magnitude of the reactions as expected?

The IStructE's *Essential Knowledge Text*: *Principles for computer analysis of structure*[110] is perhaps the best introduction to computer analysis, while *Modern structural analysis: modelling process and guidance* by MacLeod[111] and *Computational engineering* by Debney[112] are excellent resources for more detailed guidance. Probably the best introduction to qualitative analysis is *Understanding structural analysis* by Brohn[113].

10.4.3 Sizing the elements
Once the forces have been calculated the critical elements can be sized. Approximate methods should be used wherever possible, rather than full code methods. Cobb, Brooker and the IStructE's Eurocode manuals provide useful guidance on approximate methods for sizing elements, as does Morrison. Resistances of steel sections can be obtained from the 'Blue Book'[5].

Figure 10.10a: Extract from initial hand calculations for Design Museum, London

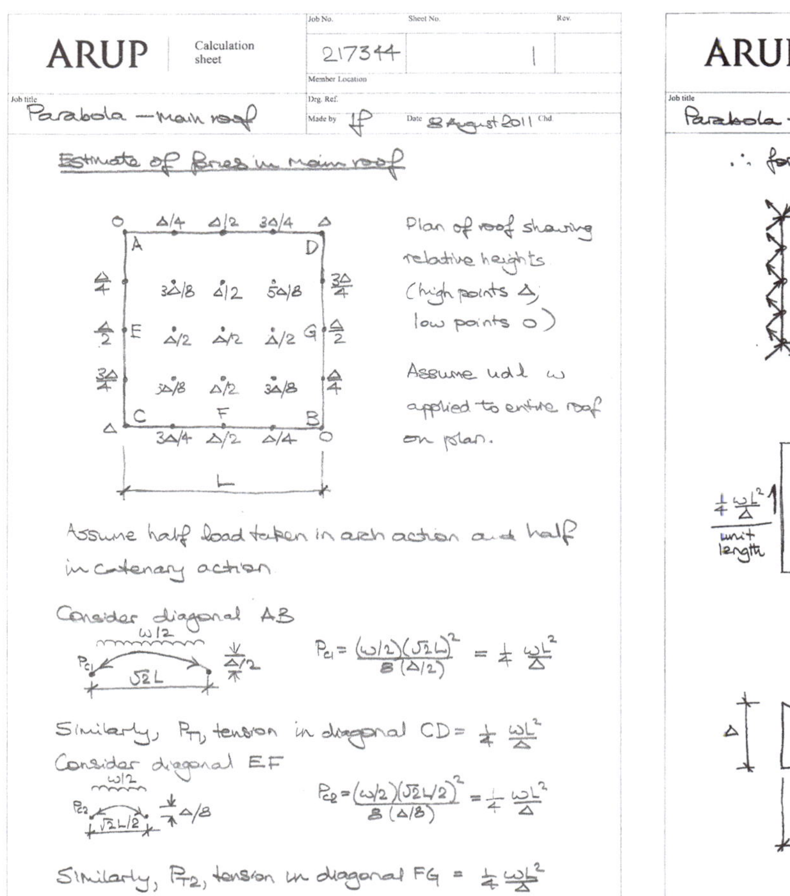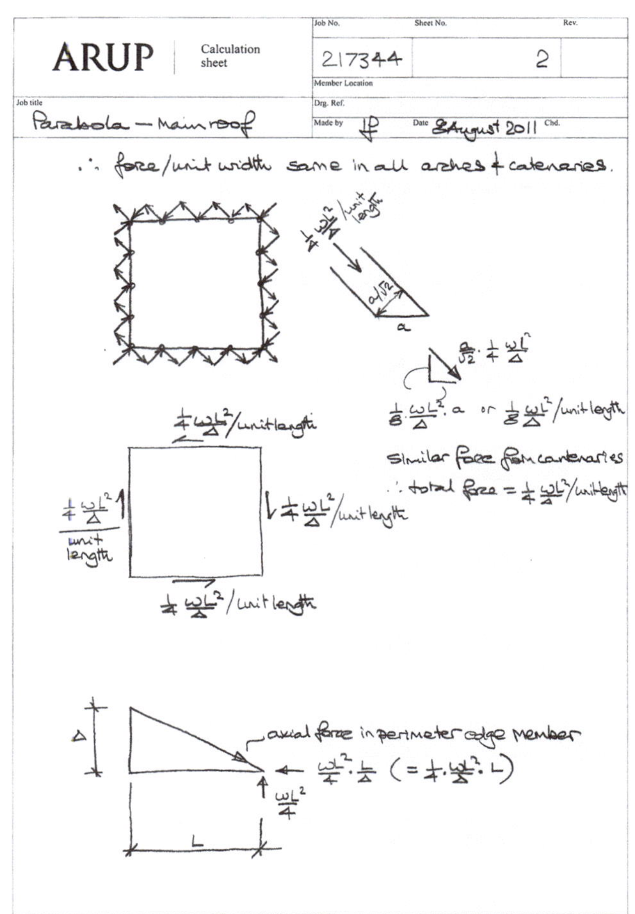

Figure 10.10b: Evolution of analysis model for Design Museum, London — from simple stick model of roof to finite element model of roof and support structure

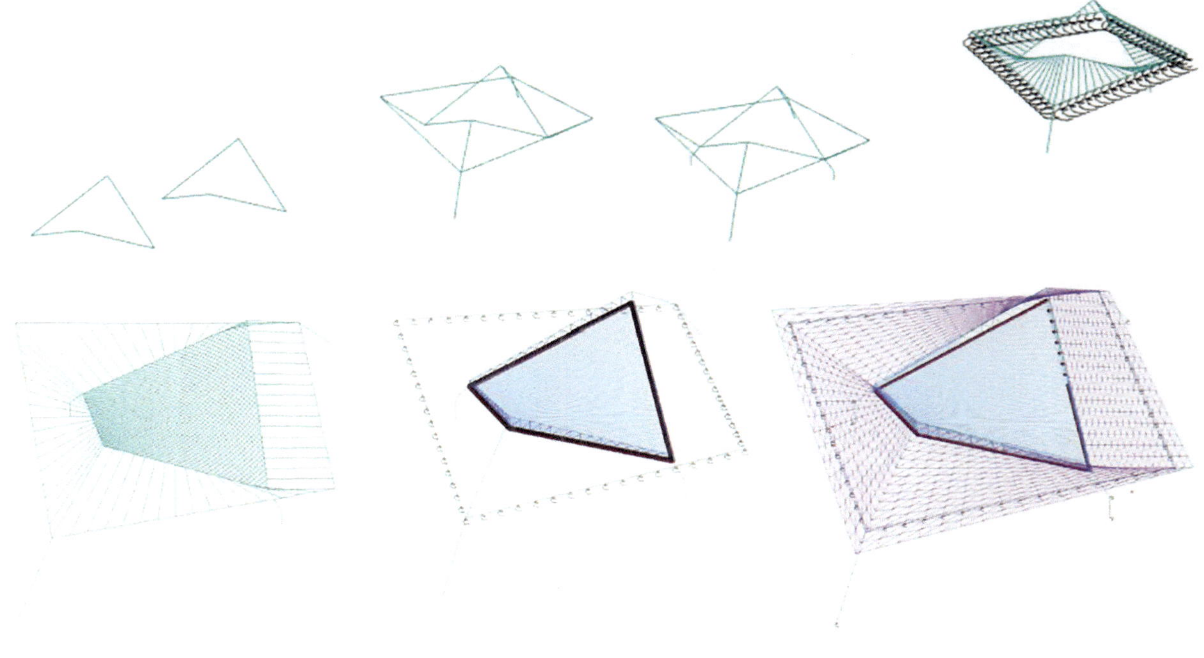

10.4.4 Typical elements
10.4.4.1 Bending elements

Bending elements are some of the most straightforward, and are generally well understood. Loads applied are resisted by the element, which will develop tension and compression in the outermost parts e.g. in a reinforced concrete beam in sagging, tension develops in the bottom part of the beam below the neutral axis (which is resisted by the reinforcement), and compression in the top part of the beam, above the neutral axis (Figure 10.11).

Figure 10.11: Simply-supported beam with compression in the top above the neutral axis and tension in the bottom

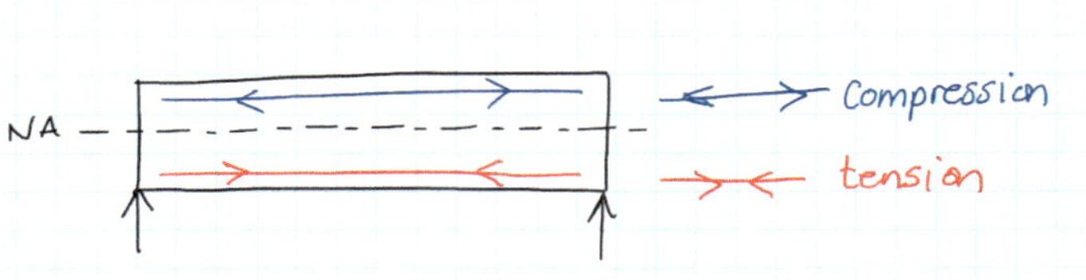

The simplest way to consider bending elements is to assume that they are simply-supported, with only reaction forces being transferred to the supporting elements. This will provide a maximum moment in the bending element for initial sizing purposes, and is often an approach adopted at concept design stage.

At a later stage, where more detailed calculations are carried out, consideration of the actual amount of fixity to the support, and how much moment will be transferred to supporting elements, can be given.

It is important, when considering bending elements as simply-supported in this way, not to underestimate or ignore the moments transferred when designing the supporting elements.

It is not necessary to size all elements at this concept design stage. Before carrying out any calculations, or sizing, of bending elements, it is worth spending time looking at the structure, to determine which are the key elements you are going to size. These may be elements with the longest span or the heaviest loading, or simply a typical floor beam which will be used extensively.

The first step when sizing bending elements is to calculate the maximum bending moment and shear force in the element. As far as possible, this should be carried out using simple hand calculations and standard formulae, such as those presented by Cobb.

When considering the loading on a bending element, it may be appropriate to simplify this for ease and speed of calculation. If there are a series of point loads along the beam, can they be grouped into a smaller number of larger point loads, or considered as an equivalent 'uniformly distributed load' (UDL) (Figure 10.12)?

Figure 10.12: Loads can be simplified to allow for simple hand calculations at concept design stage

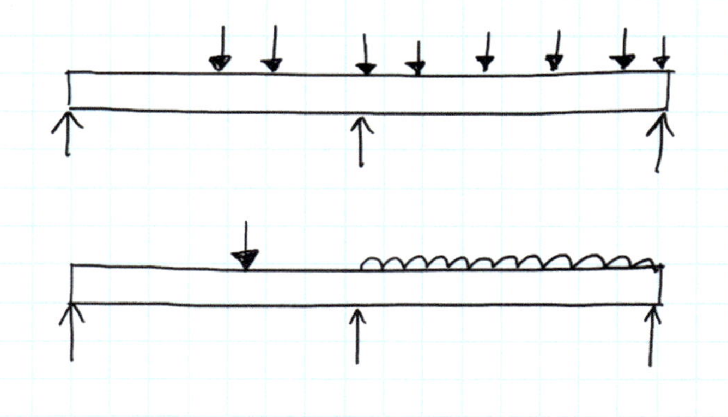

Once you have calculated the applied moments and shear forces in the element, initial sizing of element depth can be based on empirical guidance (Tables 8.4–8.6).

These initial ratios are useful to provide a starting point for element depth, which can be adopted for simplified calculations.

There are a number of calculations and considerations to remember for the various different structural materials.

Reinforced concrete

Once you have an approximate section depth and bending moment, a preliminary calculation of area of reinforcement required should be carried out to check that the section will meet the requirements of minimum and maximum reinforcement, and that the reinforcement required can be fitted into the section.

The area of reinforcement required can be calculated as:

$$A_\mathrm{s} = M_\mathrm{Ed}/0.87f_\mathrm{yk}z$$

Where
z = $[0.5 + (0.25 - 0.88K)^{0.5}]d_\mathrm{eff}$ and is limited to $0.95d_\mathrm{eff}$
K = $M_\mathrm{Ed}/f_\mathrm{ck}bd_\mathrm{eff}^2$
A_s = cross-sectional area of reinforcement
B = width of section
d_eff = effective depth
f_ck = characteristic compressive cylinder strength of concrete
f_yk = characteristic tensile strength of reinforcement
M_Ed = design moment

When considering your d_eff value you should consider likely cover to meet the requirements of both durability and fire.

At concept design stage, sizing bending elements (so there is no requirement for compression reinforcement) is generally adopted and K is often limited to 0.1.

Shear stress should also be calculated using standard formulae and, at concept design stage, should be limited to $2\mathrm{N/mm}^2$ or less.

$$v_\mathrm{Ed} = V_\mathrm{Ed}/bd_\mathrm{eff} \leq 2\mathrm{N/mm}^2$$

Where
V_Ed = design shear force
v_Ed = design shear stress
b = width of section
d_eff = effective depth

Deflection can govern in reinforced concrete design, and a greater depth than initially considered may be required. The depth may also need to increase to ensure there is adequate space to fit all the required reinforcement. Early consideration of some key reinforcement detailing should demonstrate this.

Steel

When considering steel elements, the Blue Book can be used to provide approximate sizing, based on calculated moments in the element, typically compared to buckling resistance moments of standard sections.

At concept design stage, when using the Blue Book, it should be assumed that C1 (the factor which depends on end restraint and loading) is 1.00. Consideration also needs to be given to restraint, and whether the buckling length of your element is in fact the whole length (as used to calculate your maximum bending moment), or a smaller proportion of the overall length e.g. when secondary beams are provided at mid span or third span (Figure 10.13).

Figure 10.13: Lateral torsional buckling length for long-span steel beams

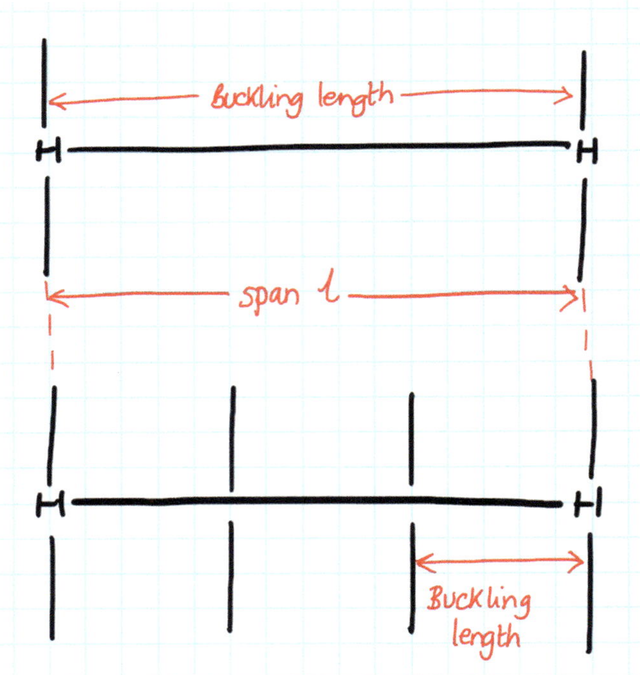

The depth of long-span steel beams will typically be governed by deflection and/or vibration so, as a minimum, deflection should be checked. If deflection is likely to govern, the quickest approach can be to rearrange standard deflection formulae, to calculate a minimum 'I value' needed to limit deflection to published limits.

e.g. for a simply supported beam with a UDL of w, $\delta = 5w_dL^4/384EI$

Therefore $I_{req} = 5w_dL^4/384\delta E$

Where
δ = deflection
w_d = design UDL (for serviceability limit state)
E = modulus of elasticity
I = second moment of area
L = span length

Deflection should be within published limits, such as L/250 for total deflection, or L/360 for deflection due to imposed load.

The Blue Book can then be used to select a section which provides both adequate resistance to the bending moment, and an I value which will satisfy deflection criteria. Using this approach, you can avoid repeatedly selecting a section which works for the bending moments, only to find it does not meet the deflection criteria.

Timber
It is important to remember that the properties of timber vary, depending on the orientation of the section, and that bending capacities, parallel and perpendicular to the grain, are different. When designing bending elements, resistance parallel to the grain is typically considered.

There are also a number of modification factors which need to be considered when designing with timber. These relate to load duration, service class, and load-sharing among others. These will need more detailed consideration at a later stage of design.

Once a preliminary section depth has been determined from published guidance, standard available section sizes should be considered. For example, if you believe that you require a 150mm deep timber element, the typical timber sections available are 150 × 38, 150 × 50, 150 × 63, 150 × 75, 150 × 100 and 150 × 150.

A standard size should be selected and the bending stress in the element calculated and checked against the allowable design stress.

The allowable design stress $f_{m,d}$, is given by:

$$f_{m,d} = (f_{m,k} \times k)/\gamma_m$$

Where
$f_{m,k}$ = characteristic bending strength parallel to grain
k = modification factor to account for service class and load duration (typically 0.8 at concept design stage)
γ_m = material partial safety factor (1.3 for solid timber)

e.g. for C16 softwood, characteristic bending parallel to the grain $f_{m,k} = 16\text{N/mm}^2$:

$$f_{m,d} = (f_{m,k} \times 0.8)/\gamma_m = (16 \times 0.8)/1.3 = 9.85\text{N/mm}^2$$

The actual bending stress in the timber element should be less than the allowable stress. This approach is similar to the permissible stress design methodology of BS 5268[72] and allows for a simplified design approach, appropriate at concept design stage. BS EN 1995-1-1[66] uses limit state design principles, and requires a more careful approach, which will be required at later stages of design.

Timber experiences deflection in both the short- and long-term. Typically, short-term deflection is elastic, while long-term deflection is commonly known as 'creep deformation' and is not elastic. Deflection can be calculated using standard formulae and adopting the E_{min} value for the relevant timber type. Historically, deflection was generally limited to 0.003 × span.

The size of timber elements can often be governed by connections, where fixings (screws, nails or bolts) are relied on to transfer loads between elements. At concept design stage, spending a lot of time designing timber connections is not appropriate. In order to ensure that the initial size of timber elements is sufficient to account for detailed design of connections (which will follow at a later stage), stress in key elements should be kept at roughly 50% of element capacity. An alternative approach is to simply design the key connections, which will inform/govern the element size.

10.4.4.2 Axially loaded elements

In buildings, the most commonly considered axially loaded elements are columns and walls. These elements typically carry the loads from the slabs and beams down to the foundations, therefore typically carrying compressive loads.

As noted in Section 10.4.4.1, when carrying out initial sizing of bending elements you may choose to consider them as simply-supported. However, in reality there is continuity of structure between bending and axial elements such that most columns will need to resist some bending moments. Eccentric loading will also result in bending moments in columns and walls. It is therefore important not to completely overlook bending moments in the design of axial elements.

When considering axial elements, such as columns and walls, it is important to first consider which are likely to be critical to your design. These may be elements with the highest compressive load, the greatest height or experience the highest bending moment along with the axial load (or a combination of these).

A preliminary load trace or run-down should be carried out for critical elements, to calculate the axial load on the column or wall.

Once you have calculated the applied axial force in the element, initial sizing of the element can be based on empirical guidance. Recommended sizes for typical elements are given in Table 10.1.

Table 10.1: Typical sizes for axially loaded elements

Element type	Ideal height-to-breadth ratio	Typical heights	Relevant sources
RC columns[a]	H/15 (braced) H/10 (unbraced)	2.5–8m	3,74
RC walls[b]	H/15–H/30 (braced)	2–4m	3,74
RC retaining walls	H/10–H/14	2–8m	3,74
Steel UC section	H/20–H/25 (single-storey) H/7–H/18 (multi-storey)	2–8m 2–4m	3 3
Steel hollow section	H/20–H/35 (single-storey) H/7–H/28 (multi-storey)	2–8m 2–4m	3 3
Timber posts[c]	Minimum dimension of 100mm		3

Notes:
[a] Larger span-to-breadth ratios are possible, but second order effects may be significant.
[b] Second order effects may be significant for larger span-to-breadth ratios.
[c] Connections may govern minimum dimensions.

These initial ratios again provide a useful starting point for element depth, which can be adopted for simplified calculations.

There are a number of calculations and considerations to remember for the various different structural materials:

Reinforced concrete
Once the axial load has been calculated, the approximate column area required can be calculated using the following ratios, or those noted previously:

$A_c = N_{Ed}/15$ assuming 1% reinforcement by area and C28/35 concrete

$A_c = N_{Ed}/18$ assuming 2% reinforcement by area and C28/35 concrete

$A_c = N_{Ed}/21$ assuming 3% reinforcement by area and C28/35 concrete

Where
A_c = cross-sectional area of concrete
N_{Ed} = design axial force

For a typical building with several floors, grade C28/35 concrete will be sufficient to carry the axial loads. The calculated area can then be converted into a section size. The minimum size required should be adjusted to provide a 'standard' column size

e.g. $A_c = 113{,}569mm^2$ would suggest a 337 × 337 column.

At concept design stage this might typically be a 350 × 350 column (grade C28/35).

If the arrangement of supported beams is approximately symmetrical (does not differ by more than 15%), and the loads on the beams are uniformly distributed, it can be considered that the moment in the element will not be significant and the element can be designed for axial load only.

The axial resistance of such a column can be considered as follows:

$N_{c,Rd} = 0.45f_{ck}A_c + 0.7f_{yk}A_s$

Where
A_c = cross-sectional area of concrete
A_s = area of reinforcement
f_{ck} = characteristic compressive cylinder strength of concrete
f_{yk} = characteristic tensile strength of reinforcement
$N_{c,Rd}$ = resistance to axial compression

Limits of minimum and maximum reinforcement in columns should be checked and adhered to, and limits on column dimensions for fire resistance should also be checked.

A simple allowance for moment transfer into the column can be made by increasing the vertical load which is applied directly to the column being considered. Sizing the column based on this increased vertical load should ensure that it is adequately sized for the applied load and moment.

For example:

- For interior columns, the load from the floor which the column is directly supporting can be multiplied by 1.25
- For edge columns, the direct load can be multiplied by 1.5
- For corner columns, the direct load can be multiplied by 2.0

This factor should only be applied to the loads from the floor which is being directly supported, not to any load in the column from higher floors.

If the column cannot be designed for axial load only, and moments need to be explicitly considered, reinforced concrete column design charts can be used for symmetrically reinforced rectangular and circular columns. These are published in the literature (e.g. Cobb). Each chart is specific to the d/h ratio of the column, where d is the effective depth and h is the overall depth, and the correct chart should be selected. The minimum moment a column should be designed for is the axial load × eccentricity (e) or $h/20$. An initial estimate of the moments transferred to the column from the slab/beams can be made using the *Manual for the design of concrete building structures to Eurocode 2*[74].

RC walls in a building are typically shear walls, which form part of the stability system. These walls are usually subject to axial load and an in-plane moment, with a small out-of-plane moment due to load eccentricity. In this case, the extreme fibre-stress at the ends of the walls should be considerec. This takes account of the axial load, distributed along its length, and the additional force (tension and compression) due to the in-plane moment.

At concept design stage, the key aspects are to ensure that there are sufficient walls in each direction for stability (if this is how the building is stabilised), and that they are sufficiently thick. Once the empirical rules mentioned previously have been considered, the wall should be checked to ensure that it is not slender. This will simplify the design at the next stage, and provide a wall of sufficient thickness for reinforcement. A check of the minimum element thickness required to satisfy fire requirements should also be made.

To ensure that vertical reinforcement bars in columns and walls do not buckle, transverse reinforcement will also be required. In columns, this will take the form of links, while in walls, horizontal bars should be provided.

Steel
Similar to bending elements, it is possible to use the 'Blue Book' to determine preliminary steel column sizes, based on the axial load and possible bending moments, and comparing them to either the axial compression capacity, or combined axial force and bending, for given standard sections.

Similar to concrete columns, if the arrangement of beams and loads supported by the column are symmetrical, it is appropriate to consider the column as designed for axial load only. At this preliminary stage, it is prudent to enhance the axial loads to take account of small out-of-balance loads if you adopt this approach.

The axial, or local, compression capacity of a section can be calculated. However, it is usually the buckling capacity which governs the design of steel elements in compression. Therefore, the compression resistance for a given buckling length in the 'Blue Book' can be identified to adequately resist the applied axial load, and the column size selected.

For preliminary design an enhanced axial load can be considered for columns subject to out-of-balance loads. Cobb gives the following guidance:

Top storey: Total axial load + 4z-z + 2y-y
Intermediate storey: Total axial load + 2z-z + y-y

Where z-z and y-y are the net axial load difference in each direction.

If both axial load and bending moments are being considered, the combined axial force and bending tables of the 'Blue Book' should be used.

In this case, the ratio of design axial force to design axial plastic resistance n is required;

$$n = N_{Ed}/N_{pl,Rd}$$

Where
N_{Ed} = design axial force
$N_{pl,Rd}$ = design plastic resistance to axial forces (given for each section in the 'Blue Book').

This value is used to identify the relevant moment resistance for a given section size, axial design load and axial plastic resistance. From this, the relevant section size can be selected. In addition, a flexural buckling check and moment check should be carried out on the section.

Columns in braced frames can be accordance with SCI P362[114] clause 6.3.4:

$$N_{Ed}/N_{min,b,Rd} + M_{y,Ed}/M_{b,Rd} + 1.5M_{z,Ed}/M_{c,z,Rd} \leq 1$$

Where
$M_{b,Rd}$ = design lateral torsional buckling resistance
$M_{y,Ed}$ = design moment, y-y axis
$M_{z,Ed}$ = design moment, z-z axis
$M_{c,z,Rd}$ = design moment resistance, z-z axis
N_{Ed} = design axial force
$N_{min,b,Rd}$ = minimum design buckling resistance (y-y, z-z axes)

Typical guidance suggests that for braced frame buildings up to three storeys, with standard floor-to-ceiling heights, a 203UC is typically sufficient, while a 254UC would usually be adopted in buildings up to five storeys.

Timber
Similar to the approach for bending elements in timber, it is possible to calculate the compressive stress in timber columns, and compare to published capacity for the given timber grade or species.

A standard size should be selected, and the compressive stress in the element calculated using the following simple formula:

$$\text{Compressive stress} = N_{Ed}/bd_{sect}$$

Where
N_{Ed} = design axial force
b = width of section
d_{sect} = depth of section

The calculated compressive stress can then be compared against the characteristic property for the appropriate grade or species of timber.

As with steel columns, it may well be necessary to consider buckling, by reducing the design resistance. For the purposes of concept design, it is often simplest to select a column size, such that buckling does not need to be considered. This is the case when relative slenderness $\lambda_{rel} < 0.3$.

For major axis (y-y) buckling:

$$\lambda_{rel,y} = \lambda_y/\pi \times (f_{c,0,k}/E_{0.05})^{0.5} \text{ and } \lambda_y = L_{cr}/i_y$$

Where
i_y = radius of gyration, y-y axis
$f_{c,0,k}$ = characteristic compressive strength parallel to grain
$E_{0.05}$ = fifth percentile value of modulus of elasticity
L_{cr} = buckling length
$\lambda_{rel,y}$ = relative slenderness, y-y axis
λ_y = slenderness, y-y axis

Corresponding values can be similarly calculated for minor axis (z-z) buckling. This can however lead to very large column sections, so may provide a starting point for sizing timber columns only.

Walls in timber construction may take the form of sawn timber stud walls or CLT panels. In timber stud walls, the centres of studs will typically vary between 300–600mm c/c, depending on span and how economic the structure is. All spans are typically governed by deflection, and require sheathing board to at least one face. Published guidance in the form of load/span tables is typically available for preliminary sizing of both stud walls and CLT panel walls[77].

10.4.4.3 Trusses
Loads applied to trusses are resisted by discrete axial forces in the truss members. Bending is resisted by tension and compression forces in the chords, while shear is resisted by tension and compression forces in the verticals and/or diagonals. The connections are only required to transmit axial forces and can be treated as pins (Figure 10.14a–b).

Loads applied to trusses between nodes will cause local bending in the supporting members (Figure 10.15). This is unlikely to have a significant effect on the member sizes in the case of trusses supporting a floor slab, or roof purlins at close centres, but every effort should be made to ensure that any large point loads are applied directly to nodes.

Figure 10.14a: Timber roof trusses, Sports Hall, Samobor, Croatia

Figure 10.14b: Storey-height truss, Watermark Place, London

Figure 10.15: Forces in trusses with loading applied at nodes only (preferred) and with loading applied between nodes

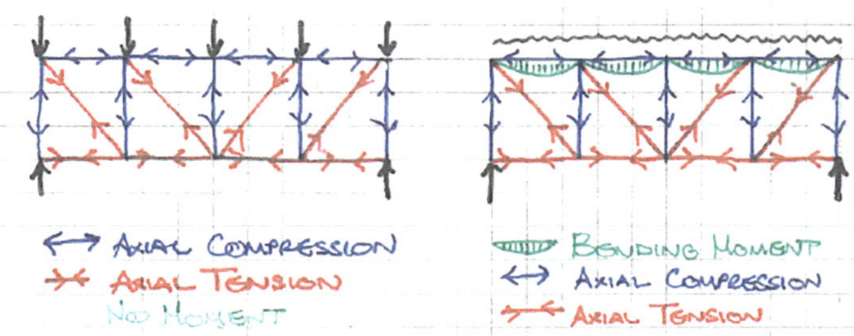

At concept design stage it is only necessary to determine the size of critical members; the size of less critical members can be refined at a later design stage as required.

Parallel chord trusses

The first step to sizing the critical members is to calculate the maximum bending moment and shear force due to the applied loads (Figure 10.16). This can generally be done using standard formulae, as presented in Cobb, the *Steel designers' manual*[115] etc. If the loading is complex, it may be more efficient to use a computer analysis package. If the loads are applied to the truss at relatively close centres, the loading can be approximated to a UDL, which simplifies the calculations.

Figure 10.16: Moment equilibrium of truss

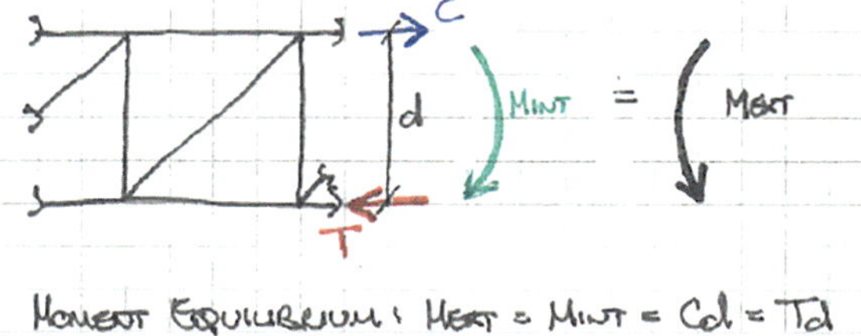

The bending moment due to the applied loads is resisted by an equal and opposite internal moment generated by the axial forces in the truss chords. The magnitude of the forces depends on the lever arm between them:

$$M_{Ed} = N_{c,Ed}d = N_{t,Ed}d$$

Therefore $N_{c,Ed} = N_{t,Ed} = M_{Ed}/d$

Where
d = depth of truss
M_{Ed} = design moment
$N_{c,Ed}$ = design axial compression
$N_{t,Ed}$ = design axial tension

Strictly, d is the distance between the centreline of the chords, but for an initial calculation it can be taken as the overall depth of the section.

The initial depth of the truss should be based on empirical guidance. Cobb recommends the span-to-depth (L/d) ratios for parallel chord trusses in Table 10.2.

Table 10.2: Span to depth (L/d) ratios for parallel chord trusses

Truss		L/d
Steel trusses subject to heavy point loads		10–15
Steel transfer trusses		10
Steel roof trusses		12–20
Parallel chord timber trusses		10–15
Reinforced/prestressed concrete trusses		10–15

The ratios for these timber trusses agree with those given in *Manual for the design of timber building structures to Eurocode 5*[77]. The ratios for timber trusses are also suitable for the initial sizing of reinforced/prestressed concrete trusses, although these are rarely used in practice.

The critical vertical/diagonal members (Figure 10.17) will be those adjacent to the supports.

For steel and timber trusses the compression elements will be critical due to the risk of buckling, although for timber trusses the element sizes may be governed by the connections between tension elements. Buckling can occur either

Figure 10.17: Critical forces in vertical (C_{VERT})/diagonal (C_{DIAG} & T_{DIAG}) truss members

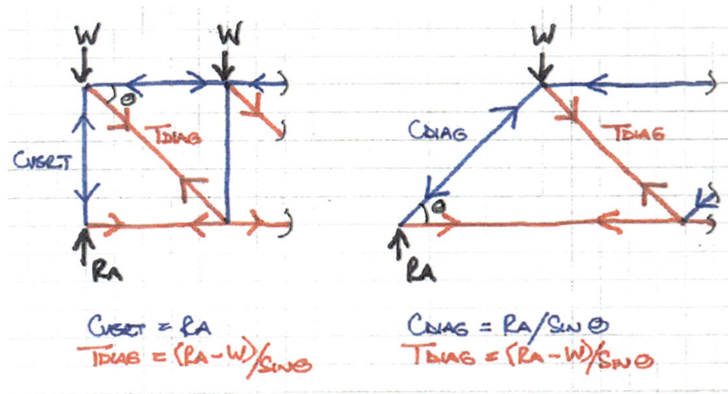

in the plane of the truss or out-of-plane. It is unlikely that there will be any intermediate restraint to verticals/ diagonals, however they will restrain the compression chord against in plane buckling. Out-of-plane restraint can be provided by cross-members (floor beams, purlins etc.) framing into the compression chord. These will generally be present if the top chord is in compression i.e. if the truss is 'sagging', but are less likely to be present if the bottom chord is in compression i.e. if the truss is 'hogging'.

This can be an issue for trusses supporting a lightweight roof. Under gravity loading, the truss will sag, with the top chord in compression, but if the wind load is sufficient to cause uplift, the truss will hog, putting the bottom chord into compression. Unless special provision has been made, the bottom chord will be entirely unrestrained between the supports. Restraint can be provided by diagonal members perpendicular to the span of the truss, triangulating the truss out-of-plane, by introducing an extra chord, or by designing the verticals/diagonals to act as cantilevers to transfer the restraint force to the roof plane (Figure 10.18).

Figure 10.18: Restraint to bottom chord of roof truss when in compression due to wind uplift

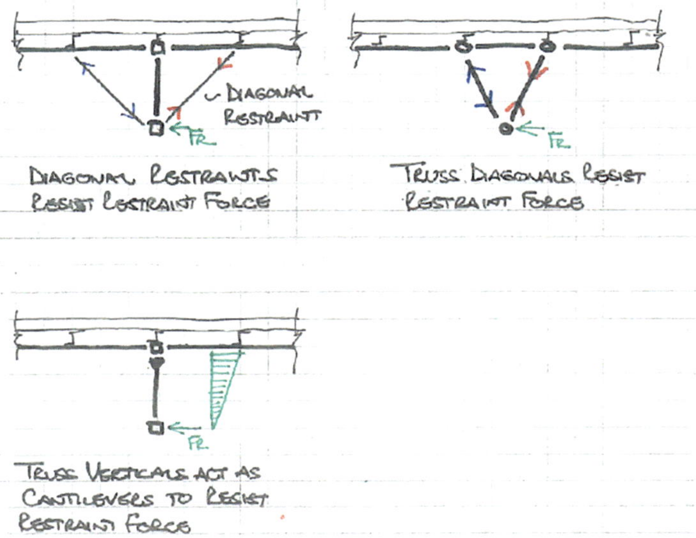

In all cases it should be verified that a load path exists e.g. floor slab providing diaphragm action, or in-plane roof bracing, for transferring the restraint forces from the members providing out-of-plane restraint to the building stability system. At concept design, the critical (effective) length for buckling should be taken as the length of the element between restraints i.e. $L_{cr} = 1.0L$ (it may be possible to justify taking a reduced effective length for detailed design).

Resistances of standard steel sections can be obtained from the 'Blue Book'. For non-standard sections, flexural buckling resistances can be calculated in accordance with Clause 6.3.1 of SCI P362. Resistances of timber sections can be calculated in accordance with *Manual for the design of steelwork building structures to Eurocode 3*[76]. Alternatively, the buckling resistance of steel and timber sections can initially be calculated using the Rankine-Gordon method:

Steel: $N_{b,Rd} = (1/N_{c,Rd} + 1/N_{cr})^{-1} = (1/f_y A_{chord} + L_{cr}^2/\pi^2 EI)^{-1}$

Timber: $F_{b,0,d} = (1/F_{c,0,k} + 1/F_{cr})^{-1}k/\gamma_m = (1/f_{c,0,k} A_{chord} + L_{cr}^2/\pi^2 EI)^{-1}k/\gamma_m$

Where
A_{chord} = cross-sectional area of chord
E = modulus of elasticity
f_y = yield strength of steel
$F_{b,0,d}$ = design buckling resistance parallel to grain
$F_{c,0,d}$ = design compressive resistance parallel to grain
F_{cr} = Euler buckling load for timber member ($\pi^2 EI/L_{cr}^2$)

I = second moment of area
k = modification factor to account for service class and moisture content
L_{cr} = buckling length
$N_{b,Rd}$ = design buckling resistance
$N_{c,Rd}$ = design resistance to axial compression
N_{cr} = Euler buckling load for timber member ($\pi^2 EI/L_{cr}^2$)
y_m = partial material factor

Compression elements in concrete trusses should be sized to ensure they are 'stocky' between restraints ($L_{cr}/b_{min} \leq 15$). The reinforcement required in the critical tension elements is given by:

$$A_{s,req} = N_{t,Ed}/0.87f_{yk}$$

The compression resistance of the concrete elements can be taken as:

$$N_{c,Rd} = 0.45f_{ck}A_c + 0.7f_{yk}A_s$$

Where
A_c = cross-sectional area of concrete
A_s = cross-sectional area of reinforcement
f_{ck} = characteristic compression cylinder strength of concrete
f_{yk} = characteristic tensile strength of reinforcement
$N_{c,Rd}$ = design resistance to axial compression
$N_{t,Ed}$ = design axial tension force

For steel and timber trusses, the same section sizes should be adopted for tension and compression members, unless there is a specific requirement to use different sections e.g. cables for tension members.

Deflection of steel and timber trusses should be checked at concept stage. The truss will deflect due to both bending and shear (Figure 10.19).

Figure 10.19: Bending and shear deflection

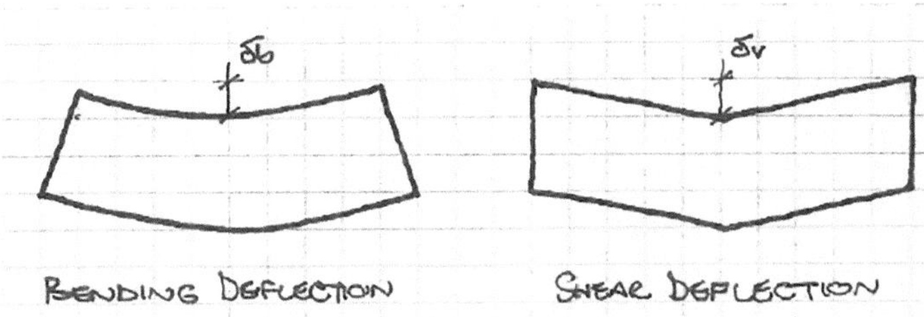

Unless the loading is very complex, the bending deflection can be approximated using standard formulae. The bending stiffness of the truss can be calculated using the parallel axis theorem:

$$I_{truss} = \Sigma I_n + A_n y_n^2$$

The verticals/diagonals contribute little to the bending stiffness, and can therefore be ignored. The 'I value' of the chords will be small in comparison to the Ay^2 term, so can also be ignored. For a planar truss with top and bottom chords of the same cross-sectional area (Figure 10.20):

$$y = d/2$$

So:

$$I_{truss} = 2A_{chord}y^2 = 2A_{chord}(d/2)^2 = A_{chord}d^2/2$$

Figure 10.20: Cross-section of planar truss

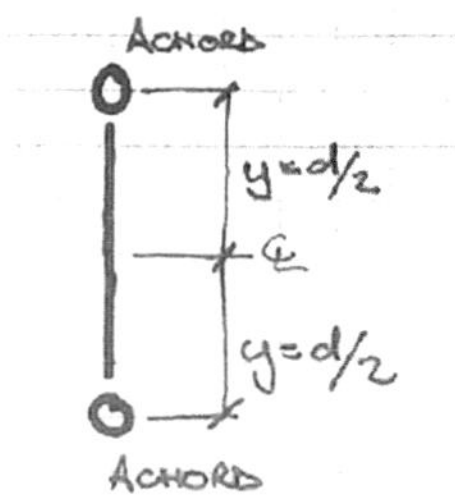

Where

A_{chord} = cross-sectional area of chord
d = depth of truss
I_{truss} = second moment of area of truss
Y = distance from neutral axis to centroid of member

The deflection due to bending should be within the limits specified in Table 7.1 of SCI P362 and the *Manual for the design of timber building structures to Eurocode 5* respectively.

Shear deflection is much more significant for trusses, than for beams with solid webs. To allow for this, the bending deflection should be limited to two-thirds of the usual value. For a case where bending deflection would normally be limited to $L/360$, allowable bending deflection of a truss would be limited to $L/360 \times \frac{2}{3} = L/540$ at concept stage. Alternatively, the bending stiffness of the truss (I_{truss}) required to limit the bending deflection to $L/360$ can be increased by 50% to allow for shear deflection.

Connection design for trusses should be investigated at an early stage of scheme design. In particular:

- For steel trusses with hollow members, are the minimum wall thicknesses governed by the connections?
- For timber trusses, are the element sizes governed by the connections?
- For concrete trusses, is it possible to place the required reinforcement at the nodes without them becoming too congested?

Vierendeel trusses

Vierendeel trusses are a special type of truss with vertical members but no diagonal members (Figure 10.21). Bending due to the applied loads is still resisted by axial tension/compression in the top and bottom chords, but the load path for transferring vertical loads to the supports is different. The vertical members can still transfer the loads in axial tension/compression but, in the absence of diagonals, vertical loads are transferred between the vertical members by shear in the horizontal members, which in turn generates local bending moments in the vertical and horizontal members. The connections need to be capable of transferring these moments between members.

The introduction of bending means that Vierendeel trusses are significantly less efficient than conventional trusses. However, they provide a viable alternative if diagonals are not acceptable (Figure 10.22).

Vierendeel trusses are statically indeterminate but, for the purposes of initial design, they can be treated as statically determinate structures by assuming a point of contraflexure, or virtual pin, at the mid-point of the members.

For the Vierendeel truss shown in Figure 10.23, the first step is to calculate the reactions:

$$R_{V,A} = R_{V,E} = V_{Ed} = 4W_d/2 = 2W_d$$

Figure 10.21: Storey-height Vierendeel truss, 4 Pancras Place, London

Figure 10.22: Comparison of forces in conventional truss (left) and Vierendeel truss (right)

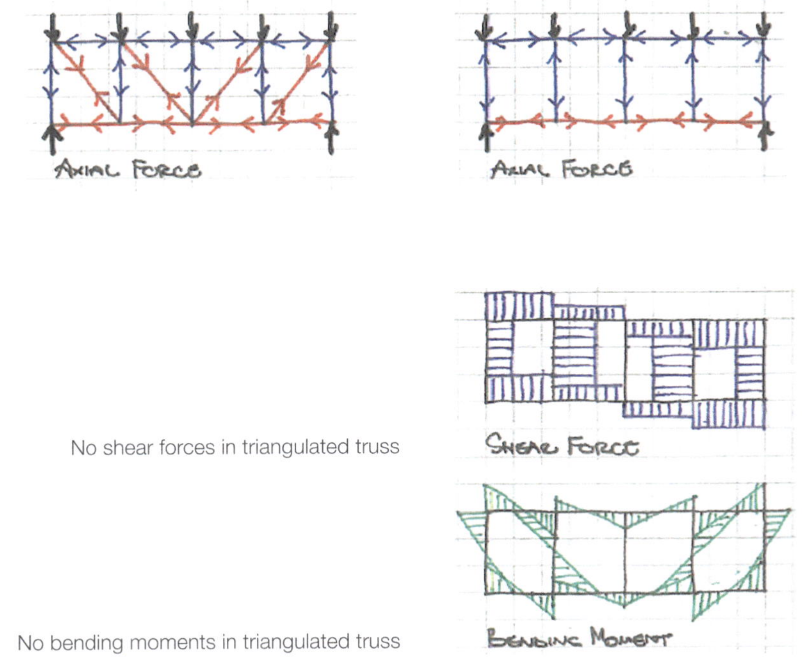

No shear forces in triangulated truss

No bending moments in triangulated truss

Figure 10.23: Location of virtual pins for static analysis of Vierendeel truss

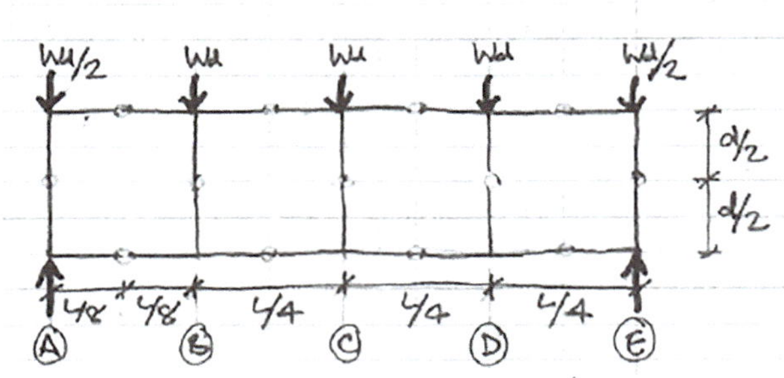

It can be assumed that the shear force due to the vertical loads is split evenly between the top and bottom chords (Figure 10.24):

$$V_{Ed,A-B} = (R_{V,A} - W_d)/2 = (2W_d - W_d/2)/2 = 3W_d/4 \text{ per chord}$$

$$V_{Ed,B-C} = (R_{V,A} - 2W_d)/2 = (2W_d/2 - 3W_d/2)/2 = W_d/4 \text{ per chord}$$

The truss is symmetrical so forces in panels C–D and D–E will be the same as those in panels B–C and A–B respectively.

Figure 10.24: Shear forces in chords

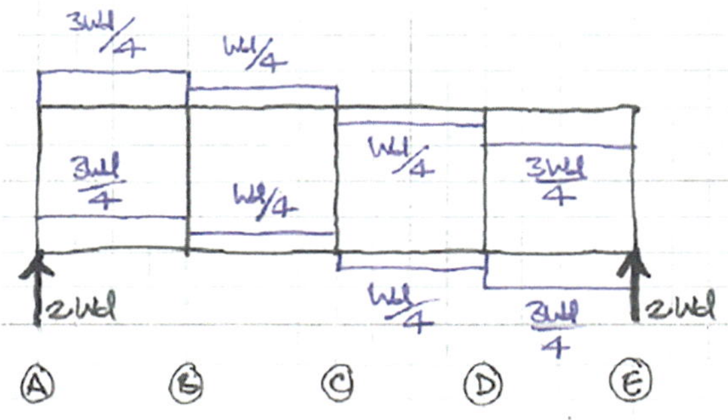

The local bending moment in the chords adjacent to the verticals can be calculated by multiplying the shear by the lever arm to the point of contraflexure i.e. half the panel length (Figure 10.25).

$$M_{Ed,A-B} = V_{Ed,A-B}(L/4)/2 = 3W_d \, 4(L/4)/2 = 3W_dL/32$$

$$M_{Ed,B-C} = V_{Ed,B-C}(L/4)/2 = W_d/4(L/4)/2 = W_dL/32$$

Figure 10.25: Moments in chords

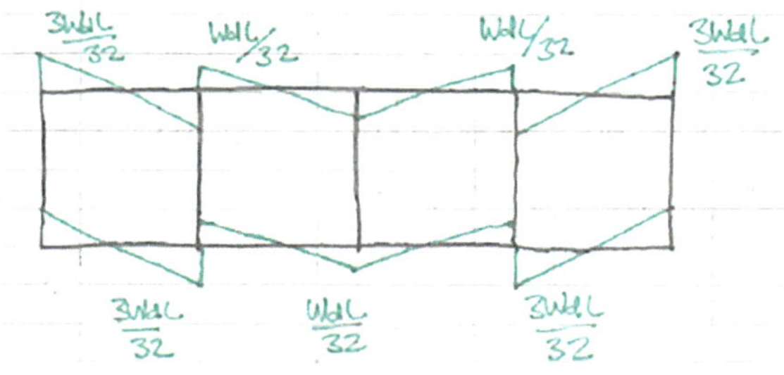

The moments in the verticals can be calculated by considering equilibrium at the connections (Figure 10.26):

$$M_{Ed,A} = M_{Ed,A-B} = 3W_dL/32$$

$$M_{Ed,B} = M_{Ed,A-B} + M_{Ed,B-C} = 3W_dL/32 + W_dL/32 = W_dL/8$$

$$M_{Ed,C} = M_{Ed,B-C} - M_{Ed,C-D} = W_dL/32 - W_dL/32 = 0$$

Figure 10.26: Moments in verticals

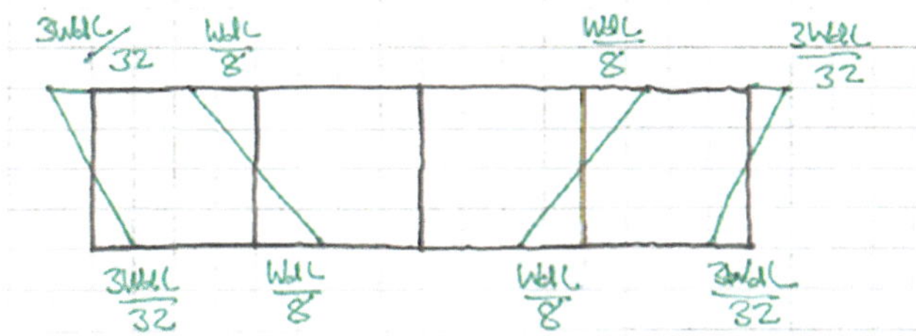

The shear in the verticals can be calculated by dividing the moment in the vertical by half their height (Figure 10.27):

$$V_{Ed,A} = M_{Ed,A}/(d/2) = 3W_dL/32(d/2) = 3W_dL/16d$$

$$V_{Ed,B} = M_{Ed,B}/(d/2) = W_dL/8(d/2) = W_dL/4d$$

$$V_{Ed,C} = M_{Ed,C}/(d/2) = 0$$

Figure 10.27: Shear forces in verticals

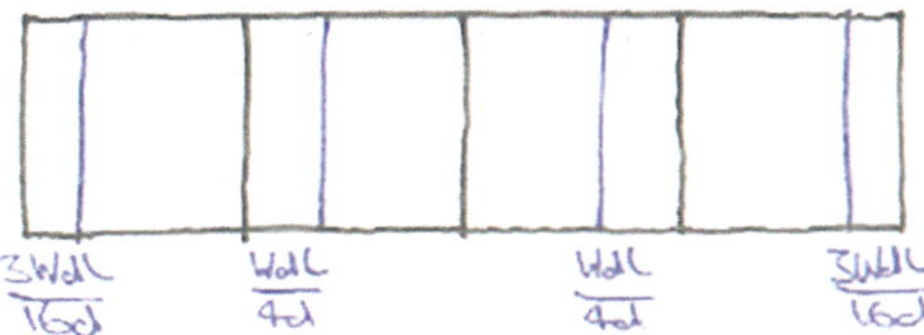

The axial force in the internal verticals is equal to half the load applied to the top of the members i.e. the load transferred to the bottom chord by the vertical.

$$N_{Ed,B} = V_{Ed,C} = W_d/2$$

The axial force in the end vertical is equal to the load applied at the top of the member ($W_d/2$) plus the shear in the top chord of the end panel (Figure 10.28):

$$N_{Ed,B} = W_d/2 + V_{Ed,A-B} = W_d/2 + 3W_d/4 = 5W_d/4$$

The axial force in the chords can be calculated by summing the shears in the vertical members:

$$N_{Ed,A-B} = V_{Ed,A} = 3W_dL/16d$$

$$N_{Ed,B-C} = V_{Ed,A} + V_{Ed,B} = 3W_dL/16d + W_dL/4d = 7W_dL/16d$$

Alternatively, the axial force in the chords can be determined by calculating the moments at the midspan of each panel, and dividing these by the lever arm between the chords (d):

$$N_{Ed,A-B} = M_{Ed,A-B}/d = (R_{V,A} - W_d)L/8d = (2W_d - W_d/2)/8d = 3W_dL/16d$$

$$N_{Ed,B-C} = M_{Ed,B-C}/d = 3(R_{V,A} - W_d)L/8d - W_dL/8d = [3(2W_d - W_d/2) - W_d]L/8d = 7W_dL/16d$$

Figure 10.28: Axial forces in chords and verticals

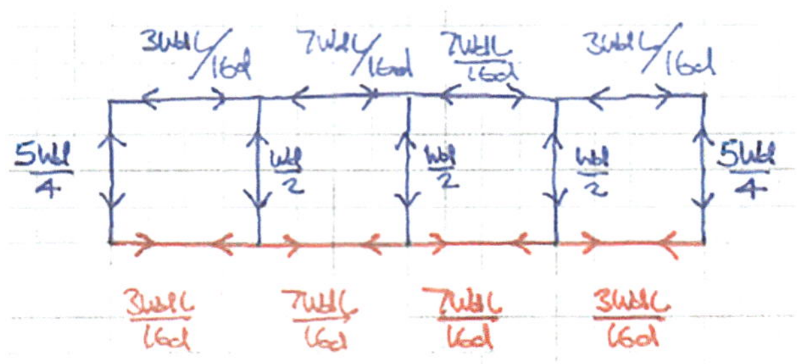

More conservatively, the maximum axial forces in the chords can be calculated by dividing the moment at midspan by the lever arm (d):

$$N_{Ed,max} = M_{Ed,max}/d = (R_{V,A}L/2 - W_d/2(L/2) - W_dL/4)/d = (2W_dL/2 - 2W_dL/4)/d = W_dL/2d$$

Where
d = depth of truss
L = span of truss
$N_{Ed,max}$ = design axial force
M_{Ed} = design moment
$R_{V,A}$ = vertical reaction
W_d = design point load

The chords need to be sized to resist axial force and moment. The compression chord will be critical for steel and timber Vierendeel trusses due to flexural buckling. Lateral torsional buckling will also need to be taken into account, unless the chords are fully restrained against lateral movement.

Steel members subject to compression and major axis bending can be checked in accordance with Clause 6.3.3 of SCI P362:

$$N_{Ed}/N_{b,y,Rd} + k_{yy}M_{y,Ed}/M_{b,Rd} \leq 1$$

$$N_{Ed}/N_{b,z,Rd} + k_{zy}M_{y,Ed}/M_{b,Rd} \leq 1$$

Where
k_{yy}, k_{zy} = interaction factors
$M_{b,Rd}$ = design lateral torsional buckling resistance
$M_{y,Ed}$ = design moment, y-y axis
$M_{z,Ed}$ = design moment, z-z axis
$M_{c,z,Rd}$ = design moment resistance, z-z axis
$N_{b,y,Rd}$ = minimum design buckling resistance, y-y axis
$N_{b,z,Rd}$ = minimum design buckling resistance z-z axis
N_{Ed} = design axial force

For initial design, k_{zy} can conservatively be taken as 1.0. Since Vierendeel members are subject to reverse curvature, with equal and opposite moments at each, k_{yy} can also be conservatively taken as 1.0.

Resistances of standard steel sections can be obtained from the 'Blue Book'. The flexural resistance of non-standard sections can be estimated using the Rankine-Gordon method or calculated using Clause 6.3 of SCI P362. The lateral torsional buckling resistance of non-standard sections can be calculated using Clause 6.3.2.3 Method 1 of SCI P362.

Resistances of timber sections subject to compression can be calculated using the *Manual for the design of timber building structures to Eurocode 5*, or approximated using the Rankine-Gordon method.

As with conventional trusses, concrete compression elements should be sized to ensure they are 'stocky' between restraints ($L_{cr}/b_{min} \leq 15$). Elements subject to compression and bending can be checked as column sections using column design charts (Figure B.4 in Brooker, Figure 15.5 in *Concise Eurocode 2*[116] etc.).

The same considerations apply for connections in Vierendeel trusses as conventional trusses, but the requirement to transfer moments means the connection design is likely to be even more critical.

The introduction of bending means a Vierendeel truss will deflect significantly more than the equivalent conventional truss. It is not easy to calculate the deflections of a Vierendeel truss by hand, but it can be done quickly using a 2D model in a computer analysis package. If deflection is critical, it is recommended that this check is carried out at concept stage. The increased deflection due to bending is reflected in the fact that Cobb recommends a span-to-depth ratio of 8–10 for the initial sizing of Vierendeel trusses.

10.4.4.4 Arches
Parabolic arch subject to UDL
A parabolic arch carrying a UDL will be purely in compression. As Robert Hooke famously noted, this is the mirror image of a cable carrying to a UDL, which will adopt a parabolic shape, purely in tension (Figure 10.29). Structures such as these, which carry loads principally in axial tension and compression, are sometimes referred to as 'funicular structures'.

Figure 10.29: Comparison of chain and parabolic arch

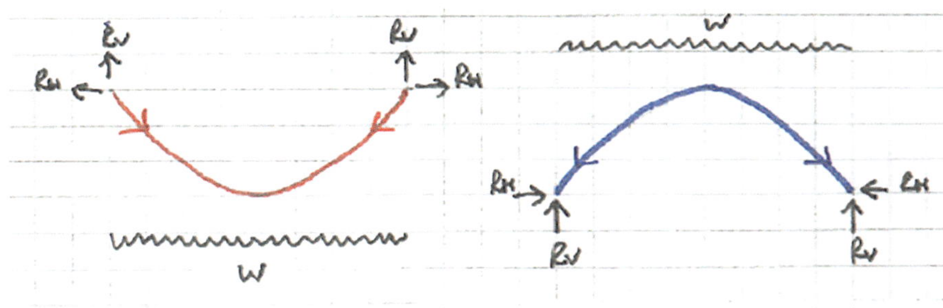

While the vertical reactions for funicular structures are identical to those for a beam with the same span and load (Figure 10.30), they also require horizontal reactions for equilibrium — inwards to resist the outward thrust for the arch, and outwards to resist the inward pull for the cable.

Figure 10.30: Parabolic arch — loads and dimensions

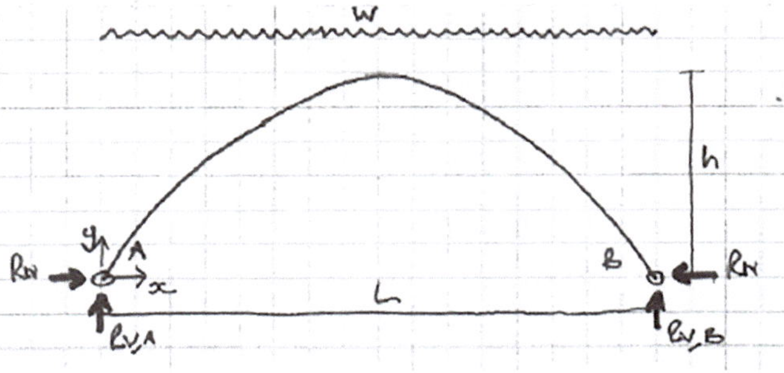

The supports can be designed 'fixed' to resist moment, or 'pinned' to allow rotation. For concept design, pinned supports should be assumed for ease of analysis. For a parabolic arch with pinned supports (sometimes referred to as a 'two-pinned arch') subject to a UDL, the reactions and compression can be calculated using statics:

$$R_{V,A} = R_{V,B} = w_d L/2$$

For a parabolic arch subject to a UDL $M = 0$ at all points (Figure 10.31).

Taking moments about the crown of the arch:

$$0 = R_{V,A} \times L/2 - w_d L/2 \times L/4 - R_{H,A} \times h$$

Substituting $R_{V,A} = w_d L/2$:

$$0 = w_d L^2/4 - w_d L^2/8 - R_{H,A} h$$

$$R_{H,A} = w_d L^2/8h$$

For horizontal equilibrium:

$$R_{H,A} = R_{H,B} = w_d L^2/8h$$

Maximum compression is at the supports:

$$N_{c,A,Ed} = N_{c,B,Ed} = (R_v^2 + R_H^2)^{0.5} = [(w_d L/2)^2 + (w_d L^2/8h)^2]^{0.5}$$

Minimum compression is at the crown:

$$N_{c,Ed} = R_H = w_d L^2/8h$$

Where
H	= height of arch
L	= span of arch
$N_{c,Ed}$	= design axial compression
R_H	= horizontal reaction
R_V	= vertical reaction
w_d	= design UDL

Figure 10.31: Free body diagram for half span of arch

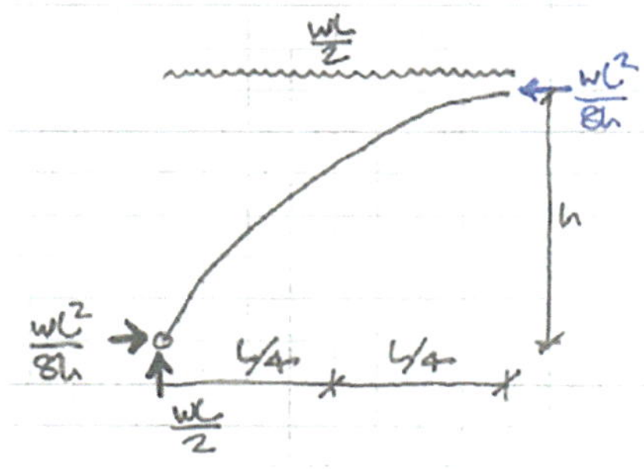

Parabolic arch subject to non-UDL

While a parabolic arch carrying a UDL will be in pure compression, parabolic arches carrying non-UDL will also be subject to bending.

The dead load on an arch can usually be approximated to a UDL (in reality, the dead load will be higher and closer to the supports due to the slope of the arch, however the resulting moments are unlikely to be significant).

Non-UDL could be due to imposed loading e.g. wind or snow loading.

For traditional masonry arches in buildings, the imposed loads are usually small in comparison to the dead load (this may not be the case for road/rail bridges), and the resulting bending is unlikely to be significant. This may also be the case for a reinforced concrete arch. However, for timber and steel arches, the imposed loads are likely to be significant compared to the dead load, and the resulting bending may dominate the design of the arch.

If the majority of the load is applied to the arch as point loads e.g. column loads on an arch acting as a transfer structure, it may be better to use an arch which reflects the funicular shape of a cable subject to a series of point loads instead of a smooth parabola (Figure 10.32).

Figure 10.32: Funicular arch with point loads, Exchange House, London

For a parabolic arch, Byfield[109] suggests a reasonably conservative load combination for concept design is to assume imposed load on one side of the arch only (Figure 10.33).

Figure 10.33: Parabolic arch with imposed loading on one side only

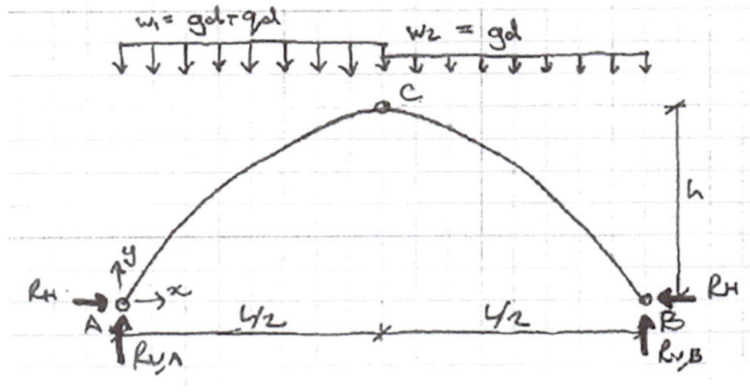

A two-pin arch (Figure 10.34) has one degree of indeterminacy, which means it is not easy to calculate the moments by hand. Introducing an extra pin at the crown transforms the structure into a statically determinate, three-pin arch. The moments can then be calculated by considering equilibrium.

Consider the three-pin arch shown in Figure 10.33:

Taking moments about B:

$$0 = R_{V,A} \times L - w_2 L/2 \times L/4 - w_1 L/2 \times 3L/4$$

$$R_{V,A}L = w_2 L^2/8 + 3w_1 L^2/8$$

$$R_{V,A} = (w_2 L + 3w_1 L)/8$$

$$R_{V,B} = (w_1 + w_2)L/2 - R_{V,A}$$

Taking moments about C:

$$0 = R_{V,A} \times L/2 - w_1 L/2 \times L/4 - R_{H,A} \times h$$

Substituting $R_{V,A} = (w_2 L + 3w_1 L)/8$:

$$0 = (w_2 L^2 + 3w_1 L^2)/16 - w_1 L^2/8 - R_{H,A}h$$

$$R_{H,A} = (w_1 + w_2)L^2/16h$$

For horizontal equilibrium:

$$R_{H,A} = R_{H,B} = (w_1 + w_2)L^2/16h$$

Taking moments at any point between A–C:

$$M_{Ed} = R_{V,A}x - w_1 x(x/2) - R_{H,A}y = R_{V,A}x - w_1 x^2/2 - R_{H,A}y$$

Similarly, at any point between C–B:

$$M_{Ed} = R_{V,A}x - w_1 L(x - L/4)/2 - w_2(x - L/2)^2/2 - R_{H,A}y$$

The geometry of the parabola is described by the equation:

$$y = -4hx^2/L^2 + 4hx/L$$

The maximum moments occur at quarter points:

At $x = L/4$

$$y = -4h(L/4)^2/L^2 + 4h(L/4) = -h/4 + h = 3h/4$$

$$M_{Ed} = R_{V,A}x - w_1 x^2/2 - R_{H,A}y = R_{V,A}L/4 - w_1 L^2/32 - 3R_{H,A}h/4$$

Where
g_d	= design UDL due to permanent loads
h	= height of arch
L	= span of arch
q_d	= design UDL due to permanent loads
R_H	= horizontal reaction
R_V	= vertical reaction
w_1, w_2	= design UDL

Figure 10.34: Moments on parabolic arch subject to asymmetrical loading

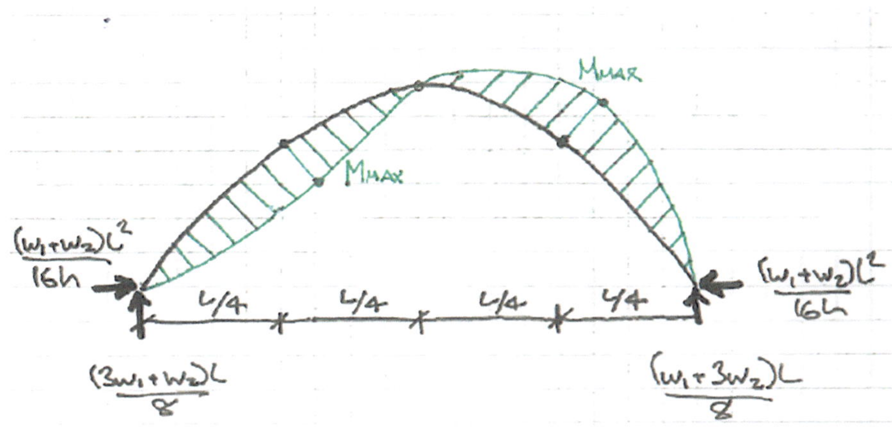

Circular arches

Some of the earliest arches constructed were semi-circular. A circular arch can also be formed by cutting an arc from the circumference of a larger circle (Figure 10.35). The geometry of a semi-circular arch and a parabolic arch with the same span and rise will be significantly different but, as the span-to-rise ratio decreases, the geometry of a circular arch approaches that of a parabolic arch.

A circular arch is not a funicular shape, so will always be subject to moments as well as axial compression. The greater the deviation from the equivalent parabolic shape, the larger the moments. Although this makes circular arches less efficient than parabolic arches they are sometimes preferred for aesthetic reasons, or to simplify construction e.g. if the arch is to be formed from precast elements, the same mould can be used for all elements of a circular arch.

The forces and moments for a three-pin circular arch can be worked out in a similar way to a three-pin parabolic arch. For a circular arch, applying the imposed load uniformly usually generates a larger moment than just applying the imposed load to one side of the arch. As with a parabolic arch, subject to a UDL:

$$R_{V,B} = R_{V,B} = w_d L/2$$

$$R_{H,A} = R_{H,B} = w_d L^2/8h$$

The moment at any point on the arch is given by:

$$M_{Ed} = R_{V,A}(L/2 + x) - w_d(L/2 + x)^2/2 - R_{H,A}y$$

The formula used to calculate y depends on whether the arch is semi-circular, or cut from the circumference of a larger circle. Unlike a parabolic arch, where x is taken as zero at the left-hand end of the arch, x is taken as zero at midspan ($L/2$). For a semi-circular arch, the span of the arch (L) will be equal to twice the radius (r), and the rise (h) will be equal to the radius (Figure 10.35).

The height of the arch at any point is given by:

$$y = (r^2 - x^2)^{0.5}$$

Figure 10.36 shows a circular arch of span L and rise h, which has been cut from the circumference of a larger circle.

The radius of the arch is given by:

$$r = (L^2/4 - h^2)/2h$$

Figure 10.35: Semi-circular arch geometry

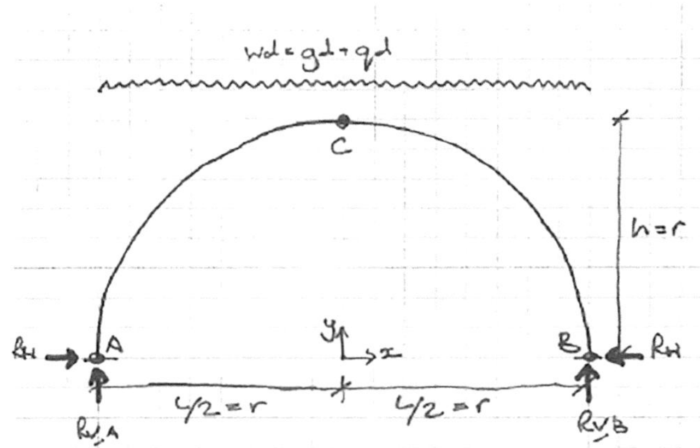

Figure 10.36: Circular arch cut from circumference of larger circle i.e. giving a non-semi-circular arch

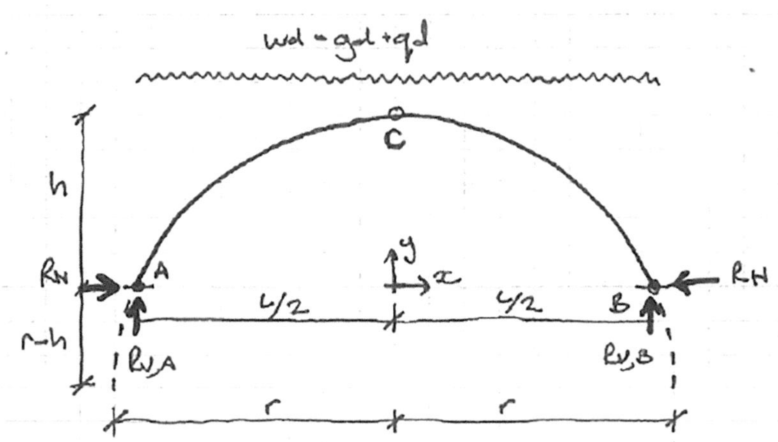

The height of the arch at any point is given by:

$$y = (r^2 - x^2)^{0.5} + h - r$$

For a circular arch subject to a UDL (Figure 10.37), the maximum moment occurs when the geometry deviates most from a parabolic arch with the same rise and span (Sandaker[117]). The maximum deviation, hence maximum moment, occurs when $y = h/2$. For a semi-circular arch this occurs at an angle of 30° to the horizontal.

Buckling
Like any compression element, arches are subject to buckling. Buckling can occur either in the plane of the arch, or out-of-plane (Figure 10.38). The in-plane buckling length (critical/effective length) of an arch with pin supports is half the swept length (the distance from one end of the arch to the other, travelling along the curve).

For a parabolic arch the swept length (S) is given by Byfield as:

$$S = L^2[ln(4h/L + (1 + 16h^2/L^2)^{0.5})]/8h + L(1 + 16h^2/L^2)^{0.5}/2$$

Swept length-to-span ratio for different span-to-rise ratios are given in Table 10.3.

Figure 10.37: Moments on circular arch subject to symmetrical loading

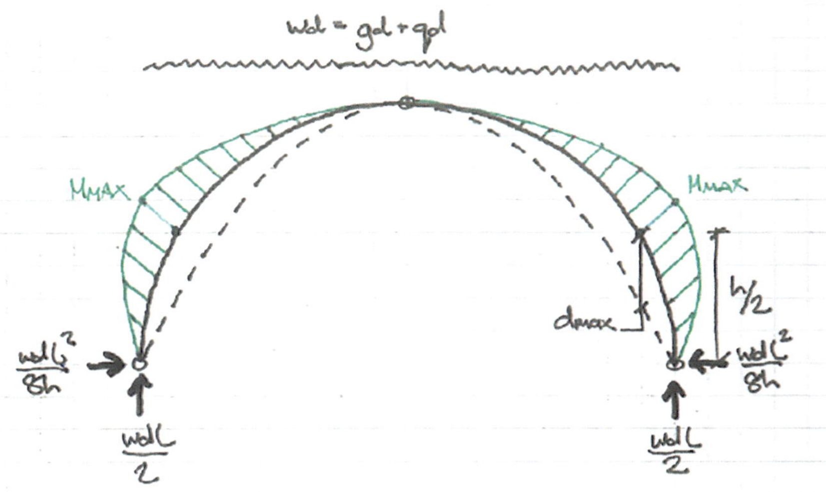

Figure 10.38: Buckling modes

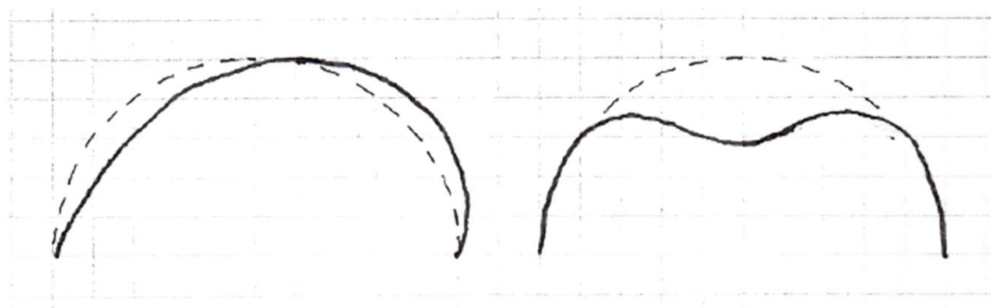

Table 10.3: Swept length-to-span ratios

Span-to-rise ratio	Swept length-to-span ratio
10	1.03
7	1.05
5	1.10
4	1.15
3	1.25
2	1.48
1	2.32

The swept length of a semi-circular arch is given by:

$$S = \pi r$$

The swept length of a circular arch cut from the circumference of a larger circle is given by:

$$S = \pi r \sin^{-1}(L/2r)/180$$

For arches with a very low span-to-rise ratio, the possibility of snap-through buckling (i.e. where the buckling length is equal, or close, to the swept length of the arch) needs to be investigated.

If no intermediate restraints are provided, the out-of-plane buckling length of an arch with pinned supports would be equal to the swept length. In building structures, intermediate restraint will generally be provided by cross-members (floor beams, purlins etc.) framing into the arch. It should be verified that a load path exists for transferring the restraint forces from the cross-members to the building stability system e.g. floor slab providing diaphragm action, or roof bracing. If the arch is subject to bending the cross-members will also provide restraint against lateral torsional buckling.

Element checks

The arch members need to be able to resist the critical combination of axial compression and bending. The maximum compression occurs at the base, rather than the point of maximum moment, but at concept stage it is reasonably conservative to check the elements based on the maximum moment and compression acting coincidentally. Both in-plane and out-of-plane buckling should be checked, unless it is obvious which axis is critical.

Steel members subject to compression and bending can be checked in accordance with Clause 6.3.5 of SCI P362:

$$N_{Ed}/N_{b,y,Rd} + k_{yy}M_{Ed,max}/M_{b,Rd} \leq 1$$

$$N_{Ed}/N_{b,z,Rd} + k_{zy}M_{Ed,max}/M_{b,Rd} \leq 1$$

Where
k_{yy}, k_{zy}	= interaction factors
$M_{b,Rd}$	= design lateral torsional buckling resistance
$M_{y,Ed}$	= design moment, y-y axis
$M_{z,Ed}$	= design moment, z-z axis
$M_{c,z,Rd}$	= design moment resistance, z-z axis
$N_{b,y,Rd}$	= minimum design buckling resistance, y-y axis
$N_{b,z,Rd}$	= minimum design buckling resistance z-z axis
N_{Ed}	= design axial force

For initial design, k_{yy} and k_{zy} can conservatively be taken as 1.8 and 1.0 respectively.

Resistances of standard steel sections can be obtained from the 'Blue Book'. For non-standard sections, flexural and lateral torsional buckling resistances can be calculated in accordance with Clauses 6.3 and 6.3.2.3 Method 1 of SCI P362 respectively.

Resistances of timber sections subject to compression can be calculated using *Manual for the design of timber building structures to Eurocode 5*. As with any timber section, the arch size may be governed by the connection design, which should be checked at an early stage of scheme design.

Rectangular reinforced concrete arches, with an effective length-to-depth ratio of less than 25, will not be subject to second order effects, and can be checked using column interaction charts. Alternatively, a more conservative estimate of the resistance can be obtained by treating the arch as a plain concrete section. The buckling resistance of a plane concrete section is given in Clause 14.4 of *Concise Eurocode 2* as:

$$N_{b,Rd} = 0.4bd_{sect}f_{ck}[1.14(1 - 2e/d_{sect}) - 0.02(L_{cr}/d_{sect})]$$

Where
b	= width of section
d_{sect}	= depth of section
e	= eccentricity = M_{Ed}/N_{Ed}
f_{ck}	= characteristic compressive cylinder strength of concrete
L_{cr}	= buckling length = swept length/2

If the effective length-to-depth ratio exceeds 25, moments due to second order effects will need to be calculated in accordance with Clause 5.6.2.2 of *Concise Eurocode 2*.

Steel or timber arches supporting a lightweight roof may experience net uplift as a result of wind loading, leading to load reversal i.e. axial tension instead of compression. While this is unlikely to be critical for the members, it may be critical for the connections. The supports will also be subject to load reversal. If it is considered likely that net uplift will occur, the implications of this should be investigated as part of the concept design.

Supports
The supports at the base of the arch need to be able to resist the horizontal thrust from the arch, as well as the vertical reactions. The most efficient way to do this is to tie the supports together. If the arch will not be subject to load reversal due to net uplift, this can be done using a cable. If load reversal can occur, the 'tie member' will go into compression, and will need to be designed accordingly. For arches terminating at ground level, it may be possible to tie the supports together using reinforcement in the ground floor slab.

If it is not possible to tie the supports together, then buttresses, braced bays or shear walls will be required to transfer the thrust to the foundations. The structural and geotechnical engineer will need to work closely together, to ensure that the proposed foundation system is suitable for transferring the thrust into the ground.

At concept stage, it is generally reasonable to assume that the support system resisting the arch thrusts is infinitely rigid. If it is considered that the thrust may lead to significant movement of the support system/foundations, the effect of this on the arch geometry and forces should be checked at an early stage of scheme design.

10.4.4.5 Fabricated steel sections (plate girders and box girders)
If it is not possible to find a suitable standard section for a member in a steel structure, a specially fabricated section can be used. Fabricated or welded sections are typically required for long-span or transfer structures, where the forces are too high for a standard section to resist, but may also be needed if the structural zone is very constrained, or if non-standard sections are required for aesthetic reasons, in the case of exposed steelwork. Fabricated open sections are often referred to as 'plate girders' (Figure 10.39a), while fabricated hollow sections are sometimes referred to as 'box girders' (Figure 10.39b).

The initial design of fabricated sections not subject to lateral torsional buckling i.e. doubly-symmetric sections or open sections, with continuous lateral restraint to the compression flange, is similar to the design of a standard beam. The section size is likely to be governed by bending or deflection, although shear may be critical for elements supporting large point loads close to the supports.

Figure 10.39a: Plate girder

Figure 10.39b: Box girder

Experience suggests that if the stresses in a fabricated section due to the ultimate factored loads are limited to $200N/mm^2$ at concept stage, the section will pass all checks required at detailed design stage.

For Class 1 or 2 sections, the moment resistance depends on the plastic modulus but, to obtain an initial section size, it can be assumed that the moment is resisted by the flanges with the web/s resisting the shear force. The flanges can then be sized in a similar way to the chords in a truss:

$$N_{c,Ed} = N_{t,Ed} = M_{Ed}/d_{sect}$$

Where
d_{sect} = depth of section
M_{Ed} = design moment
$N_{c,Ed}$ = design axial compression
$N_{t,Ed}$ = design axial tension

In a strict sense, d_{sect} is the distance between the centreline of the flanges, but for a first pass at the section size it can be taken as the overall depth of the section. If the allowable stress (f_{all}) in the flange is to be limited to $200N/mm^2$, the area of steel required to resist the forces is given by:

$$A_f = N_{c,Ed}/f_{all} = N_{t,Ed}/f_{all} = b_f t_f$$

If the stress is limited to $200N/mm^2$, deflection will usually be acceptable. If an explicit deflection check is required, the bending stiffness of the section can be conservatively taken as:

$$I = 2A_f y^2 = 2bt_f(d_{sect}/2)^2 = bt_f d_{sect}^2/2$$

Where
A_f = cross-sectional area of flange
b_f = width of flange
f_{all} = allowable stress
I = second moment of area
t_f = thickness of flange

The flanges should generally be proportioned so they meet the limits for Class 1 or 2 elements given in Table 5.1 of SCI P362. For S355 steel:

Internal compression flange:

$c_f/t_f \leq 30$

Outstand compression flange:

$c_f/t_f \leq 8$

Where
c_f = outstand length of flange
t_f = thickness of flange

For sections where these limits are exceeded e.g. wide box beams, stiffeners are likely to be required.

The minimum thickness of the web should meet the limit given in Clause 6.2.6(4) of SCI P362, to avoid shear buckling. For S355 steel:

$t_w \geq h_w/58$

If this limit is not met, stiffeners will need to be provided to prevent shear buckling. The labour costs associated with providing stiffeners will usually outweigh any savings in material costs from using a thinner web.

Shear is unlikely to be critical for elements subject to uniform loads but should be checked in accordance with Clause 6.2 of SCI P362, for elements subject to large point loads e.g. transfer beams. The shear resistance of a fabricated section is given by:

$$V_{Rd,c} = V_{pl,Rd} = A_v f_y/\sqrt{3} = h_w t_w f_y/\sqrt{3}$$

Where
A_v = shear area
h_w = height of web
t_w = thickness of web
f_y = yield strength of steel

If the applied shear is more than 50% of the shear resistance, the bending resistance of the section should be recalculated using a reduced yield strength in accordance with Clause 6.2.7 of SCI P362.

If the point loads are applied to the top flange of the element e.g. columns sitting on a transfer beam, an allowance should be made for stiffeners at the points where the loads are applied.

For sections where the compression flange is not fully restrained, the lateral torsional buckling resistance can initially be calculated in accordance with Clause 6.3.2.3 Method 3 of SCI P362. While this method is fairly straightforward, it is not strictly applicable to welded sections, and the resistance should be checked using a fully appropriate method e.g. Clause 6.3.2.3 Methods 1 or 2 of SCI P362 at an early stage in scheme design. It should be noted that lateral torsional buckling is unlikely to be critical for elements where the spacing between lateral restraints is small compared to the span.

10.5 Documenting the calculations

Concept design calculations may be the equivalent of 'back of the envelope' calculations, but they still need to be properly documented. There is often a pause after concept stage, as the client decides whether to take the project forward. If the project is put on hold for a prolonged period, it is quite possible that the original design team won't be available to work on the project when it re-starts. If the new design team are not able to understand the concept design calculations they will have to be re-done, wasting time and money. Even if the original team are available to pick up the project, it is surprising how quickly the memory of assumptions made at concept stage can fade, if not properly documented. If the project does progress, the design will evolve during the next stage and it is likely that variations to the original concept will be considered. Clearly documented concept design calculations will allow you to quickly check whether element sizing is still relevant, and update calculations quickly and efficiently to provide new structural sizes where required.

Calculations should be legible, follow a logical order and be clearly cross-referenced to the drawings and sketches (including the load paths and construction sequences). At this stage, including numerous clear sketches within the calculations is helpful, as during concept design layouts and gridlines can change. Due to this changing nature of concept design, it is important to note the revision of any drawings and sketches you refer to and on which you are basing your assumptions. Remember, the drawings should show the structure justified in the calculations, and the calculations justify the structure shown on the drawings.

Any assumptions regarding loads, restraints, construction sequence etc. should be clearly stated, and where appropriate, communicated to the rest of the design team. The calculations should conclude with a section stating any assumptions that need to be confirmed before commencing the scheme design calculations, as well as any critical checks that need to be carried out during scheme design.

10.6 Checking the calculations

One of the paradoxes of concept design is that it must often be undertaken with limited time and resources, but if the concept is flawed, it will prove difficult to justify the structural scheme, or make a profit, as the project progresses through detailed design. As such, it is essential that the concept design is thoroughly checked.

While a numerical check of the calculations is important, the main emphasis should be on checking that a suitable structural concept has been developed. Questions that need to be answered include:

- Does the proposed scheme meet the requirements of the brief?
- Have all the site/building specific constraints been identified?
- Are any assumptions that have been made, reasonable? Have all assumptions been clearly recorded in the design documents?
- Does the proposed structural system provide at least one continuous load path for transferring all the applied loads to the ground?
- Is the proposed construction sequence credible?
- Do the calculations provide sufficient confidence that it will be possible to fully demonstrate that elements forming the load paths are adequate at detailed design stage? Have the critical elements been checked?
- Is the structural design consistent with the foundation design?
- Have the calculations been coordinated with drawings/sketches i.e. do the calculations justify the structure shown on the drawings and *vice versa*?
- Has the structural design been coordinated with other aspects of the design (architecture, services, sustainability strategy etc.)?
- Have any assumptions requiring clarification/critical checks required at scheme design stage been clearly recorded?
- Has a design risk assessment been carried out and recorded?

The best way to ensure these questions have been fully considered is through a concept design review, where the design team is asked to justify their proposals to an experienced engineer (or engineers), who has not been involved in developing the proposals. Ideally, the structural design (or at least the substructure) should be reviewed in conjunction with the geotechnical design and the drainage/infrastructure design, if significant. To get the best value out of the review, the reviewers should be given an opportunity to examine as complete a set of the design documents as possible, prior to the review. The discussions should be carefully recorded and reviewed by the design team prior to issuing the design package, to ensure all the actions have been completed, or clearly highlighted as requiring action in the next design stage.

If the job has been put on hold, and the scheme design is being carried out by a new team, or the concept design has been inherited from another consultant, the new design team should carry out a similar review to satisfy themselves that the concept design is adequate before starting scheme design.

10.7 Next steps

Once you have issued your concept design information there may be a pause in the project while the client considers whether it is viable to progress. Hopefully, it will be considered a viable project and will then progress, so what are the next steps?

Before starting the next stage of design, it is important that the concept design is 'signed off'. In practice, this means holding an internal design review to ensure that everybody is satisfied that the concept design is adequate, and having agreement from the client and the rest of the design team that the structural proposals are acceptable and compatible with the project brief and the rest of the design e.g. the architectural and building services elements.

At this stage, it is also important to try and confirm any assumptions that have been made in the design, and note any particular aspects which had an impact on the initial sizing; these could include (among others) fire, vibration, exposure class and extraordinary loading. Before starting the next stage of design, you should make sure that you have been paid for the concept design, and it is worthwhile checking how much has been spent up to this point, compared to the forecast. You should also produce a list of items to be checked during the next stage, including any necessary testing etc.

The first steps of the next stage would typically be as follows:

- Review the most up to date information from the architect and the rest of the design team. Identify anything that has changed since the concept design, and consider what impact this has on the design to date. Will it be necessary to revisit any of the concept design at the start of the next stage?
- Review the concept design, and identify any assumptions which need to be verified and any outstanding items identified in the concept design review
- Consider what information you will require from the rest of the project team for the next stage of design; this might include frozen architectural layouts, structural opening sizes for key areas such as large services risers or lift shafts and specialist loading information for specific equipment
- Plan what design calculations need to be carried out during the next stage, and estimate how long this might take. At this stage, you should consider how the next stage of design will be approached. Do you intend to start building some form of computer analysis model? Will you be producing 2D or 3D information? What BIM level will the design team be working towards? What other documentation will be required, such as outline specifications, design philosophy documents, design programme or RIBA Stage 3 (Spatial Coordination) reports?
- The planned work should be related to the overall project programme, and should allow you to forecast resource requirements for the next stage
- Produce some outline dates by which you will require the information from the rest of the project team, to allow you to progress the structural design, ready to discuss at the first project meeting

When planning what calculations to undertake post-concept stage, some thought should be given to what level of structural optimisation is appropriate for the project. At concept stage, it makes sense to base member sizes on the critical elements, but carrying this approach through to detailed design would be very wasteful, in terms of cost, material usage and embodied carbon. The opposite extreme would be to size each element individually for the loads applied to it. While this would be optimal, in terms of material usage and embodied carbon, it may not produce the most efficient design in terms of cost or ease of construction. It would also place greater demands on the design fee and the way that building information is managed as it flows through the design and construction process.

In practice, the answer will probably lie somewhere between these two extremes, and will be influenced by factors such as the project team's attitude to reducing embodied carbon, budget, design fees, programme duration and the processes in place for managing the information flow. It should be recognised that the relative importance of these factors does not remain constant over time. Attitudes towards the importance of reducing embodied carbon are rapidly evolving as the scale of the climate emergency becomes evident. Advances in the use of digital tools to automate the design and fabrication process, and to better manage the information flows, open up the potential to optimise the structural design to a much greater extent than in the past. The degree of optimisation to be adopted for any project should be based on a rational assessment of the constraints and opportunities, rather than traditional, and possibly outdated, assumptions.

A recent example of a project where digital workflows have been used to optimise the structure to good effect is the Scalpel in the City of London. The structural engineers, Arup, used parametric design tools to calculate the optimal size for all the composite beams in the building, based on all the possible combinations of fabricated beam sizes, then used BIM software to successfully manage the flow of information through fabrication and construction. While it may have been possible to optimise the design process twenty years ago (albeit in a much clunkier fashion), trying to manage the information flow through to construction would have been very difficult, if not impossible.

Jon Carr
University of Sheffield and
Jon Carr Structural Design

Richard Harpin
University of Sheffield

11 Practical examples of three generic building types — with two potential solutions for each

11.1 Introduction and design philosophy

In this chapter, we will look at three relatively common types of building in the UK:

- A ten-storey office block
- A 30m span single-storey building
- A three-storey residential building

For the ten-storey office block and the 30m span single-storey building, two distinct structural concepts are presented, similar to what would be required in the IStructE Chartered Member (CM) examination. For the three-storey high residential building, presenting two solutions with the same structural form/arrangement, but constructed using different materials, would probably not be sufficiently distinct to satisfy the CM examiners.

Supporting calculations are also provided for the two ten-storey office building solutions, and the two 30m span single-storey building solutions. While no calculations are provided for the two three-storey residential building/apartment block solutions, references are provided to relevant design guides/standards etc.

While these solutions cover a range of structural forms and materials, it should be borne in mind that many other solutions are possible. The 'best' solution for any given project (if one exists, since in reality there will often be a number of potentially valid solutions) will depend on a diverse range of considerations, many of which will be project-specific.

While it may be tempting to re-use the solution adopted for an apparently similar project, every project is different in some way, so we recommend that you keep an open mind and consider all potential options, to ensure you adopt the most appropriate solution for the scheme in question.

Note the following limitations of the conceptual designs presented:

- Foundation designs and ground floor slab construction have not been considered
- The layouts shown are for illustrative proposes only, and may not comply with current UK Building Regulations requirements
- More efficient solutions may exist — the primary purpose of the calculations presented is to illustrate the concept design method/approach used
- The structural solutions shown are only for educational use, and should not be used for any other purpose

11.2 Ten-storey office building with open-plan layout

Solution 1 — Reinforced concrete flat slab

This solution (Figure 11.1 on p. 194) was chosen due to the current popularity of flat slab structures for multi-storey building in the UK.

As stated previously, many other RC options exist, and readers are directed to the Concrete Centre's *Economic concrete frame elements to Eurocode 2*[6] and the associated *RC spreadsheets*[118] for further guidance.

RC shear walls/cores or similar would be required to provide overall lateral stability.

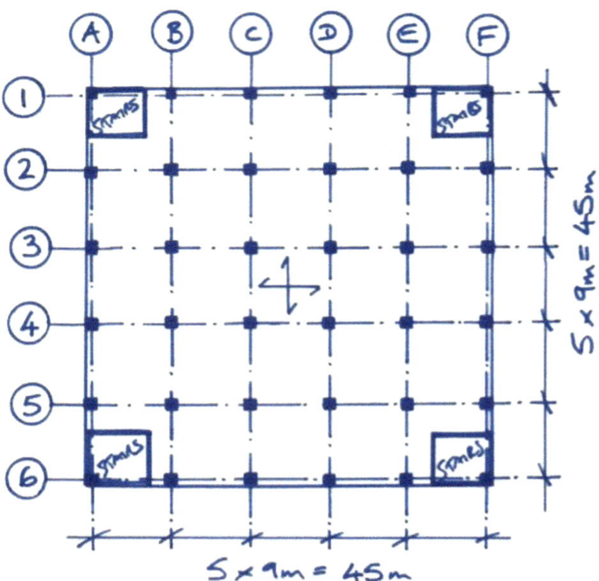

Member legend

Slab = 350mm thick (Note 6)
Internal columns = 750mm × 750mm C50/60 with 2% reinforcement
Edge columns = 550mm × 550mm C50/60 with 2% reinforcement
Corner columns = 425mm × 425mm C50/60 with 2% reinforcement

- A rectangular section may be more efficient for edge columns
- In practice, column sizes generally reduce going up the building

Notes

1. Column locations are often informed by the architect's room layouts. However, column centres of 7.2–9m generally provide a good compromise in terms of flexibility of layout and cost.
2. Lifts have not been shown, but could also contribute to lateral stability.
3. RC stair cores in each corner of building (plus lifts/service cores, if required) are deemed acceptable by inspection to provide lateral stability to a ten-storey building.
4. Location of stair cores is not ideal in terms of thermal expansion and contraction. Hence, a thermal load case analysis should be run to check for effects of 'locking-in' thermal stresses due to this and to plan dimensions of buildings exceeding 30m with no movement joints.
5. Member sizes have been determined using *Economic concrete frame elements to Eurocode 2*, assuming 120 minutes' fire resistance and mild cover requirements.
6. Slab to be 350mm thick, grade C30/37, two-way-spanning flat slab (as denoted by arrows) with no drop panels/column heads, determined in accordance with Section 3.1.10 of *Economic concrete frame elements to Eurocode 2* (be aware post-tensioning could potentially reduce slab thickness by about 50mm).
7. In some cases, edge beams may be required to reduce deflection along perimeter of building (depending on cladding system adopted).

8. Column sizes have been determined in accordance with Sections 3.3.5 and 8.4 of *Economic concrete frame elements to Eurocode 2* and/or *Manual for the design of concrete building structures to Eurocode 2*[74], based on columns between ground and first floor, 4m floor-to-floor heights, and considering live/variable load reduction. Further, an elastic reaction factor of 1.1 has been applied at upper floor levels to all internal columns i.e. giving a loaded area of $1.1 \times 9.0m \times 9.0m = 89.1m^2$ per level. Elastic reaction factors have conservatively been ignored for edge and corner columns.

9. Unfactored loads used for concept design are as follows:
   - Lightweight roof permanent load $= 0.5kN/m^2$
   - Roof variable load $= 0.6kN/m^2$
   - Floor permanent load $= 0.5kN/m^2$ + self-weight of slab
   - Floor variable load $= 2.5kN/m^2$ (office) + $1.0kN/m^2$ (lightweight partitions)
   - Cladding load $= 10kN/m$ per floor/level

Solution 2 — Braced steel frame

This solution (Figure 11.2 on p. 201) was chosen due to the current popularity of composite metal deck floor systems, combined with composite mild steel floor beams, for multi-storey building in the UK. However, various alternatives also exist e.g. precast floor units and/or cellular steel beams. Reference should be made to the Steelconstruction.info website[119], and other Steel Construction Institute (SCI) publications, for further guidance.

Either vertical steel bracing (as shown on the drawing) or RC shear walls/cores (or a combination of the two) would be required to provide overall lateral stability.

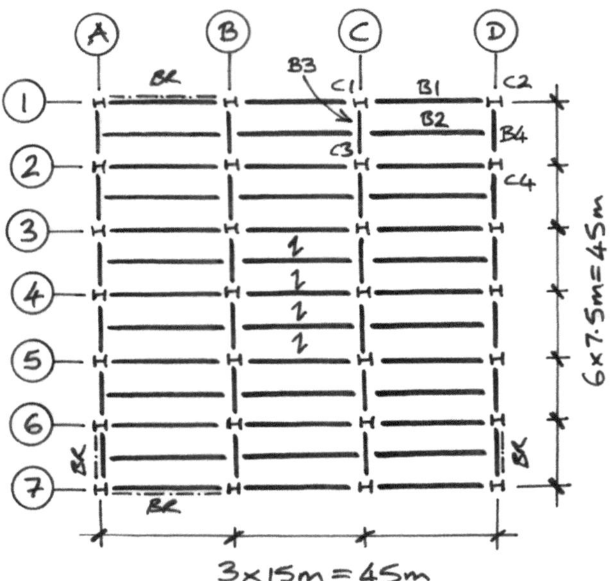

Member legend

C1 = C4 = 356 × 368 × 202UC Grade S355 column
C2 = 356 × 368 × 129UC Grade S355 column
C3 = 356 × 406 × 287UC Grade S355 column
B1 = B2 = 610 × 210 × 101UB Grade S275 composite beam
B3 = 610 × 229 × 125UB Grade S355 non-composite beam
B4 = 533 × 210 × 92UB Grade S355 non-composite beam

Notes

1. Column locations are often informed by the architect's room layouts. However, column centres of 7.2–9m generally provide a good compromise in terms of flexibility of layout and cost.
2. 'BR' denotes location of vertical steel bracing frames. Lifts, stairs and service cores are omitted for clarity, but could contribute to lateral stability.
3. Vertical bracing frames acceptable (by inspection) to provide lateral stability to a ten-storey building (span/depth ratio = 10 × 4m/7.5m = 5.33, which is reasonable for a cantilever). Location of stair cores is ideal in terms of thermal expansion and contraction. However, a thermal load case analysis may still be required to check for effects of 'locking-in' thermal stresses due to plan dimensions exceeding 30m with no movement joints.
4. Provide 120 minutes' fire resistance to structural frame in the form of applied protection (may be possible to omit this to secondary beams), and 90 minutes' to floor slabs (Table A2 of *Building Regulations Approved Document B Volume 2*[99]).
5. Assume unfactored loads used for concept design are as follows:
   - Lightweight roof permanent load = $0.5kN/m^2$
   - Roof variable load = $0.6kN/m^2$
   - Floor permanent load = $0.5kN/m^2$ + self-weight of slab and beams
   - Floor variable load = $2.5kN/m^2$ (office) + $1.0kN/m^2$ (lightweight partitions)
   - Lightweight cladding load = $1.0kN/m^2$

6. Slab to be 150mm deep (multi-span, unpropped composite), based on an 80mm deep × 0.9mm gauge trapezoidal profiled deck with normal weight concrete and A252 mesh, sized using manufacturer's load/span tables (arrows denote slab span).
7. Column sizes have been determined using the 'Blue Book', based on columns between ground and first floor, 4m floor-to-floor height at all levels, and allowing for moments due to 100mm nominal eccentricity.
8. Composite beam sizes have been determined using NCCI SN022a[120].
9. Non-composite beam sizes have been determined using the 'Blue Book' (these could potentially be designed as composite, depending on slab/beam detail).

Figure 11.1: Concrete frame option to 45m × 45m × ten-storey office building

<table>
<tr><td colspan="2">The Institution of
Structural Engineers</td><td>Job No.</td><td colspan="2">Sheet 1 of 7</td><td>Drawing No:</td></tr>
<tr><td></td><td></td><td>Made by: JFC</td><td colspan="2">Checked by:</td><td>Date: OCT 19</td></tr>
<tr><td colspan="2"></td><td colspan="4">Component:</td></tr>
<tr><td colspan="7">Project: CONCRETE FRAME OPTION TO 45m × 45m × 10 STOREY OFFICE BUILDING</td></tr>
</table>

REF	CALCULATION	OUTPUT
	<u>CONCRETE FRAME OPTION TO 45m × 45m × 10 STOREY OFFICE BUILDING</u>	
	1/ <u>TYPICAL FLOOR SLAB / FLAT SLAB</u>	
	9·0m SQUARE PANELS / BAYS THROUGHOUT, WITH 5 NO. BAYS IN EACH DIRECTION.	
BUILDING REGULATIONS APPROVED DOCUMENT 'B'	90 MINUTES FIRE RESISTANCE PERIOD ASSUMED FOR FLOOR SLABS (BUT 120 MINUTES FOR COLUMNS). SEE TABLE A2 OF BUILDING REGULATIONS APPROVED DOCUMENT 'B' FOR FURTHER DETAILS.	
	MILD COVER ASSUMED (INTERNAL)	
ECFE HANDBOOK TABLE 3.7 (SEE OVER)	FROM 'ECFE' (ECONOMIC CONCRETE FRAME ELEMENTS TO EUROCODE 2) HANDBOOK, IT CAN BE SEEN THAT;	
	RECOMMENDED SLAB THICKNESS = 347mm	
	∴ ADOPT A 350mm THICK FLAT SLAB USING GRADE C30/37 CONCRETE †	
	FOR FURTHER DETAILS, REFER TO SECTION 3.1.10 OF ECFE HANDBOOK. THIS PROVIDES A SUMMARY OF ASSUMPTIONS, REINFORCEMENT WEIGHTS, LOADS TO SUPPORTING COLUMNS, ASSUMED COLUMN SIZES (FOR PUNCHING SHEAR CHECKS ETC.) AND VARIATIONS IF REQUIRED / APPROPRIATE.	
	ALTERNATIVELY COULD USE 'CONCEPT' DESIGN SOFTWARE FROM THE CONCRETE CENTRE.	
	† CONSERVATIVELY BASED ON AN UNFACTORED IMPOSED LOAD OF 5·0 kN/m². USING 'CONCEPT' SOFTWARE WILL GIVE A THINNER SLAB, AS WOULD INTERPOLATION OF ECFE TABLE 3.7.	

Figure 11.1: *Continued*

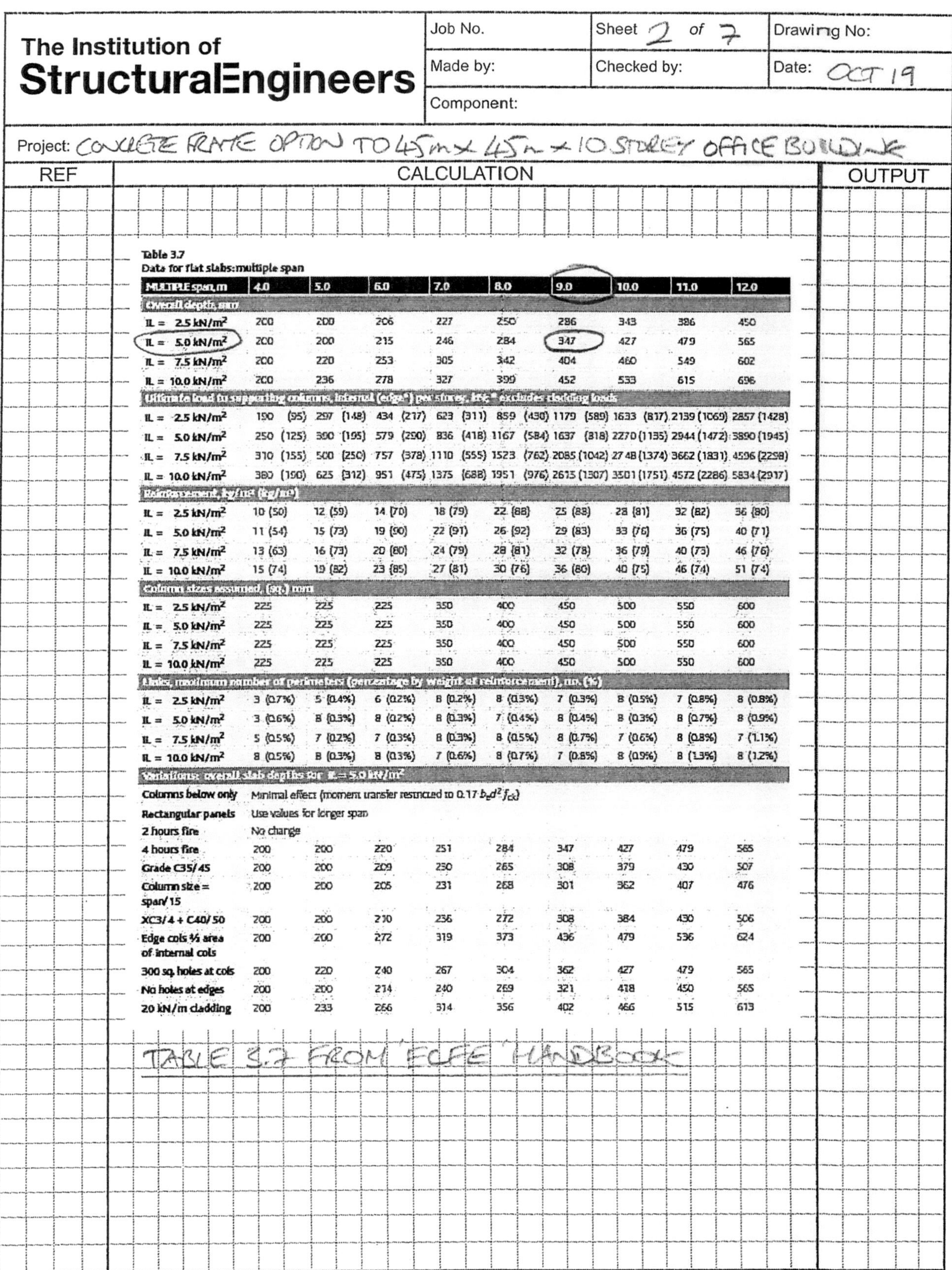

<table>
<tr><td colspan="3">The Institution of StructuralEngineers</td><td>Job No.</td><td>Sheet 2 of 7</td><td>Drawing No:</td></tr>
<tr><td colspan="3"></td><td>Made by:</td><td>Checked by:</td><td>Date: OCT 19</td></tr>
<tr><td colspan="3"></td><td>Component:</td><td></td><td></td></tr>
</table>

Project: CONCRETE FRAME OPTION TO 45m × 45m × 10 STOREY OFFICE BUILDING

REF	CALCULATION	OUTPUT

Table 3.7
Data for flat slabs: multiple span

MULTIPLE span, m	4.0	5.0	6.0	7.0	8.0	9.0	10.0	11.0	12.0
Overall depth, mm									
IL = 2.5 kN/m²	200	200	206	227	250	286	343	386	450
IL = 5.0 kN/m²	200	200	215	246	284	347	427	479	565
IL = 7.5 kN/m²	200	220	253	305	342	404	460	549	602
IL = 10.0 kN/m²	200	236	278	327	399	452	533	615	696
Ultimate load to supporting columns, internal (edge*) per storey, kN;* excludes cladding loads									
IL = 2.5 kN/m²	190 (95)	297 (148)	434 (217)	623 (311)	859 (430)	1179 (589)	1633 (817)	2139 (1069)	2857 (1428)
IL = 5.0 kN/m²	250 (125)	390 (195)	579 (290)	836 (418)	1167 (584)	1637 (818)	2270 (1135)	2944 (1472)	3890 (1945)
IL = 7.5 kN/m²	310 (155)	500 (250)	757 (378)	1110 (555)	1523 (762)	2085 (1042)	2748 (1374)	3662 (1831)	4596 (2298)
IL = 10.0 kN/m²	380 (190)	625 (312)	951 (475)	1375 (688)	1951 (976)	2615 (1307)	3501 (1751)	4572 (2286)	5834 (2917)
Reinforcement, kg/m² (kg/m³)									
IL = 2.5 kN/m²	10 (50)	12 (59)	14 (70)	18 (79)	22 (88)	25 (88)	28 (81)	32 (82)	36 (80)
IL = 5.0 kN/m²	11 (54)	15 (73)	19 (90)	22 (91)	26 (92)	29 (83)	33 (76)	36 (75)	40 (71)
IL = 7.5 kN/m²	13 (63)	16 (73)	20 (80)	24 (79)	28 (81)	32 (78)	36 (79)	40 (73)	46 (76)
IL = 10.0 kN/m²	15 (74)	19 (82)	23 (85)	27 (81)	30 (76)	36 (80)	40 (75)	46 (74)	51 (74)
Column sizes assumed, (sq.) mm									
IL = 2.5 kN/m²	225	225	225	350	400	450	500	550	600
IL = 5.0 kN/m²	225	225	225	350	400	450	500	550	600
IL = 7.5 kN/m²	225	225	225	350	400	450	500	550	600
IL = 10.0 kN/m²	225	225	225	350	400	450	500	550	600
Links, maximum number of perimeters (percentage by weight of reinforcement), no. (%)									
IL = 2.5 kN/m²	3 (0.7%)	5 (0.4%)	6 (0.2%)	8 (0.2%)	8 (0.3%)	7 (0.3%)	8 (0.5%)	7 (0.8%)	8 (0.8%)
IL = 5.0 kN/m²	3 (0.6%)	8 (0.3%)	8 (0.2%)	8 (0.3%)	7 (0.4%)	8 (0.4%)	8 (0.3%)	8 (0.7%)	8 (0.9%)
IL = 7.5 kN/m²	5 (0.5%)	7 (0.2%)	7 (0.3%)	8 (0.3%)	8 (0.5%)	8 (0.7%)	7 (0.6%)	8 (0.8%)	7 (1.1%)
IL = 10.0 kN/m²	8 (0.5%)	8 (0.3%)	8 (0.3%)	7 (0.6%)	8 (0.7%)	7 (0.8%)	8 (0.9%)	8 (1.3%)	8 (1.2%)
Variations: overall slab depths for IL = 5.0 kN/m²									
Columns below only	Minimal effect (moment transfer restricted to $0.17 \cdot b_e d^2 f_{ck}$)								
Rectangular panels	Use values for longer span								
2 hours fire	No change								
4 hours fire	200	200	220	251	284	347	427	479	565
Grade C35/45	200	200	209	230	265	308	379	430	507
Column size = span/15	200	200	205	231	268	301	362	407	476
XC3/4 + C40/50	200	200	210	236	272	308	384	430	506
Edge cols ½ area of internal cols	200	200	272	319	373	436	479	536	624
300 sq. holes at cols	200	220	240	267	304	362	427	479	565
No holes at edges	200	200	214	240	269	321	418	450	565
20 kN/m cladding	200	233	266	314	356	402	466	515	613

TABLE 3.7 FROM 'ECFE' HANDBOOK

Figure 11.1: *Continued*

<table>
<tr><td rowspan="3">The Institution of
StructuralEngineers</td><td colspan="2">Job No.</td><td>Sheet 3 of 7</td><td>Drawing No:</td></tr>
<tr><td>Made by: JFC</td><td>Checked by:</td><td>Date: OCT 19</td></tr>
<tr><td colspan="3">Component:</td></tr>
</table>

Project: CONCRETE FRAME OPTION TO 45m × 45m × 10 STOREY OFFICE BUILDING

REF	CALCULATION	OUTPUT
	2, <u>TYPICAL INTERNAL COLUMN (BETWEEN</u> <u>GROUND AND 1ST FLOOR)</u>:-	
ROOF LOADS	ASSUMING A LIGHTWEIGHT (I.E. STEEL) ROOF ; $1.0 G_k = 0.5 \text{ kN/m}^2$, AND $1.0 Q_k = 0.6 \text{ kN/m}^2$ $\therefore n = 1.35 G_k + 1.5 Q_k$ $\therefore n = 1.35 \times 0.5 + 1.5 \times 0.6 = 1.58 \text{ kN/m}^2 \text{ (ULS)}$ IGNORING CONTINUITY EFFECTS (SINCE STEEL BEAMS ARE LIKELY TO BE SIMPLY SUPPORTED) ; AXIAL LOAD FROM ROOF $= 9.0^2 \times 1.58 = 130 \text{ kN (ULS)}$	

FLOOR LOADS	ALLOWING FOR 0.5 kN/m^2 FLOOR FINISHES GIVES FOLLOWING FLOOR LOADS ;
	
	 $1.0 G_k = (0.35 \times 25 + 0.5) = 9.25 \text{ kN/m}^2 \text{ (SLS)}$
	        m kN/m³ kN/m²
	
	 $1.0 Q_k = 2.5 + 1.0 = 3.5 \text{ kN/m}^2 \text{ (SLS)}$
	       PEOPLE/ PARTITIONS
	       FURNITURE
	
	 $\therefore n = 1.35 \times 9.25 + 1.5 \times 3.5 = 17.75 \text{ kN/m}^2 \text{ (ULS)}$
	   FULL	

N.A. TO EC1, CLAUSE NA.2.7	HOWEVER, LIVE LOAD REDUCTION CAN BE TAKEN INTO ACCOUNT, IN ACCORDANCE WITH THE NATIONAL ANNEX (N.A.) TO BS EN 1991-1-1: 2002. FOR THE PROPOSED USE (OFFICES), A LIVE LOAD REDUCTION FACTOR, α_n, OF 0.6 CAN BE USED HERE, SINCE THE NUMBER OF QUALIFYING FLOORS SUPPORTED IS BETWEEN 5 AND 10.	

Figure 11.1: *Continued*

<table>
<tr><td colspan="2">The Institution of
StructuralEngineers</td><td>Job No.</td><td>Sheet 4 of 7</td><td>Drawing No:</td></tr>
<tr><td colspan="2"></td><td>Made by: JFC</td><td>Checked by:</td><td>Date: OCT 19</td></tr>
<tr><td colspan="2"></td><td colspan="3">Component:</td></tr>
<tr><td colspan="5">Project: CONCRETE FRAME OPTION TO 45 m × 45 m × 10 STOREY OFFICE BUILDING</td></tr>
<tr><td>REF.</td><td colspan="3">CALCULATION</td><td>OUTPUT</td></tr>
</table>

REF: 'ECFE' HANDBOOK SECTION 8.4.5

$$\therefore n_{REDUCED} = 1.35 \times 9.25 + 1.5 \times 0.6 \times 3.5$$

$$\text{ie} \qquad " \qquad = 15.65 \ kN/m^2 \ (ULS)$$

SECTION 8.4.5 OF 'ECFE' HANDBOOK ADVISES ELASTIC REACTION FACTORS FOR COLUMN LOADS.

FOR THE FIRST INTERIOR SUPPORTS OF CONTINUOUS SLABS OF 3 OR MORE SPANS, ELASTIC REACTION FACTOR = 1.1

$$\therefore \text{AXIAL LOAD FROM ONE FLOOR};$$

$$= 1.1 \times 9.0^2 \times 15.65 \quad = 1394 \ kN \ (ULS)$$
$$\quad ERF \qquad m^2 \qquad kN/m^2$$

HENCE, TOTAL COLUMN LOAD FROM ROOF PLUS 9 No SUSPENDED FLOORS (EXCLUDING SELF-WEIGHT OF COLUMN) IS;

$$F = 130 + 9 \times 1394 = 12,676 \ kN \ (ULS)$$

REF: ECFE HANDBOOK TABLE 3.35

HOWEVER THIS EXCEEDS THE MAXIMUM COLUMN LOAD OF 10,000 kN GIVEN IN ECFE HANDBOOK TABLE 3.35, SO USE THE EQUIVALENT STRESS VALUES GIVEN IN TABLE 4.3 OF THE ISTRUCE MANUAL FOR THE DESIGN OF CONCRETE BUILDING STRUCTURES TO EUROCODE 2 (COMMONLY KNOWN AS THE 'GREEN BOOK'), AS SHOWN OVER.

LET'S TRY 750 mm SQUARE RC COLUMN WITH f_{ck} = 50N/mm² AND 2% HIGH YIELD STEEL REINFORCEMENT (IE f_{yk} = 500 N/mm²). HENCE;

$$\text{APPLIED COMPRESSIVE STRESS} = 12,676 \times 10^3 / 750^2$$

$$\therefore \quad " \qquad " \qquad " \qquad = 22.5 \ N/mm^2$$

Figure 11.1: *Continued*

<table>
<tr><td colspan="3">The Institution of StructuralEngineers</td><td>Job No.</td><td>Sheet 5 of 7</td><td>Drawing No:</td></tr>
<tr><td></td><td>Made by: JFC</td><td>Checked by:</td><td>Date: 05 19</td></tr>
<tr><td>Component:</td></tr>
</table>

Project: CONCRETE FRAME OPTION) TO 45m × 45m × 10 STOREY OFFICE BUILDING

REF	CALCULATION	OUTPUT

Table 4.3 Equivalent 'stress' values

Reinforcement (500MPa) percentage ρ	Equivalent stresses (MPa) for concrete strength classes		
	C25/30	C30/37	C35/45
$\rho = 1\%$	14	17	19
$\rho = 2\%$	18	20	22
$\rho = 3\%$	21	23	25
$\rho = 4\%$	24	27	29

The equivalent 'stresses' given in Table 4.3 are derived from the expression:

$$stress = 0.44\, f_{ck} + \frac{\rho}{100}(0.67\, f_{yk} - 0.44\, f_{ck})$$

Where: f_{ck} is the characteristic concrete strength in MPa

f_y the characteristic strength of reinforcement in MPa

ρ the percentage of reinforcement.

HENCE, 'ALLOWABLE' STRESS S EQUAL TO;

$$= 0.44 \times 50 + \frac{2}{100} \times (0.67 \times 500 - 0.44 \times 50)$$

$$= 22.0 + \frac{2}{100} \times 313 = 22.0 + 6.26 \simeq 28.3 \text{ N/mm}^2$$

$$> 22.5 \text{ ok}$$

∴ 750 mm SQUARE C50 INTERNAL COLUMNS

WITH 2% REINFORCEMENT SHOULD BE OK (AS

LONG AS THEY'RE NOT SLENDER)

3, TYPICAL EDGE COLUMN (BETWEEN GROUND

AND 1ST FLOOR) —

DATA IS GENERALLY AS PER INTERNAL COLUMN

BUT EDGE COLUMNS ARE ALSO SUBJECT TO

AN UNFACTORED CLADDING LOAD OF 10kN/m

AT EACH FLOOR LEVEL.

AXIAL LOAD FROM ROOF = 50% OF INTERNAL COLUMN

"   "   "   "   " $= 0.5 \times 130 = 65$ kN/ULS

Figure 11.1: *Continued*

<table>
<tr><td colspan="2">The Institution of StructuralEngineers</td><td>Job No.</td><td colspan="2">Sheet 6 of 7</td><td>Drawing No:</td></tr>
<tr><td colspan="2"></td><td>Made by: JFC</td><td colspan="2">Checked by:</td><td>Date: OCT 19</td></tr>
<tr><td colspan="2"></td><td colspan="4">Component:</td></tr>
<tr><td colspan="6">Project: CONCRETE FRAME OPTION TO 45m × 45m × 10 STOREY OFFICE BUILDING</td></tr>
</table>

REF	CALCULATION	OUTPUT
ECFE HANDBOOK SECTION 8.4.5	FOR PERIMETER COLUMNS, AN ELASTIC REACTION FACTOR OF 0.5 SHOULD BE CONSIDERED FOR IN-SITU RC FLOORS. ∴ AXIAL LOAD FROM ONE FLOOR IS; $= 0.5 \times 9.0^2 \times 15.65 = 634$ kN (ULS) ERF m² kN/m² CLADDING LOAD $= 1.35 \times 9.0 \times 10.0 = 122$ kN (ULS) F.O.S.† m kN/m PER FLOOR	
'GREEN BOOK' SECTION 4.8.4	WE DO HOWEVER NEED TO CONSIDER ECCENTRICITIES FOR EDGE COLUMNS, IN ACCORDANCE WITH SECTION 4.8.4 OF THE iSTRUCTE 'GREEN BOOK'. THIS REQUIRES A FACTOR OF 1.5 TO BE APPLIED TO THE FLOOR LOADS FROM THE FLOOR LEVEL IMMEDIATELY ABOVE THE COLUMN BEING DESIGNED (IE – FROM 1ST FLOOR LEVEL IN THIS CASE). NOTE THAT THE CLADDING LOADS WILL BALANCE OUT HERE SO HAVE BEEN IGNORED WHEN CALCULATING THE ADDITIONAL LOAD DUE TO ECCENTRICITIES ∴ TOTAL COLUMN LOAD EQUALS; $F = 65 + 8 \times (634 + 122) + (1 \times (1.5 \times 634 + 122))$ ROOF 2ND FLOOR AND ABOVE 1ST FLOOR ∴ $F = 7186$ kN (ULS) WHILST THE CHARTS IN SECTION 3.3.8 OF THE ECFE HANDBOOK COULD BE USED HERE, WE'LL USE THE SAME METHOD AS FOR INTERNAL COLUMNS. † F.O.S. = FACTOR OF SAFETY	

Figure 11.1: *Continued*

<table>
<tr><td colspan="2">The Institution of StructuralEngineers</td><td>Job No.</td><td>Sheet 7 of 7</td><td>Drawing No:</td></tr>
<tr><td></td><td></td><td>Made by: JfC</td><td>Checked by:</td><td>Date: OCT 19</td></tr>
<tr><td></td><td></td><td colspan="3">Component:</td></tr>
</table>

Project: CONCRETE FRAME OPTION TO 45m × 45m × 10 STOREY OFFICE BUILDING

REF	CALCULATION	OUTPUT
	TRY 550mm SQUARE R.C. COLUMNS (f_{ck} = 50 N/mm² AND f_{yk} = 500 N/mm², AS BEFORE):	

APPLIED COMPRESSIVE STRESS = $1186 \times 10^3 / 550^2$

∴ " " " = 23.8 N/mm²

ALLOWABLE STRESS = 28.3 N/mm² > 23.8 ✓

∴ ADOPT 550mm SQUARE C50 EDGE COLUMNS WITH 2% REINFORCEMENT SHOULD BE OK (AS LONG AS THEY'RE NOT SLENDER)

4, TYPICAL CORNER COLUMN (BETWEEN GROUND AND 1ST FLOOR) –

ADOPT A SIMILAR PROCESS AS USED ABOVE FOR SIZING EDGE COLUMNS, EXCEPT THAT A FACTOR OF 2.0 SHOULD BE APPLIED TO THE FLOOR LOADS FROM 1ST FLOOR LEVEL WHEN TAKING ECCENTRICITIES INTO ACCOUNT. ALSO NOTE THAT CLADDING LOADS WON'T BALANCE OUT FOR A CORNER COLUMN, SO SHOULD BE CONSIDERED WHEN CALCULATING ECCENTRICITY EFFECTS (IN OTHER WORDS, THE 2.0 FACTOR SHOULD BE APPLIED TO CLADDING LOADS AS WELL AS FLOOR LOADS).

Figure 11.2: Steel frame option to 45m × 45m × ten-storey office building

<table>
<tr><td colspan="2" rowspan="2">The Institution of
StructuralEngineers</td><td>Job No.</td><td colspan="2">Sheet 1 of 12</td><td>Drawing No:</td></tr>
<tr><td>Made by: JFC</td><td colspan="2">Checked by:</td><td>Date: OCT 19</td></tr>
<tr><td colspan="2">Component:</td><td colspan="2"></td></tr>
</table>

Project: STEEL FRAME OPTION TO 45m × 45m × 10 STOREY OFFICE BUILDING

REF	CALCULATION	OUTPUT
	STEEL FRAME OPTION TO 45m × 45m × 10 STOREY OFFICE BUILDING —	
	1/ TYPICAL FLOOR SLAB (COMPOSITE METAL DECK) —	
	SLAB SPANS 3.75m WITH 2 OR MORE SPANS (IE — GIVING CONTINUITY) AND IS UNPROPPED.	
BUILDING REGS. APPROVED DOCUMENT B, TBL A2	90 MINUTES FIRE RESISTANCE ASSUMED FOR FLOOR SLAB (BUT 120 MINUTES FOR FLOOR BEAMS AND COLUMNS). REFER TO TABLE A2 OF BUILDING REGULATIONS APPROVED DOCUMENT B FOR FURTHER DETAILS.	
SPAN/LOAD TABLES	USING MANUFACTURER'S SPAN/LOAD TABLES AND FIRE DESIGN TABLES, AS SHOWN OVER, IT CAN BE SEEN THAT;	
	A 150mm THICK COMPOSITE METAL DECK SLAB, WITH A 0.9mm GAUGE 80mm DEEP DECK, WITH A252 MESH REINFORCEMENT, IS ADEQUATE.	
	2, BEAM 'B2' : 15m SPAN COMPOSITE MILD STEEL INTERNAL BEAM (NB: BEAM 'B1' IS SIMILAR) —	
	THESE 15m SPAN, SIMPLY SUPPORTED, SECONDARY BEAMS, OCCUR AT 3.75m CENTRES AND ARE SUBJECT TO AN IMPOSED LOAD OF 3.5 kN/m² (UNFACTORED). THEY ARE SIZED USING NCCI SN022a, THE RELEVANT CHART FROM WHICH IS SHOWN ON CALC. SHEET 3. NOTE THAT FOR SUCH LONG SPAN BEAMS SUPPORTING MODEST UDLS, DEFLECTION OR VIBRATION ARE LIKELY TO BE CRITICAL.	
	∴ ADOPT 610 × 210 × 101 UB GRADE S275 MILD STEEL COMPOSITE BEAMS	

Figure 11.2: *Continued*

The Institution of **StructuralEngineers**			
Job No.	Sheet 2 of 12		Drawing No:
Made by: J.FC	Checked by:		Date: OCT 19
Component:			

Project: STEEL FRAME OPTION TO 45m × 45m ×10 STOREY OFFICE BUILDING

REF	CALCULATION	OUTPUT

Span/load table — Normal weight concrete

Support Condition	Slab Depth (mm)	Concrete Volume (m³/m²)	0.9 Gauge Imposed Load				1.0 Gauge Imposed Load				1.2 Gauge Imposed Load			
			FW	5.0	6.7	10.0	FW	5.0	6.7	10.0	FW	5.0	6.7	10.0
Single - Unpropped	140	0.098	3.73	3.73	3.73	3.51	4.11	4.11	4.11	3.56	4.28	4.28	4.28	3.67
	150	0.108	3.63	3.63	3.63	3.63	4.01	4.01	4.01	5.77	4.20	4.20	4.20	3.90
	160	0.118	3.55	3.55	3.55	3.55	3.92	3.92	3.92	3.92	4.12	4.12	4.12	4.12
	170	0.128	3.47	3.47	3.47	3.47	3.83	3.83	3.83	3.83	4.05	4.05	4.05	4.05
	175	0.133	3.43	3.43	3.43	3.43	3.79	3.79	3.79	3.79	4.02	4.02	4.02	4.02
	200	0.158	3.26	3.26	3.26	3.26	3.61	3.61	3.61	3.61	3.86	3.86	3.86	3.86
	250	0.208	2.97	2.97	2.97	2.97	3.30	3.30	3.30	3.30	3.57	3.57	3.57	3.57
Multiple - Unpropped	140	0.098	4.01	4.01	4.01	3.51	4.47	4.47	4.47	3.56	4.90	4.90	4.68	3.67
	150	0.108	3.91	3.91	3.91	3.73	4.36	4.36	4.36	3.79	5.09	5.09	4.99	3.90
	160	0.118	3.82	3.82	3.82	3.82	4.26	4.26	4.26	4.03	4.98	4.98	4.98	4.14
	170	0.128	3.73	3.73	3.73	3.73	4.17	4.17	4.17	4.17	4.87	4.87	4.87	4.38
	175	0.133	3.69	3.69	3.69	3.69	4.12	4.12	4.12	4.12	4.82	4.82	4.82	4.50
	200	0.158	3.50	3.50	3.50	3.50	3.92	3.92	3.92	3.92	4.60	4.60	4.60	4.60
	250	0.208	3.20	3.20	3.20	3.20	3.58	3.58	3.58	3.58	4.23	4.23	4.23	4.23
Multiple - Propped	140	0.098	4.90	4.74	4.33	3.51	4.90	4.81	4.39	3.56	4.90	4.90	4.51	3.67
	150	0.108	5.25	5.04	4.61	3.73	5.25	5.11	4.67	3.79	5.25	5.24	4.79	3.90
	160	0.118	5.60	5.34	4.88	3.97	5.60	5.41	4.95	4.03	5.60	5.55	5.08	4.14
	170	0.128	5.95	5.64	5.16	4.20	5.95	5.72	5.23	4.26	5.95	5.86	5.36	4.38
	175	0.133	6.13	5.79	5.26	4.31	6.13	5.87	5.37	4.38	6.13	6.02	5.50	4.50
	200	0.158	6.50	6.17	5.64	4.90	6.50	6.46	5.91	4.97	6.50	6.50	6.22	5.10
	250	0.208	6.37	6.37	6.22	5.48	6.50	6.50	6.50	5.75	6.50	6.50	6.50	6.15

SPAN/LOAD TABLE FOR COMPOSITE METAL DECK SLAB

Simplified fire design table — Normal weight concrete

Fire Rating (Hrs)	Slab Depth (mm)	Span (m) for given Imposed Load (kN/m²)											
		A142			A193			A252			A393		
		5.0	6.7	10.0	5.0	6.7	10.0	5.0	6.7	10.0	5.0	6.7	10.0
1.0	140	3.65	3.31	2.86	3.91	3.55	3.06	4.01	3.79	3.27	4.01	4.01	3.51
	150	3.84	3.49	3.02	3.91	3.75	3.25	3.91	3.91	3.48	3.91	3.91	3.73
	160	3.82	3.64	3.16	3.82	3.82	3.41	3.82	3.82	3.65	3.82	3.82	3.82
	170	3.73	3.71	3.23	3.73	3.73	3.49	3.73	3.73	3.73	3.73	3.73	3.73
	175	3.69	3.69	3.26	3.69	3.69	3.52	3.69	3.69	3.69	3.69	3.69	3.69
	200	-	-	-	3.50	3.50	3.50	3.50	3.50	3.50	3.50	3.50	3.50
	250	-	-	-	-	-	-	3.20	3.20	3.20	3.20	3.20	3.20
1.5	150	3.31	3.01	2.61	3.60	3.27	2.83	3.89	3.54	3.06	3.91	3.91	3.51
	160	3.47	3.17	2.75	3.79	3.45	3.00	3.82	3.74	3.25	3.82	3.82	3.74
	170	3.61	3.30	2.87	3.73	3.60	3.14	3.73	3.73	3.41	3.73	3.73	3.73
	175	3.65	3.34	2.91	3.69	3.65	3.18	3.69	3.69	3.46	3.69	3.69	3.69
	200	-	-	-	3.50	3.50	3.31	3.50	3.50	3.50	3.50	3.50	3.50
	250	-	-	-	-	-	-	3.20	3.20	3.20	3.20	3.20	3.20
2.0	160	3.04	2.77	2.42	3.37	3.07	2.67	3.70	3.37	2.93	3.82	3.82	3.42
	170	3.18	2.91	2.54	3.53	3.23	2.81	3.73	3.55	3.09	3.73	3.73	3.62
	175	3.24	2.97	2.59	3.60	3.30	2.88	3.69	3.62	3.16	3.69	3.69	3.69
	200	-	-	-	3.50	3.43	3.02	3.50	3.50	3.32	3.50	3.50	3.50
	250	-	-	-	-	-	-	3.20	3.20	3.20	3.20	3.20	3.20

FIRE DESIGN TABLE FOR COMPOSITE METAL DECK SLAB

HENCE, THE FIRE CONDITION IS CRITICAL SINCE MAXIMUM ALLOWABLE SPAN = 3.89m, WHICH EXCEEDS ACTUAL SPAN OF 3.75m (WHILST MAX. ALLOWABLE SPAN FROM FIRST TABLE = 3.91m, SO NOT CRITICAL).

Figure 11.2: *Continued*

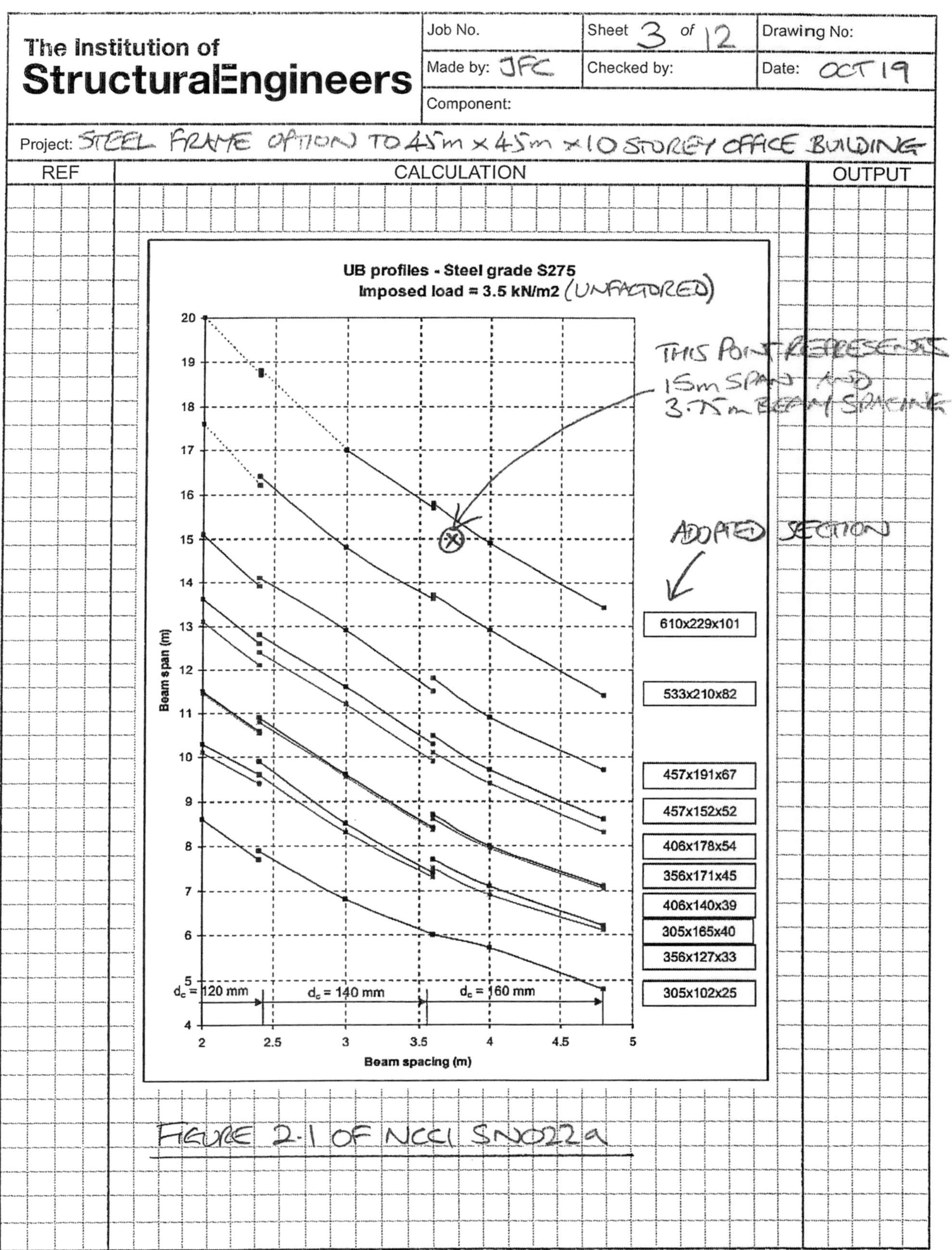

Figure 11.2: *Continued*

The Institution of **StructuralEngineers**	Job No.	Sheet 4 of 12	Drawing No:
	Made by: JFC	Checked by:	Date: OCT 19
	Component:		

Project: STEEL FRAME OPTION TO 45m × 45m × 10 STOREY OFFICE BUILDING

REF	CALCULATION	OUTPUT

3, BEAM 'B3' : 7.5m SPAN SIMPLY SUPPORTED

NON-COMPOSITE MILD STEEL INTERNAL BEAM —

THESE 7.5m SPAN NON-COMPOSITE MILD STEEL PRIMARY BEAMS OCCUR AT 15m CENTRES AND ARE SUBJECT TO A SIGNIFICANT CENTRAL POINT LOAD, P. HENCE, STRENGTH (IE — BENDING OR SHEAR) WILL ALMOST CERTAINLY BE CRITICAL HERE.

REF: STEEL DESIGNER'S MANUAL

FROM THE STEEL DESIGNER'S MANUAL, OR SIMILAR PUBLICATIONS;

$$\text{MAX BENDING MOMENT} = \frac{PL}{4} \quad (\text{AT MIDSPAN})$$

$$\text{MAX SHEAR FORCE} = P/2 \quad (\text{ALONG FULL LENGTH OF BEAM})$$

$$\text{MAX DEFLECTION} = \frac{PL^3}{48EI} \quad (\text{AT MIDSPAN})$$

AREA OF FLOOR CONTRIBUTING TO POINT LOAD, P;

$$A = 3.75\,(m) \times 15.0\,(m) = 56.25\,m^2$$

USING FIGURES / DATA FROM THE COMPOSITE METAL DECK MANUFACTURER'S LITERATURE / WEBSITE;

$$\text{SLAB SELF-WEIGHT} = \underset{\substack{(kN/m^2) \\ \text{METAL DECKING}}}{0.11} + \underset{\substack{(m^3/m^2) \quad (kN/m^3) \\ \text{CONCRETE}}}{0.108^\dagger \times 25}$$

$$'' \quad '' \quad '' \quad '' = 2.81\,kN/m^2 \ (\text{UNFACTORED})$$

$\dagger$ 0.108 m^3/m^2 IS THE AVERAGE VOLUME OF CONCRETE PER SQUARE METRE OF FLOOR SLAB.

Figure 11.2: *Continued*

The Institution of **StructuralEngineers**	Job No.	Sheet 5 of 12	Drawing No:
	Made by: JFC	Checked by:	Date: OCT 19
	Component:		

Project: STEEL FRAME OPTION TO 45m × 45m × 10 STOREY OFFICE BUILDING

REF	CALCULATION	OUTPUT
	ALSO INCLUDE AN ALLOWANCE OF $0.5 kN/m^2$ FOR STEEL FLOOR BEAMS (NOTE THAT THIS FIGURE SHOULD BE REVIEWED ONCE THE ACTUAL BEAM SIZES/WEIGHTS HAVE BEEN DETERMINED, AND REVISED IF NECESSARY). HENCE;	

TOTAL DEAD LOAD $= 2.81 + 0.5 + 0.5$
(SLAB) (BEAMS) (FINISHES)

ie " " " $= 3.81 kN/m^2$ (UNFACTORED)

$\therefore n = 1.35 G_k + 1.5 Q_k$

ie $n = 1.35 \times 3.81 + 1.5 \times 3.5 = 10.4 kN/m^2$ (ULS)

$\therefore$ POINT LOAD, $P = 10.4 \times 56.25 = 585 kN$ (ULS)

$\therefore M_{MAX} = \dfrac{PL}{4} = \dfrac{585 \times 7.5}{4} = 1097 kNm$ (ULS)

$\therefore R_{MAX} = \dfrac{P}{2} = \dfrac{585}{2} = 293 kN$ (ULS)

TRY A 610 × 229 × 125 UB GRADE S355 MILD STEEL BEAM;

<u>BENDING CHECKS;</u>

WE NEED TO CAREFULLY CONSIDER LATERAL TORSIONAL BUCKLING (LTB) EFFECTS HERE, SINCE WE CAN'T ASSUME FULL/CONTINUOUS LATERAL RESTRAINT TO THE COMPRESSION (IE TOP) FLANGE OF THIS BEAM. INDEED, IT IS ONLY RESTRAINED AT MIDSPAN (BY THE SECONDARY BEAM BEING SUPPORTED) AS WELL AS AT SUPPORTS.

A CONSERVATIVE APPROACH WOULD BE TO ASSUME A VALUE FOR C_1 (A FACTOR WHICH DEPENDS ON THE SHAPE OF THE BENDING MOMENT DIAGRAM) $= 1.0$, WHICH WOULD RESULT IN A 610 × 229 × 140 UB GRADE S355 SECTION BEING REQUIRED.

Figure 11.2: *Continued*

The Institution of **StructuralEngineers**	Job No.	Sheet 6 of 12	Drawing No:
	Made by: JFC	Checked by:	Date: OCT 19
	Component:		

Project: STEEL FRAME OPTION TO 45m × 45m × 10 STOREY OFFICE BUILDING

REF	CALCULATION	OUTPUT
SCI P362 TABLE 6.4	FROM TABLE 6.4 OF SCI DESIGN GUIDE P362, IT CAN BE SEEN THAT $C_1 = 1.35$ FOR THE OVERALL BENDING MOMENT DIAGRAM GIVEN BY A CENTRAL POINT LOAD ON A SIMPLY SUPPORTED BEAM, AS SHOWN BELOW;	

P

R 3.75m 3.75m R

HOWEVER, THE ABOVE IGNORES THE (BENEFICIAL) LATERAL RESTRAINT PROVIDED AT MIDSPAN BY THE SECONDARY BEAM. A MORE ACCURATE APPROACH WOULD BE TO CONSIDER THE SHAPE OF THE BENDING MOMENT DIAGRAM BETWEEN THE POINTS OF LATERAL RESTRAINT (IE — FOR EACH 3.75m LONG HALF OF THE BEAM), AS SHOWN BELOW;

0 kNm   1097 kNm

R   ℄ BEAM
3.75m

REF	CALCULATION	OUTPUT
DITTO	FROM TABLE 6.4 OF SCI P362 (SHOWN OVER), IT CAN BE SEEN THAT THIS ARRANGEMENT EQUATES TO $\psi = 0.00$ AND, HENCE, $C_1 = 1.77$.	
SCI P363 (BLUE BOOK) ↑ SEE OVER	CONSERVATIVELY TAKING $L_e = 4.0m$, FROM THE 'BLUE BOOK', IT CAN BE SEEN THAT; $M_b = 1150$ kNm FOR $C_1 = 1.5$ & $M_b = 1270$ kNm FOR $C_1 = 2.0$	

∴ $M_b = 1215$ kNm FOR $C_1 = 1.77$ WHICH > 1097 kNm SO OK (BY INTERPOLATION).

Figure 11.2: *Continued*

<table>
<tr><td colspan="2" rowspan="2">The Institution of
StructuralEngineers</td><td>Job No.</td><td colspan="2">Sheet 7 of 12</td><td>Drawing No:</td></tr>
<tr><td>Made by: JFC</td><td>Checked by:</td><td></td><td>Date: OCT 19</td></tr>
<tr><td></td><td></td><td colspan="4">Component:</td></tr>
</table>

Project: STEEL FRAME OPTION TO 45m × 45m ×10 STOREY OFFICE BUILDING

REF	CALCULATION	OUTPUT

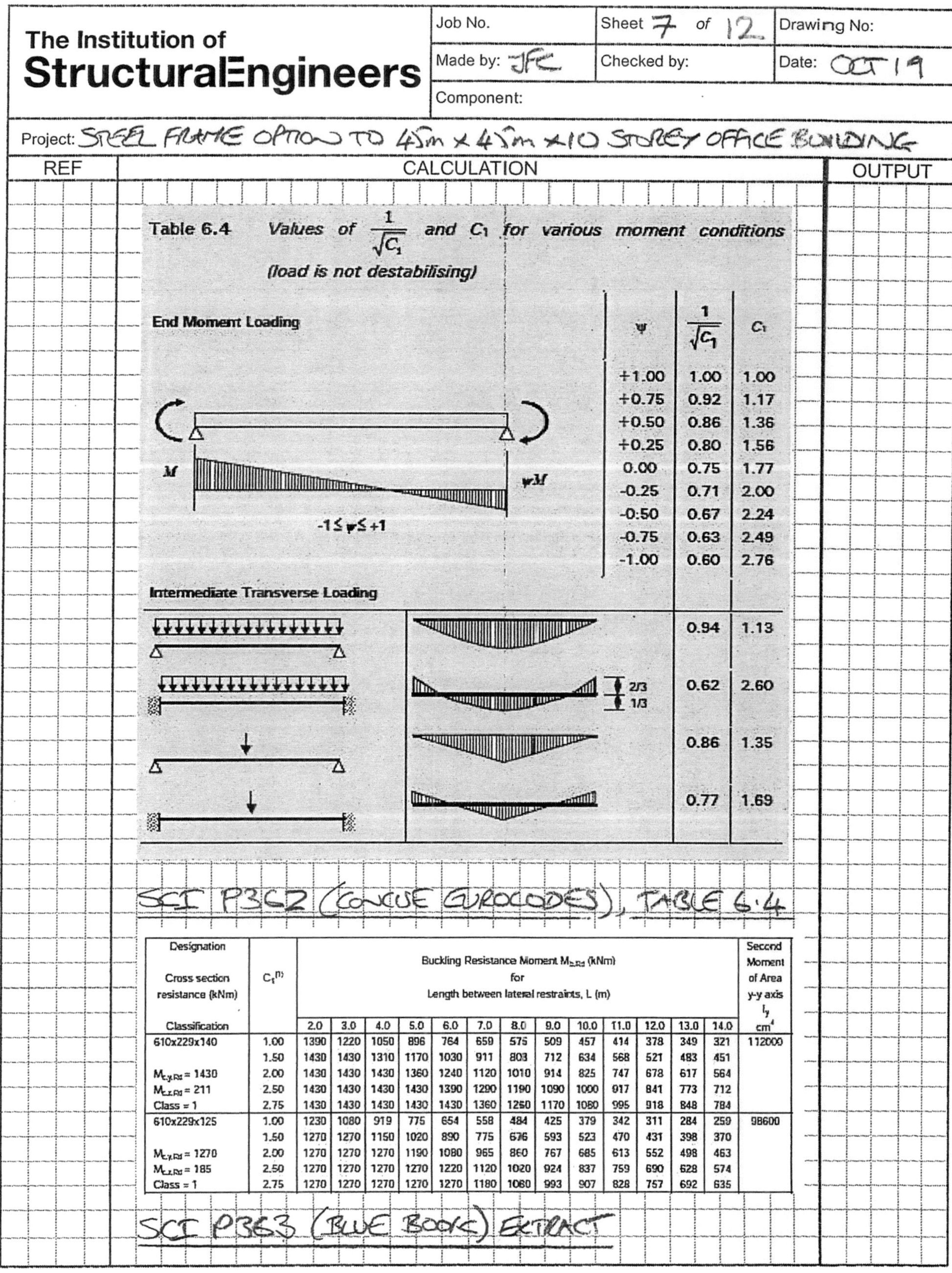

SCI P362 (CONCISE EUROCODES), TABLE 6.4

Designation	C_t [1]	Buckling Resistance Moment $M_{b,Rd}$ (kNm) for Length between lateral restraints, L (m)												Second Moment of Area y-y axis I_y
Cross section resistance (kNm) Classification		2.0	3.0	4.0	5.0	6.0	7.0	8.0	9.0	10.0	11.0	12.0	13.0	14.0	cm⁴
610x229x140	1.00	1390	1220	1050	896	764	659	575	509	457	414	378	349	321	112000
	1.50	1430	1430	1310	1170	1030	911	803	712	634	568	521	483	451	
$M_{c,y,Rd}$ = 1430	2.00	1430	1430	1430	1360	1240	1120	1010	914	825	747	678	617	564	
$M_{c,z,Rd}$ = 211	2.50	1430	1430	1430	1430	1390	1290	1190	1090	1000	917	841	773	712	
Class = 1	2.75	1430	1430	1430	1430	1430	1360	1260	1170	1080	995	918	848	784	
610x229x125	1.00	1230	1080	919	775	654	558	484	425	379	342	311	284	259	98600
	1.50	1270	1270	1150	1020	890	775	676	593	523	470	431	398	370	
$M_{c,y,Rd}$ = 1270	2.00	1270	1270	1270	1190	1080	965	860	767	685	613	552	498	463	
$M_{c,z,Rd}$ = 185	2.50	1270	1270	1270	1270	1220	1120	1020	924	837	759	690	628	574	
Class = 1	2.75	1270	1270	1270	1270	1270	1180	1060	993	907	828	757	692	635	

SCI P363 (BLUE BOOK) EXTRACT

Figure 11.2: *Continued*

The Institution of StructuralEngineers	Job No.		Sheet **8** of **12**	Drawing No:
	Made by: **JFC**	Checked by:		Date: **OCT 19**
	Component:			

Project: STEEL FRAME OPTION TO 45m × 45m × 10 STOREY OFFICE BUILDING

REF	CALCULATION	OUTPUT

SHEAR CHECKS;

WHILST SHEAR ISN'T USUALLY CRITICAL FOR FLOOR BEAMS IN OFFICE BUILDINGS, IT COULD BE HERE DUE TO THE HIGH POINT LOAD APPLIED AT MIDSPAN.

SECTION 6.2.6 of SCI P362

PLASTIC SHEAR CAPACITY $= V_{pl,Rd} = A_v(f_y/3^{0.5})/\gamma_{M0}$

CONSERVATIVELY TAKE A_v AS AREA OF BEAM WEB;

SCI P363

$\therefore A_v = (612.2 - 2 \times 19.6) \times 11.9 = 6819 \text{ mm}^2$

$h \qquad t_f \qquad t_w$

SECTION 6.2.6 of SCI P362

$f_y = 345 \text{ N/mm}^2$ (REDUCED FROM 355 N/mm² DUE TO FLANGE THICKNESS, t_f, BEING GREATER THAN 16mm)

γ_{M0} = MATERIAL FACTOR OF SAFETY FOR CROSS-SECTIONS

$\therefore \gamma_{M0} = 1.0$ (AS SECTION 6.1 OF SCI P362)

HENCE, $V_{pl,Rd} = 6819 \times (345/3^{0.5})/1.0 = 1,358,189 \text{ N}$ (ULS)

$\therefore \qquad " \qquad = 1358 \text{ kN (ULS)} \gg 293 \text{ kN}$

SINCE THE ACTUAL SHEAR FORCE $\not> 0.5 V_{pl,Rd}$, THE EFFECT OF SHEAR ON BENDING RESISTANCE CAN BE IGNORED.

DEFLECTION CHECKS;

$\delta_{MAX} = PL^3/48EI$

WHERE P IS THE UNFACTORED LIVE OR IMPOSED LOAD.

$\therefore P = 3.5 \times 56.25 = 197 \text{ kN (SLS)}$

$\quad \text{kN/m}^2 \qquad \text{m}^2$

SCI P363 (BLUE BOOK)

$I_{yy} = 98,600 \text{ cm}^4$

$\therefore \delta_{MAX} = 197 \times 10^3 \times 7500^3/(48 \times 210,000 \times 98,600 \times 10^4)$

Figure 11.2: *Continued*

<table>
<tr><td colspan="3">The Institution of StructuralEngineers</td><td>Job No.</td><td>Sheet 9 of 12</td><td>Drawing No:</td></tr>
<tr><td></td><td></td><td></td><td>Made by: JFC</td><td>Checked by:</td><td>Date: OCT 19</td></tr>
<tr><td></td><td></td><td></td><td colspan="3">Component:</td></tr>
</table>

Project: STEEL FRAME OPTION TO 45m × 45m × 10 STOREY OFFICE BUILDING

REF	CALCULATION	OUTPUT
	ie δmax = 8.4 mm = SPAN/893	
	ASSUMING NO BRITTLE FINISHES, LIMIT = SPAN/200	
	OR " SOME " " , " = SPAN/360	
	IN EITHER CASE δmax IS WELL WITHIN ALLOWABLE LIMITS.	
	NOTE THAT VIBRATION SHOULD BE CHECKED AT A LATER STAGE, ALTHOUGH ITS UNLIKELY TO BE CRITICAL HERE (IE – DUE TO THE LOW SPAN TO DEPTH RATIO OF THE SECTION ADOPTED, WHICH IS EQUIVALENT TO 7500/612.2 = 12.25, WHICH IS CONSISTENT WITH THE GUIDANCE PROVIDED IN CHAPTER 7 OF THIS BOOK).	
	∴ ADOPT A 610 × 229 × 125 UB GRADE S355 SECTION FOR BEAM TYPE 'B3'	
	4/ BEAM 'B4' : 7.5m SPAN SIMPLY SUPPORTED	
	NON – COMPOSITE MILD STEEL INTERNAL BEAM –	
	DESIGN METHOD IS SIMILAR TO BEAM 'B3' ABOVE SO WON'T BE REPEATED HERE.	
	NOTE THAT BEAM B4 IS ALSO SUBJECT TO A UDL, DUE TO CLADDING.	
	5/ TYPICAL INTERNAL COLUMN, REF. 'C3' (BETWEEN GROUND AND 1ST FLOOR) –	
	ROOF LOAD = 1.58 kN/m² (IE – AS RC FRAME).	
	∴ AXIAL LOAD FROM ROOF = 1.58 × 15.0 × 7.5 = 178 kN (ULS)	
	kN/m² m m	

Figure 11.2: *Continued*

<table>
<tr><td colspan="3">The Institution of StructuralEngineers</td><td>Job No.</td><td>Sheet 10 of 12</td><td>Drawing No:</td></tr>
<tr><td>Made by: JFC</td><td>Checked by:</td><td>Date: OCT 19</td></tr>
<tr><td colspan="3">Component:</td></tr>
</table>

Project: STEEL FRAME OPTION TO 45m × 45m × 10 STOREY OFFICE BUILDING

REF	CALCULATION	OUTPUT

REF: N.A. TO BS EN 1991-1-1: 2002

LIVE LOAD REDUCTION CAN BE TAKEN INTO ACCOUNT FOR ANY FLOORS SUPPORTED, IN ACCORDANCE WITH CLAUSE N.A. 2.7 OF THE NATIONAL ANNEX (N.A.) TO BS EN 1991-1-1: 2002

$$\therefore \alpha_n = 0.6 \ (\text{FOR } 6 \leq n \leq 10)$$

∴ LOAD FROM ONE FLOOR LEVEL EQUALS;

$$= 15.0 \times 7.5 \times (1.35 \times 3.81 + 1.5 \times 0.6 \times 3.5)$$
$$\quad (m) \quad\quad (m) \quad\quad \text{F.O.S.}^\dagger \ (kN/m^2) \ \text{F.O.S.} \ \alpha_n \ (kN/m^2)$$
$$= 933 \ kN / FLOOR \ (ULS)$$

HENCE, TOTAL COLUMN LOAD FROM ROOF PLUS 9 NO. SUSPENDED FLOORS IS EQUAL TO;

$$= 178 + 9 \times 933 = 8575 \ kN \ (ULS)$$

(NB — COLUMN SELF-WEIGHT WON'T BE SIGNIFICANT BUT SHOULD BE INCLUDED AT DETAILED DESIGN STAGE).

TRY A 356 × 406 × 287 UC GRS355 SECTION;

REF: SCI P363 (BLUE BOOK) ↑ NO EXTRACT PROVIDED

FOR Le = 1.0 L₀ = 4.0m (IE — ASSUME A 'SIMPLE' OR NOMINALLY PIN JOINTED STEEL FRAME);

$$N_{b,z,Rd} = 10,600 \ kN > 8575 \ kN \ (OK)$$

$$(NB - \text{COLUMN S.W.} = 1.35 \times 10 \times 4.0 \times 2.87$$
$$\quad\quad\quad \text{F.O.S. STOREYS (m)} \ (kN/m)$$

$$ie \quad\quad\quad \text{''} \quad\quad \text{''} = 155 \ kN \ (SLS)$$

$$\therefore 8575 + 155 = 8730 \ kN, \ STILL < 10,600 \ kN).$$

∴ ADOPT 356 × 406 × 287 UC GRS355 COLUMNS

$$\dagger \ F.O.S. = FACTOR \ OF \ SAFETY$$

Figure 11.2: *Continued*

The Institution of StructuralEngineers	Job No.	Sheet 11 of 12	Drawing No:
	Made by: JFC	Checked by:	Date: OCT 19
	Component:		

Project: STEEL FRAME OPTION TO 45m × 45m × 10 STOREY OFFICE BUILDING

REF	CALCULATION	OUTPUT
	6, TYPICAL EDGE COLUMN, REF 'C1' (BETWEEN GROUND AND 1ST FLOOR) – KEY DATA IS GENERALLY AS PER THE INTERNAL COLUMN CONSIDERED EARLIER, BUT EDGE COLUMNS ARE ALSO SUBJECT TO A CLADDING LOAD OF 10 kN/m (UNFACTORED) AT EVERY FLOOR LEVEL. ∴ AXIAL LOAD FROM ROOF AND FLOORS (EXCLUDING CLADDING) = 50% OF AN INTERNAL COLUMN. ∴ $F = 0.5 \times (178 + 9 \times 933) = 4288$ kN (ULS) CLADDING LOAD = $\underset{FoS}{1.35} \times \underset{m}{15.0} \times \underset{kN/m}{10} = 203$ kN (ULS) PER FLOOR ∴ TOTAL LOAD = $4288 + 9 \times 203 = 6115$ kN (ULS) EXCL. SWT TRY A $356 \times 368 \times 202$ UC GRADE S355 SECTION; SELF – WEIGHT = $1.35 \times 10 \times 4.0 \times 2.02 = 109$ kN (ULS) ∴ TOTAL LOAD = $6115 + 109 = 6224$ kN (ULS) INCL. SWT.	
SCI P362 SECTION 6.3.4 & APPENDIX A.5	BUT WE ALSO NEED TO CONSIDER NOMINAL BENDING MOMENTS ABOUT ONE AXIS OF EDGE COLUMNS (AND ABOUT BOTH AXES OF CORNER COLUMNS), IN ACCORDANCE WITH SECTION 6.3.4 AND APPENDIX A.5 OF SCI P362. WHILST THIS CHECK WON'T BE CARRIED OUT HERE, THE LARGE SECTION ADOPTED AND THE RESIDUAL CAPACITY TO RESIST SUCH A MOMENT SHOULD ENSURE THE SECTION IS ADEQUATE. HENCE, QUICK CHECK ON AXIAL COMPRESSION;	
SCI P363 (BLUE BOOK) ↑ NO EXTRACT PROVIDED HERE	FOR $L_{cr} = 1.0 L_0 = 4.0$m, $N_{b,z,Rd} = 7290$ kN > 6224 ✓ ∴ ADOPT A $356 \times 368 \times 202$ UC GR S355	

Figure 11.2: *Continued*

<table>
<tr><td colspan="2">The Institution of StructuralEngineers</td><td>Job No.</td><td>Sheet 12 of 12</td><td>Drawing No:</td></tr>
<tr><td colspan="2"></td><td>Made by: JFc</td><td>Checked by:</td><td>Date: OCT 19</td></tr>
<tr><td colspan="2"></td><td colspan="3">Component:</td></tr>
<tr><td colspan="5">Project: STEEL FRAME OPTION TO 45m × 45m ×10 STOREY OFFICE BUILDING</td></tr>
<tr><td>REF</td><td colspan="3">CALCULATION</td><td>OUTPUT</td></tr>
</table>

7, EDGE COLUMN, REF 'C4' (BETWEEN GROUND AND 1ST FLOOR) —

A SIMILAR PROCESS AS USED ABOVE FOR COLUMN C1 CAN BE USED HERE, THE ONLY DIFFERENCE BEING THAT COLUMN C4 SUPPORTS A SMALLER CLADDING LOAD.

8, TYPICAL CORNER COLUMN, REF 'C2' (BETWEEN GROUND AND 1ST FLOOR —

AGAIN, A SIMILAR PROCESS AS USED FOR COLUMN C1 CAN BE USED HERE, THE MAIN DIFFERENCE BEING THAT CORNER COLUMNS HAVE NOMINAL BENDING MOMENTS ABOUT BOTH AXES.

SCI P362 (CONCISE EUROCODE) — NB: FOR FURTHER GUIDANCE ON HOW TO CHECK COLUMNS SUBJECT TO AN AXIAL COMPRESSION AND BENDING ABOUT ONE OR MORE AXES, REFER TO SECTIONS 6.3.3 AND 6.3.4 OF SCI P362, AS WELL AS APPENDICES A.4 AND A.5 OF THE SAME DOCUMENT.

INTENTIONALLY BLANK

11.3 30m span single-storey building with open-plan layout

Solution 1 — Steel frame with roof trusses

This solution (Figure 11.3 on p. 218) represents 30m span × 3m deep simply-supported steel roof trusses at 7.5m centres i.e. with a span-to-depth ratio of 10, spanning onto 15m high UB section columns (for which the wind-induced bending moments are significant, hence the use of UB rather than UC sections).

For overall lateral stability, the frame is braced both on-plan i.e. in the plane of the roof and vertically, along all four elevations.

In practice, the 30m length of the roof trusses is such that splices would be required for transport purposes, although the 15m-long columns should not need splicing.

Alternatives might include replacing the trusses with cellular steel beams, or using conventional steel portal frames, possibly with light gauge Z or C-section purlins at 1.8m centres (although such purlins would not be capable of acting as plan-bracing members in the way that the UB section purlins adopted here do).

Typical cross-section and roof plan (not to scale):

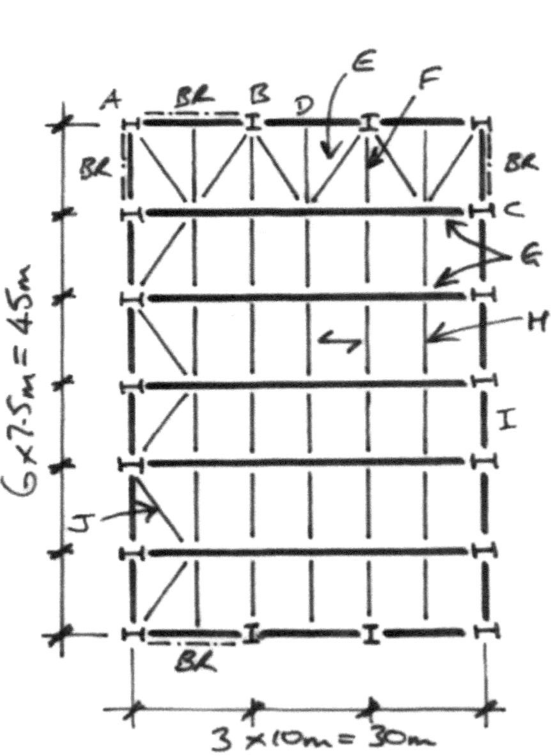

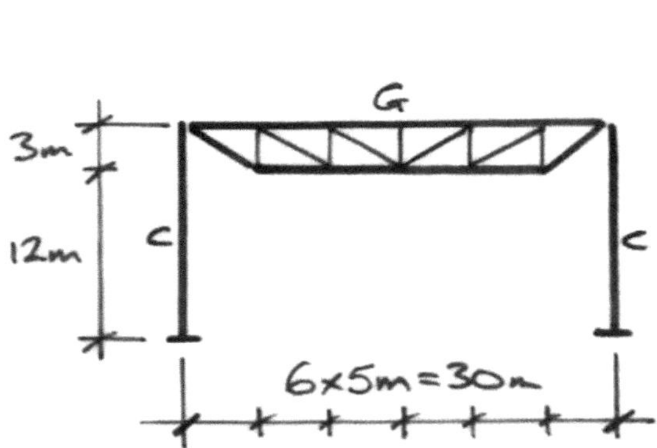

Member legend (Grade S275 mild steel unless noted otherwise)

A = Similar to C but stiffened/braced about minor axis (or possibly use a UC, SHS or RHS, subject to appropriate strength and deflection checks)

B = 610 × 229 × 113UB mild steel column

C = 610 × 229 × 101UB mild steel column

D = 356 × 127 × 39UB or Grade S355 300 × 200 × 6.3 RHS beam

E = J = 168.3 × 5 CHS Grade S355 (slenderness limited to 180)

F = H = 254 × 146 × 37UB

G = Truss with 140 × 140 × 5 SHS top and bottom chords, 120 × 120 × 8 SHS diagonals and 120 × 120 × 5 SHS verticals (all in Grade S355 mild steel)

I = 254 × 146 × 31UB or Grade S355 250 × 250 SHS

Notes

1. Frame centres of 7.2–9m generally provide a good compromise in terms of flexibility of layout and cost.
2. 'BR' denotes location of vertical steel bracing frames.
3. Vertical bracing frames acceptable (by inspection) to provide lateral stability to a 15m high single-storey building (worst case span/depth ratio = 15m/7.5m = 2.0, which is very reasonable for a cantilever). Movement joints are not normally required to single-storey steel framed buildings of this size.
4. Fire protection is not generally required to single-storey buildings, but is dependent on their proximity to other buildings and the site boundary.
5. Assume unfactored loads used for concept design are as follows:
   - Lightweight roof permanent load = $0.5kN/m^2$
   - Roof variable load = $0.6kN/m^2$

   Also need to check the wind uplift condition, and provide lateral restraint to the compression chords of trusses and compression flanges of beams and columns, as required.
6. Steel member sizes have been determined using the 'Blue Book'. Truss chord sizes are based on mid-span moment, while vertical and diagonal sizes are based on forces at end of truss.
7. Truss depth based on a rule of thumb = span/10 (although you could probably reduce depth to 2.5m, although the section sizes and the total truss weight would increase).
8. Beam and column depths are based on span/25–span/30 (approx.).

Solution 2 — Glulam arches

This solution (Figure 11.4 on p. 226) represents 30m span × 0.76m deep three-pinned glulam arches at 7.5m centres i.e. with a span-to-depth ratio of approx. 40, spanning directly onto foundations. Columns are required along the two gable walls, to support vertical loads from the gable roof beams, as well as horizontal wind loads acting on the gables. Alternatively, glulam arches could also be provided along the gables, with timber 'windposts' (vertical members with vertically-slotted head connections, to prevent them resisting any vertical roof loads) provided to resist winds loads acting on the gables.

The arches themselves provide overall lateral stability in the 'side-to-side' direction, although bracing in the plane of the roof is required for stability in the other direction.

The use of three-pinned arches has benefits in terms of ease of hand-analysis (as they are statically determinate) and transport (the arches can be delivered in two halves, which are connected together on-site).

Alternative shapes/geometries could be used e.g. a parabolic arch, which would avoid the presence of bending moments in the arch, when subject to uniform loads.

Note that due to the relative complexities of designing in timber, the calculations produced are essentially in accordance with BS EN 1995-1-1, even at concept design stage.

Typical cross-section and roof plan (not to scale):

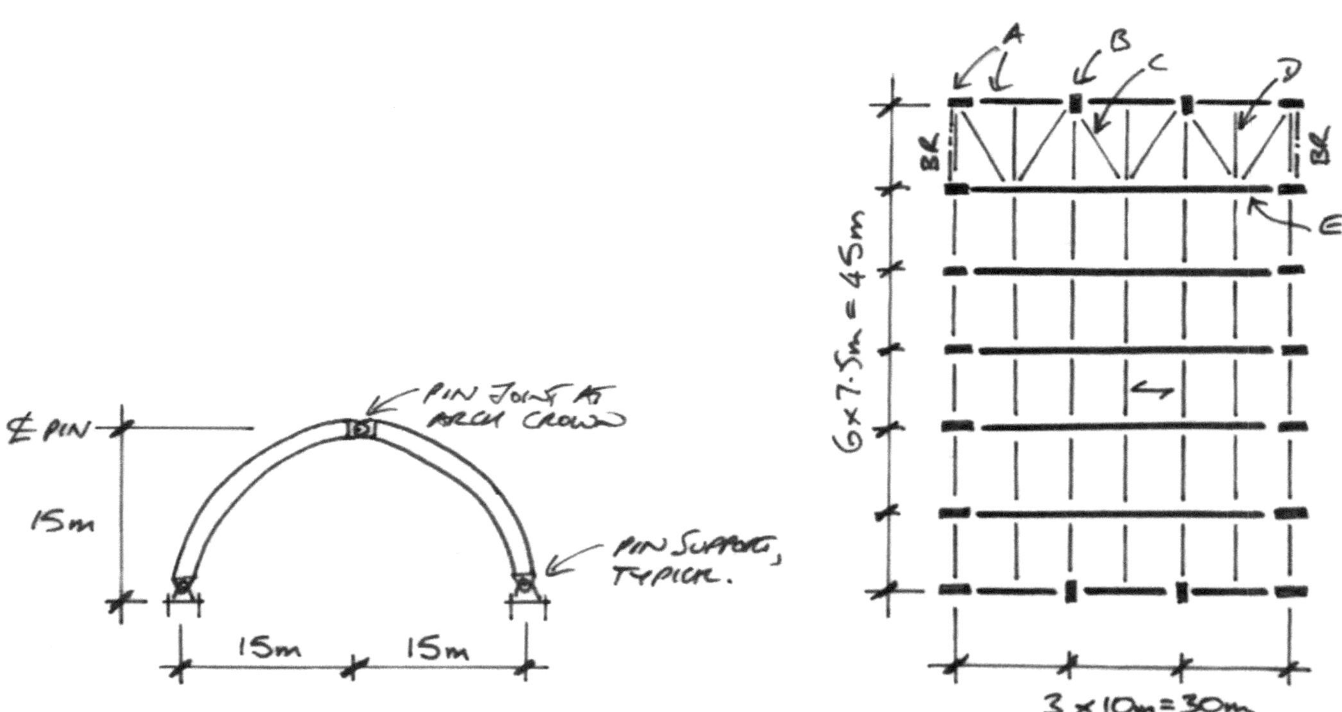

Member legend (all sizes are based on Grade GL28h glulam)
A = 760mm deep × 240mm wide arch member (could potentially reduce size if member type B can resist vertical as well as horizontal loads)
B = 760mm × 240mm column, with lateral restraints at 5m centres
C = 200mm square bracing member
D = 375mm deep × 125mm wide purlin/bracing member
E = 760mm deep × 240mm wide arch member

Notes

1. Frame centres of 7.2–9m generally provide a good compromise in terms of flexibility of layout and cost.
2. 'BR' denotes location of vertical steel bracing frames.
3. Vertical bracing frames acceptable (by inspection) to provide lateral stability to a 15m high single-storey building (worst case span/depth ratio = 15m/7.5m = 2.0, which is very reasonable for a cantilever). Movement joints are not normally required to single-storey timber framed buildings of this size.
4. Fire protection is not generally required to single-storey buildings, but is dependent on their proximity to other buildings and the site boundary.
5. Assume unfactored loads used for concept design are as follows:
   - Lightweight roof permanent load = $0.5kN/m^2$
   - Roof variable load = $0.6kN/m^2$

   (giving maximum thrust of 143kN at the feet of the arch, and maximum bending moment of 276kNm half way between the feet and the crown). Also need to check wind uplift condition, and provide lateral restraint to the compression flanges of the glulam arches and beams as required.
6. Glulam arch member to be 760mm deep × 240mm wide Grade GL28h, with lateral restraints provided at 5m centres, determined using *Manual for the design of timber building structures to Eurocode 5*[77]. The manual suggests an arch depth of approx. span/40 i.e. 30m/40 = 0.75m), and a maximum depth to breadth ratio of 3:1 to avoid lateral torsional buckling issues, based on the ends being restrained against rotation i.e. giving a depth of 0.75/3 = 0.25m.
7. Members B and D were sized as members in bending, with a span-to-depth ratio of 20, and with a maximum depth-to-breadth ratio of 3:1 (approx.).
8. Member C was sized by limiting its slenderness ratio to 50, but also need to check self-weight deflection.
9. Members A–D can be designed using a similar method to member E.

Figure 11.3: Steel option for 30m span single-storey building

The Institution of StructuralEngineers		
Job No.	Sheet 1 of 8	Drawing No:
Made by: JFC	Checked by:	Date: OCT 19
Component:		

Project: STEEL OPTION FOR 30m SPAN SINGLE STOREY BUILDING

REF	CALCULATION	OUTPUT
	STEEL OPTION FOR 30m SPAN SINGLE STOREY BUILDING -	

STEEL OPTION FOR 30m SPAN SINGLE STOREY BUILDING -

1/ ROOF TRUSSES (MEMBER REF. 'G') -

30m SPAN, SIMPLY SUPPORTED STEEL ROOF TRUSSES PROVIDED AT 7.5m c/c

LIGHTWEIGHT ROOF, WITH FOLLOWING DOWNWARD LOADS ASSUMED;

$1.0 G_k = 0.5 \, kN/m^2$ AND $1.0 Q_k = 0.6 \, kN/m^2$

∴ UDL ON A TYPICAL INTERNAL TRUSS EQUALS;

$$UDL = 7.5 \times (1.35 \times 0.5 + 1.5 \times 0.6) = 11.8 \, kN/m \, (ULS)$$

m   FOS kN/m² FOS kN/m²

IN PRACTICE, ROOF LOADS WILL PROBABLY BE APPLIED AS A SERIES OF POINT LOADS AT NODES OF TRUSSES (BY PURLINS), BUT FOR CONCEPT STAGE WILL CONSIDER A UDL APPLIED TO TRUSS FOR CALCULATING AXIAL FORCES IN ALL MEMBERS (AND IGNORING ANY ASSOCIATED BENDING MOMENTS). HENCE;

MAXIMUM BENDING MOMENT $= WL^2/8$

ie   "   " $= 11.8 \times 30^2/8 = 1328 \, kNm \, (ULS)$

∴ AXIAL FORCE IN TOP AND BOTTOM CHORDS IS;

$F_{CHORDS} = \pm 1328 / TRUSS \, DEPTH = \pm 1328/3.0$

ie   " $= \pm 443 \, kN \, (ULS)$

IN PRACTICE, WIND UPLIFT LOADS WOULD OF COURSE NEED TO BE CONSIDERED. HOWEVER;

Figure 11.3: *Continued*

<table>
<tr><td colspan="2">The Institution of StructuralEngineers</td><td>Job No.</td><td>Sheet 2 of 8</td><td>Drawing No:</td></tr>
<tr><td></td><td></td><td>Made by: JFC</td><td>Checked by:</td><td>Date: OCT 19</td></tr>
<tr><td></td><td></td><td>Component:</td><td></td><td></td></tr>
</table>

Project: STEEL OPTION FOR 30m SPAN SINGLE STOREY BUILDING

REF	CALCULATION	OUTPUT
	THIS IS UNLIKELY TO GIVE HIGHER MEMBER FORCES THAN THOSE ASSOCIATED WITH THE MAXIMUM DOWNWARD LOAD CASE CONSIDERED ABOVE. A KEY ISSUE IS TO ENSURE THAT ANY TRUSS MEMBERS WHICH ARE IN TENSION FOR THE DOWNWARD LOAD CASE ARE CHECKED AS COMPRESSION MEMBERS FOR THE WIND UPLIFT CASE. FURTHER, OUT OF PLANE LATERAL RESTRAINT SHOULD BE PROVIDED TO THE BOTTOM CHORD OF THE TRUSS (AS DESCRIBED BELOW IN MORE DETAIL).	
	1.1, CHORD MEMBERS: STRENGTH CHECK –	
	ASSUME EFFECTIVE LENGTH, $L_{cr} = 1.0 L_0$ (ABOUT BOTH AXES)	
	THE ABOVE ASSUMPTION IS BASED ON OUT OF PLANE LATERAL RESTRAINT BEING PROVIDED TO THE TOP CHORD BY PURLINS AT EACH NODE POSITION (IE – AT 5m c/c) WHILST DIAGONAL KNEE-BRACES PROVIDE LATERAL RESTRAINT TO THE BOTTOM CHORD, ALSO AT EACH NODE POSITION. NOTE THAT THE UB SECTION PURLINS ADOPTED FOR MEMBER TYPES 'F' AND 'H' SHOULD BE CAPABLE OF RESISTING THE RESULTANT FORCES FROM THE KNEE BRACES.	
	TRY $140 \times 140 \times 5$ SHS GRADE S355 HOT-FINISHED MILD STEEL TOP AND BOTTOM CHORDS.	
SCI P363 (BLUE BOOK)	FOR $L_{cr} = 5.0m$, $N_{b,Rd} = 508 kN > 443$ OK	
	1.2, DIAGONAL MEMBERS: STRENGTH CHECK –	
	VERTICAL REACTION AT END OF TRUSS;	
	$R = WL/2 = 11.8 \times 30.0/2 = 177 kN$ (ULS)	
	LENGTH OF DIAGONALS $= (3.0^2 + 5.0^2)^{1/2} = 5.83m$	
	$\therefore$ BY TRIGONOMETRY, FORCE IN DIAGONALS IS;	
	$F = 5.83/3.0 \times 177 = 344 kN$ (ULS) TENSION	

Figure 11.3: *Continued*

The Institution of **StructuralEngineers**	Job No.	Sheet 3 of 8	Drawing No:
	Made by: JFC	Checked by:	Date: OCT 19
	Component:		

Project: STEEL OPTION FOR 30m SPAN SINGLE STOREY BUILDING

REF	CALCULATION	OUTPUT
	AT CONCEPT STAGE, THIS FORCE CAN CONSERVATIVELY BE CONSIDERED AS A COMPRESSION FORCE (IE — SO THAT THE WIND UPLIFT CASE IS TAKEN INTO ACCOUNT)	
	TRY 120 × 120 × 8 SHS GRADE S355 HOT-FINISHED MILD STEEL DIAGONAL MEMBERS;	
SCI P363 (BLUE BOOK)	FOR L_{cr} = 6.0m (> 5.83m), $N_{b,Rd}$ = 368kN > 344 ✓	
	1.3, <u>VERTICAL MEMBERS: STRENGTH CHECK</u> —	
	FROM SECTION 1.2 ABOVE, IT CAN BE SHOWN THAT;	
	VERTICAL REACTION AT ENDS OF TRUSS = 177 kN (UDS)	
	∴ DESIGN VERTICAL MEMBER FOR THIS COMPRESSION FORCE;	
	L_{cr} = 1.0 l_0 = 3.0m	
SCI P363 (BLUE BOOK)	TRY A 120 × 120 × 5 SHS GRADE S355 HOT-FINISHED SECTION;	
	$N_{b,Rd}$ = 623 kN > 177 kN ✓	
	(NOTE THAT A SMALLER / LIGHTER SECTION COULD BE JUSTIFIED HERE).	
	1.4, <u>ROOF TRUSS DEFLECTION CHECK</u>;	
	BENDING DEFORMATION = δ = $5\omega L^4 / 384 EI$	
	THE SECOND MOMENT OF AREA IS CALCULATED USING THE PARALLEL AXIS THEOREM, CONSERVATIVELY IGNORING THE RELATIVELY INSIGNIFICANT SECOND MOMENTS OF AREA OF THE INDIVIDUAL CHORDS.	

Figure 11.3: *Continued*

<table>
<tr><td colspan="2">The Institution of StructuralEngineers</td><td>Job No.</td><td>Sheet 4 of 8</td><td>Drawing No:</td></tr>
<tr><td></td><td></td><td>Made by: JFC</td><td>Checked by:</td><td>Date: OCT 19</td></tr>
<tr><td></td><td></td><td colspan="3">Component:</td></tr>
</table>

Project: STEEL OPTION FOR 30m SPAN SINGLE STOREY BUILDING

REF	CALCULATION	OUTPUT

HENCE, $I_{TRUSS} = A.\bar{y}^2$

WHERE $\bar{y}$ = DISTANCE FROM $\mathcal{L}$ OF CHORDS TO NEUTRAL AXIS AND A = CROSS-SECTIONAL AREA OF CHORDS.

IE $A = 26.7 \, cm^2$ FOR $140 \times 140 \times 5$ SHS SECTIONS

SECTION (NTS)

$\mathcal{L}$ TOP CHORD
$\bar{y} = 1500 \, mm$
NEUTRAL AXIS (N.A.)
$\bar{y} = 1500 \, mm$
$\mathcal{L}$ BOTTOM CHORD

$\therefore I_{TRUSS} = 2 \times (26.7 \times 10^2) \times 1500^2 = 1.2015 \times 10^{10} \, mm^4$

(COMPARED TO A SECOND MOMENT OF AREA EQUAL TO $807 \, cm^4 \equiv 0.000081 \times 10^{10} \, mm^4$, SO NOT SIGNIFICANT, AS ASSUMED).

$\therefore \delta_{TRUSS} = \dfrac{5 \times (7.5 \times 0.6) \times 30{,}000^4}{384 \times 210 \times 10^3 \times 1.2015 \times 10^{10}} = 18.8 \, mm$

1.0 QK

IE " = SPAN/1596 << SPAN/200 (NO BRITTLE FINISHES)

2, PRIMARY COLUMNS (MEMBER REF. 'C') –

AGAIN AS CALCULATED IN SECTION 1.2 ABOVE, AXIAL COMPRESSION IN COLUMNS = 177 kN (ULS).

(NOTE THAT SOME COLUMNS WILL ALSO BE SUBJECT TO AXIAL FORCES WHERE THEY FORM PART OF VERTICAL BRACING FRAMES, BUT THEY HAVEN'T BEEN DESIGNED, ALTHOUGH THE METHOD IS THE SAME).

FOR CALCULATING BENDING MOMENT IN COLUMNS DUE TO WIND (IE. WITH COLUMNS SPANNING TOP TO BOTTOM WITH PINNED ENDS) A WIND LOAD = 1.0 kN/m² (SLS) HAS BEEN ASSUMED.

$\therefore$ UDL WIND = 1.0 $\times$ 7.5 = 7.5 kN/m (SLS)

kN/m² ... m

WIND LOAD ... FRAME CENTRES

Figure 11.3: *Continued*

The Institution of StructuralEngineers	Job No.	Sheet 5 of 8	Drawing No:
	Made by: JFC	Checked by:	Date: OCT 19
	Component:		

Project: STEEL OPTION FOR 30m SPAN SINGLE STOREY BUILDING

REF	CALCULATION	OUTPUT
	$\therefore$ UDL WIND = 1.5 × 7.5 = 11.25 kN/m (ULS)	
	$\therefore$ M WIND = $WL^2/8$ = 11.25 × 15.0²/8 = 316 kNm (ULS)	
	kN/m m	
	TRY 610 × 229 × 101 UB GR.S275 HOT-FINISHED MILD STEEL COLUMNS;	
	2.1 STRENGTH CHECKS –	
	ASSUMING LATERAL RESTRAINT IS PROVIDED TO THE COLUMNS MINOR AXIS (e.g. AS ON SEC - USING DIAGONAL KNEE-BRACING FIXED BACK TO SIDE RAILS, OR SIMILAR, SUBJECT TO THE NECESSARY DESIGN CHECKS BEING CARRIED OUT);	
SCI P363 (BLUE BOOK)	$N_{b,z,Rd}$ = 1670 kN	
	(NOTE THAT, WHILST THE MAXIMUM EFFECTIVE LENGTH CONSIDERED IN THE 'BLUE BOOK' IS 14m AS OPPOSED TO THE ACTUAL VALUE OF 15m, THE VALUE OF $N_{b,y,Rd}$ = 3370 kN FOR L_{cr} = 14m, SO MINOR AXIS FLEXURAL BUCKLING IS CLEARLY THE CRITICAL CONDITION).	
	$M_{b,Rd}$ = 579 kNm (CONSERVATIVELY TAKING C_1 = 1.0)	
	HENCE, THE INTERACTION EQUATION IS;	
	$\dfrac{177}{1670}$ + $\dfrac{316}{579}$ = 0.11 + 0.55 = 0.66 < 1.0 ✓	
	2.2 DEFLECTION CHECK –	
	$\delta = \dfrac{5wL^4}{384EI} = \dfrac{5 × 7.5 × 15,000^4}{384 × 210,000 × 75,800 × 10^4}$ = 31.1 mm	
	i.e. δ = SPAN/482 < SPAN/200 ✓	
	(NB: A SMALLER/LIGHTER SECTION COULD PROBABLY BE JUSTIFIED HERE).	

Figure 11.3: *Continued*

<table>
<tr><td colspan="2">The Institution of
Structural Engineers</td><td>Job No.</td><td colspan="2">Sheet 6 of 8</td><td>Drawing No:</td></tr>
<tr><td colspan="2"></td><td>Made by: JFC</td><td colspan="2">Checked by:</td><td>Date: OCT 19</td></tr>
<tr><td colspan="2"></td><td colspan="4">Component:</td></tr>
<tr><td colspan="7">Project: STEEL OPTION FOR 30m SPAN SINGLE STOREY BUILDING</td></tr>
<tr><td>REF</td><td colspan="5">CALCULATION</td><td>OUTPUT</td></tr>
</table>

3, CORNER COLUMNS AND GABLE COLUMNS
(MEMBER REFS 'A' AND 'B') –

THESE CAN BE DESIGNED USING A SIMILAR METHOD TO THE HAUNCH COLUMNS (IE AS SECTION 2, ABOVE), ALTHOUGH CORNER COLUMNS WILL BE SUBJECT TO WIND LOADS, AND HENCE BENDING MOMENTS, ABOUT BOTH AXES. SOME OF THESE COLUMNS WILL ALSO BE SUBJECT TO AXIAL FORCES FROM VERTICAL BRACING FRAMES.

4, GABLE BEAMS (MEMBER REF 'D') –

WHILST NOT DESIGNED HERE, THESE MEMBERS ARE SUBJECT TO MAJOR AXIS BENDING FROM THE DEAD AND IMPOSED (SNOW + WIND) LOADS ON THE ROOF, AS WELL AS AXIAL FORCES FROM THE PLAN BRACING IN THE PLANE OF THE ROOF TO THE END BAY.

DEPENDING ON THE CLADDING SYSTEM ADOPTED TO THE GABLE WALL, GABLE BEAMS MAY ALSO BE SUBJECT TO MINOR AXIS BENDING.

BOTH STRENGTH AND SERVICEABILITY CHECKS + SHOULD BE CARRIED OUT HERE.

5, PLAN BRACING (MEMBER REFS 'E' AND 'J') –

THESE MEMBERS WERE SIZED BY LIMITING THEIR SLENDERNESS TO A MAXIMUM OF 180, BUT WOULD NEED CHECKING FOR ANY TENSION AND COMPRESSION FORCES PRESENT (THE LATTER WOULD ALMOST ALWAYS BE CRITICAL). THIS SHOULD AVOID ISSUES DUE TO BENDING MOMENTS AND

+ IN THIS CASE, DEFLECTION (IE NO NEED FOR VIBRATION CHECKS ON ROOF BEAMS).

Figure 11.3: *Continued*

The Institution of **StructuralEngineers**	Job No.	Sheet 7 of 8	Drawing No:
	Made by: JFC	Checked by:	Date: OCT 19
	Component:		

Project: STEEL OPTIONS FOR 30m SPAN SINGLE STOREY BUILDING

REF	CALCULATION	OUTPUT
	DEFLECTION ASSOCIATED WITH THE MEMBER'S SELF-WEIGHT.	

6, PURLINS TO END BAYS (MEMBER REF. 'F')–

SIMILAR TO MEMBER REF. D, DISCUSSED IN SECTION 4, ABOVE, THESE MEMBERS ARE SUBJECT TO MAJOR AXIS BENDING FROM THE ROOF LOADS, AS WELL AS AXIAL FORCES FROM THE PLAN BRACING TO THE END BAY OF THE ROOF. BOTH STRENGTH AND DEFLECTION CHECKS SHOULD BE CARRIED OUT ON THESE MEMBERS.

7, PURLINS TO INTERNAL BAYS (MEMBER REF. 'H')–

PURLINS ARE AT 5.0m c/c, AND SPAN 7.5m BETWEEN ROOF TRUSSES. THEY ARE DESIGNED AS SIMPLY SUPPORTED.

7.1, STRENGTH CHECKS–

CONSIDERING DOWNWARD (IE – DEAD + SNOW) LOADS;

$$UDL = 5.0 \times (1.35 \times 0.5 + 1.5 \times 0.6) = 7.9 \text{ kN/m (ULS)}$$

$$\therefore M = WL^2/8 = 7.9 \times 7.5^2/8 = 56 \text{ kNm (ULS)}$$

TRY 254 × 146 × 37 UB GRADE S355 HOT-FINISHED MILD STEEL PURLINS;

THE PURLINS TOP FLANGES SHOULD BE LATERALLY RESTRAINED BY THE METAL ROOF DECKING BEING SUPPORTED (ALTHOUGH IT CANNOT BE ASSUMED THIS IS ALWAYS THE CASE, SO APPROPRIATE CHECKS SHOULD BE CARRIED OUT – IT MAY BE AS SIMPLE AS TALKING TO THE DECK MANUFACTURER'S TECHNICAL DEPARTMENT). THIS WILL PREVENT LATERAL

Figure 11.3: *Continued*

The Institution of **StructuralEngineers**	Job No.	Sheet 8 of 8	Drawing No:
	Made by: JFC	Checked by:	Date: OCT 19
	Component:		

Project: STEEL OPTION FOR 30m SPAN SINGLE STOREY BUILDING

REF	CALCULATION	OUTPUT

TORSIONAL BUCKLING (LTB) IN THE DOWNWIND LOADING CASE. HOWEVER, THE PURLINS' BOTTOM FLANGES WILL BE IN COMPRESSION FOR THE WIND UPLIFT CASE, SO NEED TO CONSIDER LTB HERE.

See P363 (BLUE BOOK)

$$M_{c,y,Rd} = 133 \text{ kNm} \gg 56 \text{ ok}$$

FOR $L_{cr} = 1.0 L_b = 7.5 \text{m};$

$$M_{b,Rd} = (59.0 + 52.2)/2 = 55.6 \text{ kNm} \dagger$$

HENCE, AS LONG AS THE MAXIMUM BENDING MOMENT DUE TO THE WIND UPLIFT DOESN'T EXCEED 55.6 kNm (WHICH IS CONSIDERED HIGHLY UNLIKELY) THEN SECTION IS ADEQUATE IN BENDING.

WHILST NOT CALCULATED HERE, IT SHOULD BE NOTED THAT (RELATIVELY MODEST) AXIAL LOADS WILL ALSO BE PRESENT IN THESE PURLINS. HOWEVER, THERE IS LIKELY TO BE SUFFICIENT CAPACITY REMAINING IN THE CHOSEN SECTION TO RESIST SUCH AXIAL FORCES AND THE MOMENT DUE TO WIND UPLIFT, ALTHOUGH A COMBINED (INTERACTION EQUATION) CHECK SHOULD BE CARRIED OUT.

FOR $L_{cr} = 7.5 \text{m}, \quad N_{b,z,Rd} = \dfrac{(208 + 163)}{2} = 186 \text{ kN (ULS)}$

7.2, <u>DEFLECTION CHECK</u> –

$$\delta = \frac{5WL^4}{384 EI} = \frac{5 \times (5 \times 0.6) \times 7500^4}{384 \times 210,000 \times 5540 \times 10^4} = 10.6 \text{ mm}$$

ie $\delta = \text{SPAN}/707 < \text{SPAN}/200 \text{ ok}$

$\dagger$ CALCULATED BY INTERPOLATING VALUES FOR L_{cr} OF 7m AND 8m, AND ASSUMING $C_1 = 1.0$ (CONSERVATIVE)

Figure 11.4: Timber option for 30m span single-storey building

<table>
<tr><td colspan="3">The Institution of StructuralEngineers</td><td>Job No.</td><td>Sheet 1 of 6</td><td>Drawing No:</td></tr>
<tr><td></td><td></td><td></td><td>Made by: JE</td><td>Checked by:</td><td>Date: OCT 19</td></tr>
<tr><td></td><td></td><td></td><td colspan="3">Component:</td></tr>
</table>

Project: TIMBER OPTION FOR 30m SPAN SINGLE STOREY BUILDING

REF	CALCULATION	OUTPUT

<u>3-PINNED SEMI-CIRCULAR GLULAM ARCH</u>

OPTION FOR 30m SPAN SINGLE STOREY BUILDING

1/ <u>GEOMETRY AND UNIFORM LOADS</u>

POSITION OF MAX. MOMENT, TYPICAL

PIN

x, y

$h = 15.0m$

$30°$

R_H — PIN — R_H

15.0m — 15.0m

R_V — R_V

POSITION OF MAXIMUM COMPRESSION, TYPICAL

AGAIN, BASE CONCEPT DESIGN ON UDL DUE TO DOWNWARD LOADS;

$\therefore \ 1.0G_k = 0.5 \ kN/m^2$ AND $1.0Q_k = 0.6 \ kN/m^2$

ASSUME ARCHES OCCUR AT 7.5m c/c

$\therefore$ UDL $= (1.35 \times 0.5 + 1.5 \times 0.6) \times 7.5 = 11.8 \ kN/m \ (ULS)$

$R_V = \dfrac{WL}{2} = 11.8 \times 30.0 / 2 = 177 \ kN \ (ULS)$

$\therefore R_H = \dfrac{WL^2}{8h} = \dfrac{11.8 \times 30.0^2}{8 \times 15.0} = 89 \ kN \ (ULS)$

$\therefore$ MAXIMUM COMPRESSION IN ARCH (AT BASE) IS;

$F_{c,0,d} = (177^2 + 89^2)^{0.5} = 198 \ kN \ (ULS)$

MAXIMUM MOMENT OCCURS AT 30° FROM THE HORIZONTAL (AS SHOWN ABOVE).

$x = r \cos\theta = 15.0 \times \cos 30° = 13.0m \ (APPROX)$

Figure 11.4: *Continued*

<table>
<tr><td colspan="2" rowspan="2">The Institution of
StructuralEngineers</td><td>Job No.</td><td colspan="2">Sheet 2 of 6</td><td>Drawing No:</td></tr>
<tr><td>Made by: JFC</td><td colspan="2">Checked by:</td><td>Date: OCT 19</td></tr>
<tr><td colspan="2">Component:</td></tr>
<tr><td colspan="4">Project: TIMBER OPTION FOR 30m SPAN SINGLE STOREY BUILDING</td></tr>
<tr><td>REF</td><td colspan="2">CALCULATION</td><td>OUTPUT</td></tr>
</table>

$$y = r \sin\theta = 15.0 \times \sin 30° = 7.5m$$

∴ MAXIMUM MOMENT CAN BE CALCULATED BY TAKING MOMENT DUE TO BOTH VERTICAL AND HORIZONTAL SUPPORT REACTIONS, AS WELL AS THE APPLIED UDL, ABOUT THE POINT CALCULATED ABOVE);

$$\therefore M_{y,d} = R_V(r-x) - R_H \cdot y - \frac{W(r-x)^2}{2}$$

$$ic \quad '' \quad = 177 \cdot (15-13) - 89 \times 7.5 - \frac{11.8(15-13)^2}{2}$$

$$ic \quad '' \quad = 354 - 664 - 24 = -334\ kNm\ (ULS)$$

2, INITIAL SIZING –

NB: SUBSEQUENT REFERENCES FOR THIS SET OF CALCULATIONS REFER TO THE INSTITUTE'S MANUAL FOR THE DESIGN OF TIMBER BUILDING STRUCTURES TO EUROCODE 5.

TABLE 4.6
$$DEPTH \simeq \frac{L}{40} = \frac{30}{40} = 0.75m = 750mm$$

SAY 760mm TO SUIT STANDARD GLULAM DEPTHS.

TABLE 4.3
WIDTH IS SIZED TO AVOID LATERAL TORSIONAL BUCKLING (LTB) EFFECTS.

ASSUMING ENDS OF GLULAM ARCH ARE RESTRAINED AGAINST ROTATION, AND MEMBER IS HELD IN LINE BY PURLINS AT CENTRES NOT MORE THAN 30 × MEMBER BREADTH;

$$\therefore WIDTH_{(MIN)} = MAXIMUM\ OF\ \frac{h}{4}\ OR\ \frac{PURLIN\ SPACING}{300}$$

$$ic \quad '' \quad = MAXIMUM\ OF\ 760/4\ OR\ 5000/30$$

† ie – GIVING A MAXIMUM DEPTH TO BREADTH RATIO OF 4:1.

Figure 11.4: *Continued*

The Institution of **StructuralEngineers**	Job No.	Sheet 3 of 6	Drawing No:
	Made by: JFC	Checked by:	Date: OCT 19
	Component:		

Project: TIMBER OPTIONS FOR 30m SPAN SINGLE STOREY BUILDING

REF	CALCULATION	OUTPUT
	IC — MINIMUM WIDTH = MAXIMUM OF 190mm OR 167mm	
	∴ " " " = 190mm †	
	BUT ROUND UP TO 200mm	
	∴ TRY A 760 × 200 GRADE GL28h GLULAM ARCH	
TABLE 3.15	∴ CHARACTERISTIC STRENGTHS ARE ;	
	$f_{c,0,k}$ = 26.5 N/mm² (COMPRESSION PARALLEL TO GRAIN)	
	$f_{m,k}$ = 28 N/mm² (BENDING PARALLEL TO GRAIN)	
	3, <u>DESIGN RESISTANCES —</u>	
TABLE 2.1	ASSUME SERVICE CLASS 2 (EG — COLD ROOFS AND TIMBER FRAME WALLS)	
TABLE 2.2	ASSUME SHORT-TERM LOADING DURATION CLASS (IE — LESS THAN ONE WEEK), DUE TO SNOW.	
TABLE 2.3	∴ k_{mod} = 0.9	
	SINCE ARCHES ARE AT 7.5m c/c, LOAD CAN'T BE EFFICIENTLY TRANSFERRED BETWEEN THEM.	
TABLE 3.20	∴ k_{sys} = 1.0	
DITTO	SINCE h (DEPTH) > 150mm k_h = 1.0	
SECTION 5.3.1	3.1, <u>COMPRESSION RESISTANCE —</u>	
	$f_{c,0,d} = \dfrac{k_{sys} \cdot k_{mod} \cdot f_{c,0,k}}{\gamma_m}$	
TABLE 3.19	γ_m = MATERIAL FACTOR OF SAFETY = 1.25 FOR GLULAM (FOR FUNDAMENTAL LOAD COMBINATIONS)	

† HENCE, k_{crit} = 1.0, AS DISCUSSED LATER ON,

Figure 11.4: *Continued*

The Institution of **StructuralEngineers**	Job No.	Sheet 4 of 6	Drawing No:
	Made by: JFC	Checked by:	Date: OCT 19
	Component:		

Project: TIMBER OPTION FOR 30m SPAN SINGLE STOREY BUILDING

REF	CALCULATION	OUTPUT

$$\therefore f_{c,0,d} = \frac{1.0 \times 0.9 \times 26.5}{1.25} = 19.1 \, N/mm^2$$

SECTION 5.2.1 †

3.2, BENDING RESISTANCE —

$$f_{m,y,d} = \frac{k_h \cdot k_{crit} \cdot k_{sys} \cdot k_{mod} \cdot f_{m,k}}{\gamma_m}$$

TABLE 3.20

$k_{crit} = 1.0$ (LATERAL TORSIONAL BUCKLING NOT AN ISSUE, AS PER CALCULATION SHEETS 2 & 3, BASED ON TABLE 4.3).

$$\therefore f_{m,y,d} = \frac{1.0 \times 1.0 \times 1.0 \times 0.9 \times 28}{1.25} = 20.1 \, N/mm^2$$

3.3, SECTION PROPERTIES —

$$A = b.h = 200 \times 760 = 152 \times 10^3 \, mm^2$$

$$W_y = \frac{bh^2}{6} = \frac{200 \times 760^2}{6} = 19.3 \times 10^6 \, mm^3$$

3.4, DESIGN STRESSES —

SECTION 5.3

$$\sigma_{c,0,d} = F_{c,0,d} / A = \frac{198 \times 10^3}{152 \times 10^3} = 1.3 \, N/mm^2$$

$$\therefore \sigma_{m,y,d} = \frac{M_{y,d}}{W_y} = \frac{334 \times 10^6}{19.3 \times 10^6} = 17.3 \, N/mm^2$$

† IT MAY BE NECESSARY TO REFER TO SECTION 5.2.2 AT DETAILED DESIGN STAGE, SINCE THIS COVERS CURVED GLULAM BEAMS.

Figure 11.4: *Continued*

<table>
<tr><td rowspan="3">The Institution of StructuralEngineers</td><td>Job No.</td><td>Sheet 5 of 6</td><td>Drawing No:</td></tr>
<tr><td>Made by: JFC</td><td>Checked by:</td><td>Date: OCT 19</td></tr>
<tr><td colspan="3">Component:</td></tr>
</table>

Project: TIMBER OPTION FOR 30m SPAN SINGLE STOREY BUILDING

REF	CALCULATION	OUTPUT

REF: SECTION 3.5

3.5, <u>COMBINED AXIAL COMPRESSION AND BENDING</u>

CONSERVATIVELY CONSIDER MAXIMUM COMPRESSION AND BENDING FORCES AS BEING CO-EXISTENT.

a, <u>LOCAL CAPACITY CHECK</u> [†]

$$\left(\frac{\sigma_{c,0,d}}{f_{c,0,d}}\right)^2 + \frac{\sigma_{m,y,d}}{f_{m,y,d}} \leq 1$$

$$\therefore \left(\frac{1.30}{19.1}\right)^2 + \frac{17.3}{20.1} = 0.07^2 + 0.86 = 0.87$$

$$< 1.0 \quad \text{ok}$$

b, <u>FLEXURAL STABILITY CHECK : MAJOR AXIS</u>

$$\frac{\sigma_{c,0,d}}{k_{c,y} \cdot f_{c,0,d}} + \frac{\sigma_{m,y,d}}{f_{m,y,d}} \leq 1$$

REF: LENGTH LIKE / HALF ARCH = $L_{ef,y}$

EFFECTIVE LENGTH AS IN MAJOR AXIS OF ARCH (IE — IN VERTICAL DIRECTION);

$$L_{ef,y} = \frac{L_{ARCH}}{2} = \frac{2\pi r}{4} = \frac{2\pi \times 15}{4} = 23.6 \text{m}$$

REF: FIGURE 3.6

$$\therefore \frac{L_{ef,y}}{h} = \frac{23.6}{0.76} = 31, \quad \therefore k_{c,y} = 0.33$$

HENCE, INTERACTION EQUATION BECOMES;

$$\frac{1.3}{0.33 \times 19.1} + \frac{17.3}{20.1} = 0.21 + 0.86 = 1.07 > 1.0$$

[†] DONE FOR COMPLETENESS, SINCE BUCKLING EFFECTS WILL ALMOST CERTAINLY BE CRITICAL.

Figure 11.4: *Continued*

<table>
<tr><td colspan="2">The Institution of StructuraIEngineers</td><td>Job No.</td><td>Sheet 6 of 6</td><td>Drawing No:</td></tr>
<tr><td colspan="2"></td><td>Made by: JFC</td><td>Checked by:</td><td>Date: OCT 19</td></tr>
<tr><td colspan="2"></td><td colspan="3">Component:</td></tr>
<tr><td colspan="5">Project: TIMBER OPTION FOR 30m SPAN SINGLE STOREY BUILDING</td></tr>
<tr><td>REF</td><td colspan="3">CALCULATION</td><td>OUTPUT</td></tr>
</table>

∴ INCREASE WIDTH OF SECTION FROM 200mm TO 240mm. THIS WILL REDUCE BOTH AXIAL COMPRESSIVE STRESS AND MAJOR AXIS BENDING STRESS BY A FACTOR OF $200/240 = 0.83$

∴ $\sigma_{c,0,d} = 0.83 \times 1.3 = 1.08 \text{ N/mm}^2$

& $\sigma_{m,y,d} = 0.83 \times 17.3 = 14.4 \text{ N/mm}^2$

HENCE, INTERACTION EQUATION BECOMES;

$$\frac{1.08}{0.33 \times 19.1} + \frac{14.4}{20.1} = 0.17 + 0.72 = 0.89 < 1 \text{ OK}$$

C, FLEXURAL STABILITY CHECK : MINOR AXIS –

PURLINS PROVIDE LATERAL RESTRAINT SINGLE.

∴ $L_{ef,z} = 5.0 \text{m}$

∴ $\dfrac{L_{ef,z}}{b} = \dfrac{5000}{240} = 21$

FIGURE 3.6

∴ $k_{c,z} = 0.62$ (APPROX.)

SINCE THIS FACTOR IS HIGHER THAN $k_{c,y}$ (WHICH = 0.33, AS CALCULATED ON PREVIOUS PAGE), THIS CHECK ISN'T CRITICAL.

∴ <u>760mm DEEP × 240mm WIDE GRADE GL28h GLULAM ARCH OK</u>

NB: CONNECTION DESIGN MAY IN FACT GOVERN THE SIZE OF THE GLULAM ARCH, SO SHOULD IDEALLY BE CHECKED AT CONCEPT DESIGN STAGE.

11.4 Three-storey residential building/apartment block with cellular layout

Solution 1 — Cross-laminated timber (CLT) walls and floors

The significant increase in use of engineered timber products warrants the inclusion of this solution, in which CLT panels are used for upper floors, the roof and loadbearing walls, both internally and to the building perimeter.

As well as resisting vertical loads, the CLT floor and roof panels act as diaphragms, which transfer lateral loads to the CLT wall panels which, in turn, act as shear walls to transmit the lateral loads down to the foundations.

The floor and wall thicknesses shown have all been determined using the pre-scheme design guide published by TRADA[4].

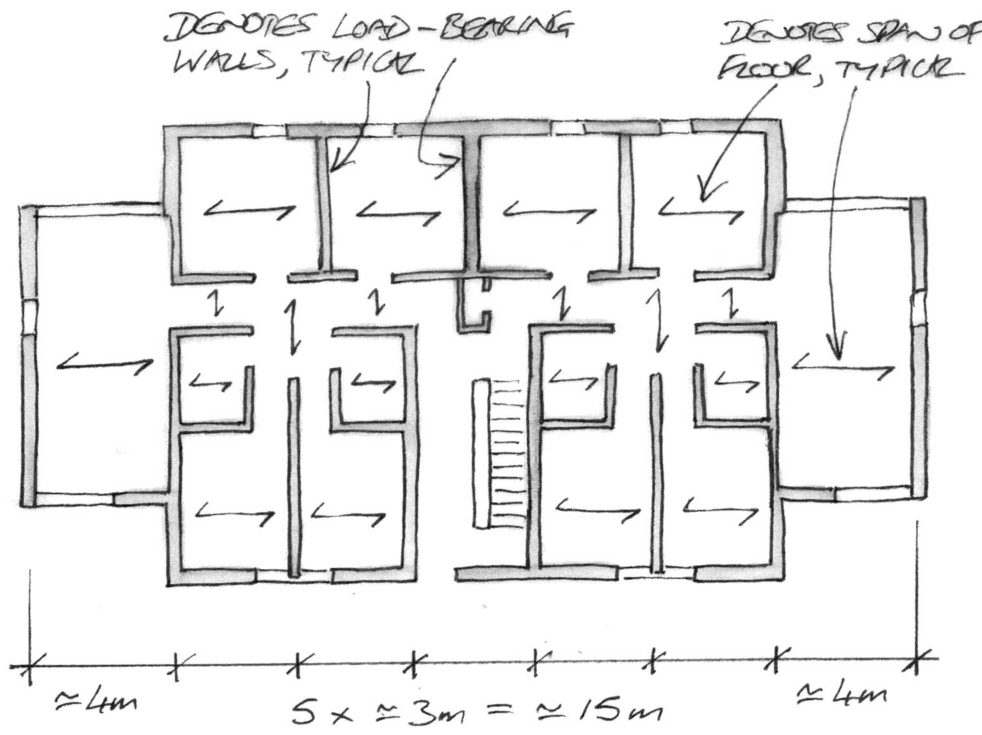

Notes

1. Assuming a three-storey building, with up to 3.0m floor-to-floor heights, 60mm thick CLT wall panels are adequate for all loadbearing walls i.e. perimeter and internal walls, and have been sized in accordance with Section 5.3.4 of the TRADA's pre-scheme design guide.
2. Loadbearing CLT wall panels are deemed to be acceptable (by inspection), to provide lateral stability to a three-storey building. See Section 3.2.2 of the TRADA guide.
3. Suspended upper floors to comprise 160mm thick CLT timber panels where span = 4m, potentially reducing to 100mm thick CLT panels, where span = 3m, sized in accordance with Table 5.1.5 of the TRADA guide.
4. 100mm thick CLT timber panels could be used for the roof, if a flat roof was adopted.
5. All CLT sizes quoted are based on Grade GL24 timber.
6. Unfactored loads used for concept design are as follows:
   - Roof permanent load = $0.5kN/m^2$ + self-weight of timber roof trusses, or similar (assuming a duo-pitch roof)
   - Roof variable load = $0.6kN/m^2$
   - Floor permanent load = $0.5kN/m^2$ + self-weight of CLT floor
   - Floor variable load = $1.5kN/m^2$ (residential) typically but potentially increasing to $3.0kN/m^2$ in communal areas
   - Cladding load = 10kN/m per floor/level (conservatively)

Solution 2 — Masonry walls with precast unit floors

This solution represents a more traditional approach, in which precast unit floors with an *in situ* concrete topping are used for suspended floors (and potentially for the roof as well, if a flat roof is required), supported by loadbearing masonry walls, both internally and to the building perimeter.

The precast unit floors will act as diaphragms which transfer lateral loads to the masonry walls, which in turn act as shear walls, to transmit the lateral loads down to the foundations.

The precast unit floor depth shown has been determined using span/load tables published by the manufacturer, while the wall specification has been determined using *Building Regulations Approved Document A*[80].

An alternative solution could be to use sawn timber joists for upper floors, although this form of construction could have acoustic implications i.e in terms of noise-transfer between adjacent flats/apartments. Timber joists could also potentially be used for the roof, if a flat roof is required, or trussed rafters could be used to create a pitched roof. Timber joist sizes can quickly be determined by span tables published by TRADA[74].

Disproportionate collapse is a potential concern with loadbearing masonry buildings — further guidance is provided in the accompanying drawing notes.

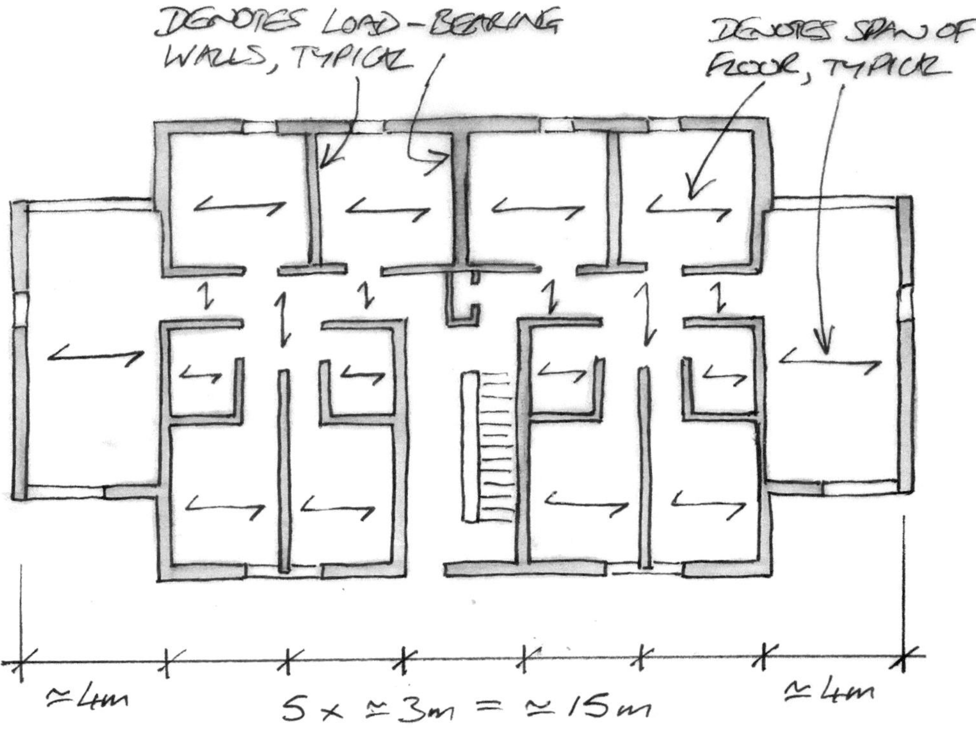

Notes

1. For the layout shown, the presence of several significant lengths of masonry wall in the two orthogonal directions are deemed to be acceptable (by inspection) to provide lateral stability to a three-storey building (assuming maximum 2.7m floor-to-floor heights). This assumes 100mm block inner leaf and 103mm brick or 100mm block outer leaf cavity walls to the building perimeter, and 140mm blockwork walls internally from ground to first, potentially reducing to 100mm at upper storeys (subject to the bearing requirements for the beam and block floor being satisfied).
2. The internal wall thicknesses quoted could increase for any walls acting as compartment or separating walls.
3. Preliminary loadbearing wall thicknesses have been determined using Section 2C of *Building Regulations Approved Document A*, which also provides guidance on the strength requirements of masonry units and mortar. For loadbearing masonry buildings which do not comply with the limitations on which this guidance is based, refer to note 4.

4. For further guidance on lateral stability and wall thicknesses of loadbearing masonry buildings, refer to *Manual for the design of plain masonry in building structures to Eurocode 6*[75].
5. Suspended upper floors to comprise a 150mm deep precast unit floor system (sized using manufacturer's load/span tables) with *in situ* concrete topping as required.
6. Unfactored loads used for concept design are as follows:
   - Roof permanent load = $0.5kN/m^2$ + self-weight of timber roof trusses, or similar (assuming a duo-pitch roof)
   - Roof variable load = $0.6kN/m^2$
   - Floor permanent load = $0.5kN/m^2$ + self-weight of beam and block floor slab
   - Floor variable load = $1.5kN/m^2$ typically, but potentially increasing to $3.0kN/m^2$ in communal areas
   - Cladding load = 10kN/m at each floor level
7. This is a 'Consequence Class 2A' i.e. lower risk building, for disproportionate collapse/robustness purposes, as defined in Section 5.1 of *Building Regulations Approved Document A*. In addition to complying with requirements A1 and A2, it is necessary to provide effective horizontal ties, or effective anchorage of suspended floors to walls. See Section 5.2 of BS EN 1996-1-1[67] and Section 6 of PD 6697[121] for further details.

11.5 Suggested reading

- Brettle, M.E. *et al. Steel building design: concise Eurocodes. In accordance with Eurocodes and the UK National Annexes (Revised edition) (Publication P362)*. Ascot: SCI, 2017
- Brooker, O. *Concrete buildings scheme design manual: a handbook for the IStructE chartered membership examination, based on Eurocode 2*. Camberley: MPA – The Concrete Centre, 2009
- Goodchild, C.H., Webster, R.M. and Elliott, K.S. *Economic concrete frame elements to Eurocode 2: a pre-scheme handbook for the rapid sizing and selection of reinforced concrete frame elements in multi-storey buildings designed to Eurocode 2*. Camberley: The Concrete Centre, 2009
- HM Government (2013). *The Building Regulations 2010. Approved Document A: Structure*. Available at: https://www.gov.uk/government/publications/structure-approved-document-a [Accessed: March 2020]
- HM Government (2019). *The Building Regulations 2010. Approved Document B: Fire safety. Volume 1: Dwellings; Volume 2: Buildings other than dwellings*. Available at: https://www.gov.uk/government/publications/fire-safety-approved-document-b [Accessed: March 2020]
- Davison, B. and Owens, G.W. eds. *Steel designers' manual (7th edition)*. Chichester: Wiley-Blackwell, 2012
- The Institution of Structural Engineers. *Manual for the design of concrete building structures to Eurocode 2*. London: IStructE Ltd, 2006
- The Institution of Structural Engineers. *Manual for the design of plain masonry in building structures to Eurocode 6 (2nd edition)*. London: IStructE Ltd, 2018
- The Institution of Structural Engineers. *Manual for the design of steelwork building structures to Eurocode 3*. London: IStructE Ltd, 2010
- The Institution of Structural Engineers. *Manual for the design of timber building structures to Eurocode 5 (2nd edition)*. London: IStructE Ltd, 2020
- *NA to BS EN 1991-1-1:2002: UK National Annex to Eurocode 1. Actions on structures. General actions. Densities, self-weight, imposed loads for buildings*. London: BSI, 2002
- Ranasinghe, K. *Eurocode 5 span tables: for solid timber members in floors, ceilings and roofs for dwellings (4th edition)*. High Wycombe: BM TRADA, 2014
- Steelconstruction.info website. Available at: https://www.steelconstruction.info [Accessed: March 2020]
- Steel Construction Institute. *Interactive 'Blue book'*. Available at: https://www.steelforlifebluebook.co.uk [Accessed: March 2020] [Interactive version of: Steel Construction Institute. *Steel building design: design data (SCI Publication P363)*. Ascot: SCI, 2015]
- Steel Construction Institute. *NCCI: Initial design of composite beams (SN022a-EN-GB)*. Ascot: SCI, 2006

James Norman
University of Bristol

12 What to produce at the end of the conceptual design process

At the end of the conceptual design process, you will have produced a large number of sketches, calculations and had many conversations. Much of this information will have been formative in the development of the design, but won't be used in the final concept or going forward. So, at the end of the conceptual design process, it is useful to pull everything together. Reflect on the process and articulate not just the final outcome, but the reason that you have ended up there and the route you have taken. Each project is different, but most will require at the end of the process a report with some drawings appended. As the design solidifies, this may also be a good time to look forward and finalise fees, programme and any critical specification items. Of course, these may all change again at the next stage, especially as you go through the planning permission process, but it is important to start the conversations.

12.1 Stage 2 report — the only output our client looks at

In the design office you will have generated a large amount of information. But let's be honest, the client is never going to see that. They will, however, look at the Stage 2 report. From our perspective, the report is simply a summary of the work to date. From their perspective, they have paid thousands of pounds for you to carry out the concept design work — there is no hole in the ground, no structure to be seen, all that they get is the report. It is important to realise that while for you the report may not be of great significance, for the client it is the only tangible outcome at this stage of their financial investment. It is important that the report is tailored to them. It should contain all the information they need, in a way that is accessible for them. It should avoid being overly technical but at the same time, when key decisions are required, it should plainly spell them out. It is also worth noting that at this stage, many aspects may still be unresolved e.g. the quantity surveyor (QS) may be expecting two schemes, so that they can cost them both. This may be the first time the client has seen the loading requirements. Ideally, they should be given the right information to decide, and these decisions should then be adhered to throughout the rest of the design process.

12.1.1 The importance of communicating the design
Several parties have an interest in the Stage 2 report:

- The client will want to know what they are getting
- The architect will want to make sure the design is coordinated and considered
- The QS will want enough information to cost the project, and possibly a number of different options
- The building services engineer will want to ensure that adequate allowance has been made for plant rooms and loading, and that the design accounts for services penetrations. They may also want to ensure adequate thermal mass is available
- You will want to capture the design process, the reason for the decisions you have made e.g. why a movement joint was placed on the gridline that was chosen and not somewhere else, and the agreed way forward with the client

There are two options to this situation:

1. We write a single detailed report for everyone. For some client's this is not just a suitable approach, it is what they will be expecting. For other clients, the wealth of information may make it difficult to understand what the report is saying, and they may not understand the decisions they are being asked to make.
2. Produce a simple report. This still covers the major information, and outlines the decisions and choices the client needs to make, but does not contain all the technical information. We would still recommend you capture all this information, but it may well be in the form of a 'design philosophy', which can either be included as an appendix, or shared with key members of the design team, and not included in the official report.

Either way, following the completion of Stage 2, you will want to consider creating a design philosophy. This document will contain much of the same information as in your Stage 2 report in the first instance. However, its purpose is slightly different. The aim is to inform anyone picking up your calculations of the underlying approach, assumptions and information. It should be a live document, that is updated as and when new information comes to light, or when a change is made in the approach. Ultimately, it should be sent, along with your calculations, to Building Control (in the UK) for their review, and should provide them with a thorough overview of your approach to the design.

Alongside the design philosophy (which can be a rather weighty document), you may want to consider including a face sheet. This is a single page summary of the design approach, and outlines the key information e.g. structural form, ground type, foundations and method of stability. This may be particularly helpful for quality assurance reviews where, in a single page, the reviewer can see what you are doing and where the design challenges lie.

12.1.2 How to make reports accessible and professional — a 'style guide'

Before considering anything else you will need to know whether:

- you are working to a set format for the project (often dictated by the architect)
- you are working to a set format dictated by your own firm who may have a series of agreed work templates
- you can create your own template

If you can create your own template, you have a number of simple decisions to make:

Portrait or landscape?

This one is simple. Design reports look and read better in landscape. I don't know why, almost everything else in life works better in portrait (books, magazines, manuals — although Lego instructions work better in landscape). It will also enable easier integration with the drawings in the appendix. Frustratingly, a design philosophy will work better in portrait, as it will later sit at the front of the pages of calculation. So, if you include a design philosophy, you will need to consider whether to create it now in landscape, and later convert to portrait, or whether to print it two pages to a page so it still fits the report format.

A3 or A4?

A Stage 2 report should be A3 so that the drawings can be appended while still at a sensible scale (A1 drawings scale to A3). However, the report will often be printed at A4 so it needs to be readable at this smaller scale.

Of course, on very large projects where A0 drawings are required, the drawings don't naturally scale to A3. However, A2 is an impractical size for a report and at A4 the drawings will be so small (four times smaller than the original scale), that it is hard to imagine anyone will be able to read them. We would therefore suggest that, while it may feel clumsy, A3 should be used in this situation as well.

One, two, three or four columns of text?

This is a matter of opinion. These are my views but you are welcome to disagree.

- Assuming you are working in landscape, there is no situation where you should use one column of text. If you have a table which takes up almost the whole page and need a couple of sentences above/below it, they should still be in multiple columns or on a separate page. A single column of text looks unprofessional and is exhausting to read

- For most of my professional career, I produced reports with two columns of text. Looking back at them now, I think they look a little lazy. Two columns of text look a lot like two pages printed side by side. While it is better than one column, it is still not the preferred approach
- Three columns of text works. It looks good, feels professional and provides flexibility for formatting. I would, given the choice, always use three columns of text
- Four columns of text are too busy. It looks cramped and a little messy. Lines of text become short, which leads to either messy paragraphs or strange gaps in the text if you justify it

In order of preference I would say three, two, four, one but, in reality, I don't see a situation when either four or one would be required (Figure 12.1).

Figure 12.1: Columns of text

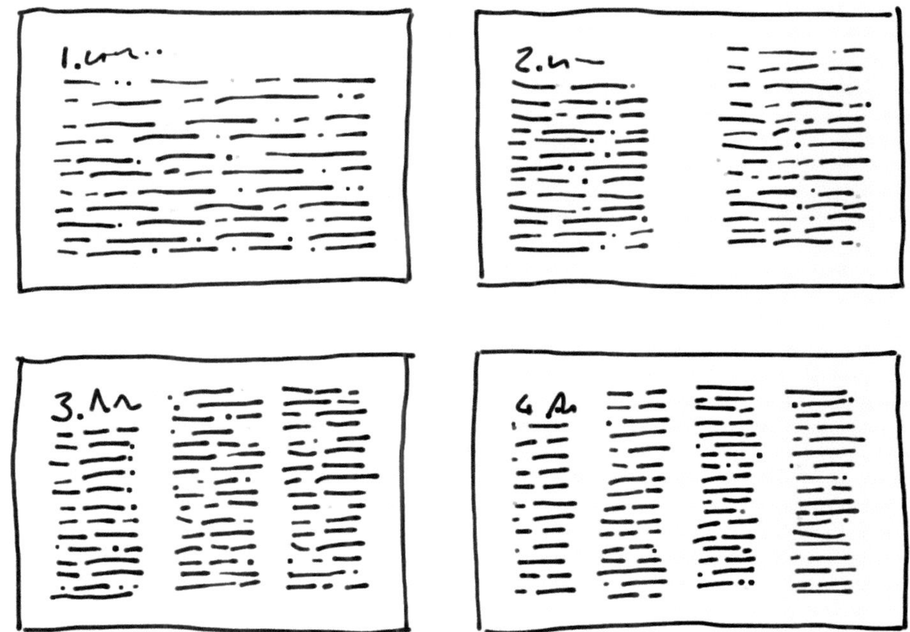

The use of imagery

A report absolutely must contain images. In fact, I would be so bold to say that every page must have an image of some sort. For many clients, the first thing they will do when they pick up a report is flick through it and look at the images. Occasionally (and frustratingly), they may only look at the images. The images need to tell the story of the design. They don't all have to be professional renders, and they don't all have to show the final design. Particularly difficult is finding good images of foundations or drainage. But, even here, we can draw the client in. We can either use photos from other projects e.g. highlighting the use of piles versus strip footings, or include hand-drawn and annotated sketches that reinforce the design. Likewise, for health and safety, an image may seem hard to come by, but a process diagram showing how the risks were reviewed or even images of your ISO 9001 and 14001 certificates will help break up the text, and make the report seem more inviting. When planning the report, plan what images will go where. Have some sketches of the building from different vantage points that you can add on pages where there really isn't a great image otherwise.

Tables can also be used — they may not feel like images but the careful use of colour in a table can have a similar effect on the reader (Figure 12.2).

Figure 12.2: Use of imagery

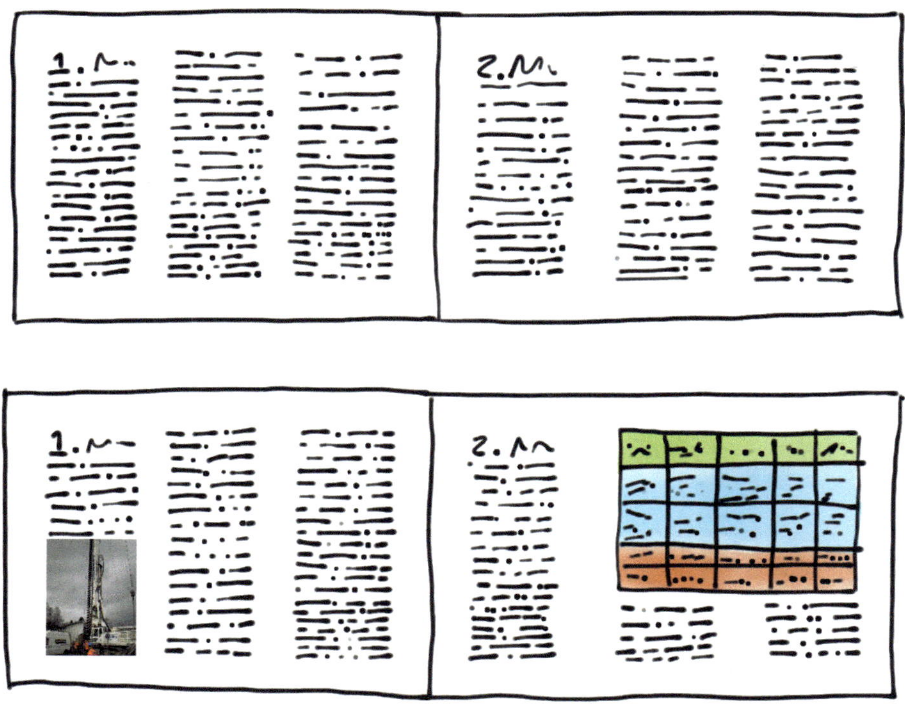

The position of an image

Images can either be interspersed through the text, or have their own page — or you can do a bit of both.
My personal preference is to always have images on the left and text on the right, with additional images within
the text when necessary (Figure 12.3).

Figure 12.3: Position of images

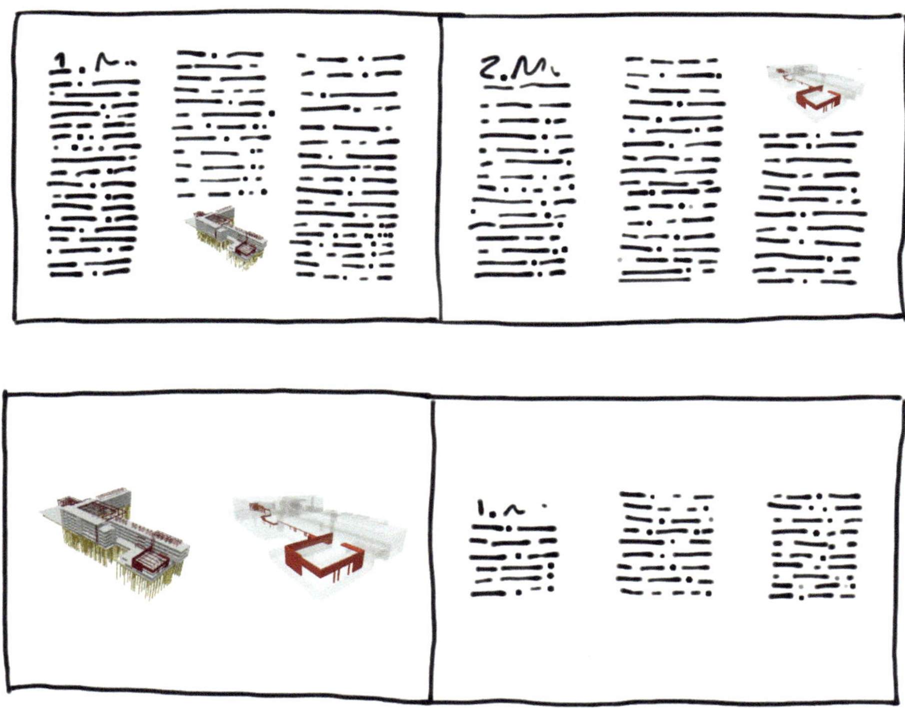

Formatting

When it comes to formatting, less is definitely more. Avoid congested headers and footers. A page number is useful. I am not convinced any other information is — the client already knows it's a Stage 2 report, that you are the engineer, the project name and are probably not too concerned about what your internal admin project number is. Avoid too many font types and sizes, and having too many heading levels — this book uses four (1, 1.1, 1.1.1 and level four has no number, and generally that should be adequate. Above all else, avoid using text boxes and other fussy devices, especially the automatic shadow feature. It may be useful once or twice to highlight something really important e.g. a key decision the client needs to make, but overuse makes the report look less professional.

White space

Don't be afraid of white space. White space used well will enhance the report, not make it seem lightweight (Figure 12.4). Have a look in a few modern photography books to see how they use white space to increase impact. This is important for both text and imagery. For images, think about the size it needs to be, and don't be afraid to make it that size and have white space around it. An A3 image will need to be of very high quality and resolution to work at that size. You may have a nice render of the project that you want to use that works, but you may have a couple of hand-drawn sketches, which when blown up lose much of their aesthetic charm. Generally, hand-drawn images look better shrunk down or at the same size rather than blown up. With text, allow each page to hold a few ideas rather than ram it full of information. If the section on foundations only requires half a page, space it out so that it is in three approximately equal columns with white space around it. Don't try and cram in half the next section as well.

Figure 12.4: White space

At university, you are often required to produce work to a set page length, encouraging you to fit as much information as you can on each page. Now you are producing professional reports, there is no page limit. A 40-page report, which is 50% white space, will read much better than a 20-page report with 0% white space.

12.1.3 Content — what's in and what's out?

In simple terms, the report should cover three things:

- The brief
- The design process
- The proposed solution/s

The brief — what the client asked for, what they are getting and why they differ
As outlined in Chapter 5, there are occasions when the engineer is required to produce or co-create the brief. If you have produced the brief, it is essential that it is articulated at the start of the Stage 2 report, as this will act both as a measure of the success of the proposed design (remembering that the brief and the design have developed in tandem), and act as a benchmark for the design going forward, which can be especially useful when faced with cost escalation. It may be that the brief is deterministic, or that it has some softer requirements which cannot be summarised in numbers. In this situation, don't be afraid to use exemplars to try and capture the ethereal nature of what the client is after.

If you have co-created the brief, it is likely that the architect will outline it in their Stage 2 report. It may then be useful to either input into their report or, where technical decisions have been made or need to be made, make it clear to the client that this is the case, and that they are effectively signing up to a brief by agreeing these elements. Examples include loading, design life and limits of deflection and vibration. Often, the client will look to our technical judgement on these issues, so helping them make sense of the information can be useful e.g. showing how the number of people in a m^2 equates to different variable actions (Figure 12.5), or suggesting they visit some offices to experience the bounciness of a CLT timber and precast concrete floor.

Figure 12.5: Differing numbers of people in 1m^2

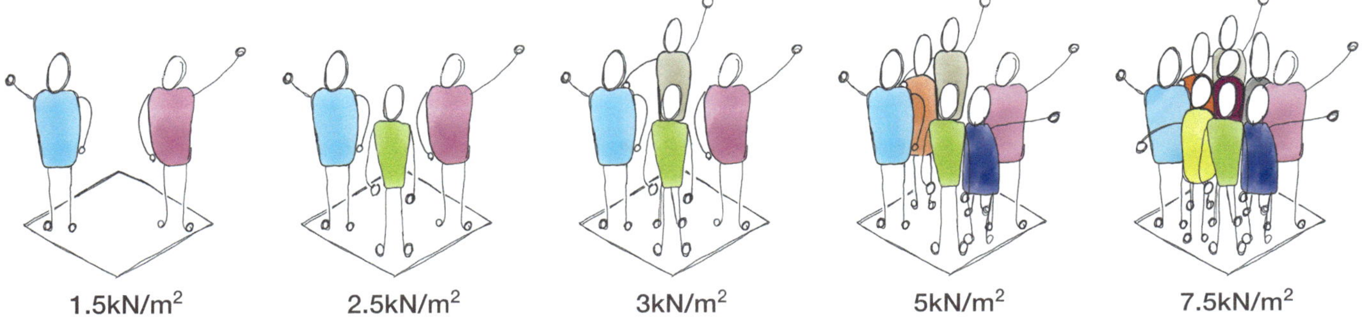

Documenting the design process to provide a rational response to the brief
Whether you have created the brief, co-created it or are simply working to a pre-defined brief, it is important to try and capture the design development. This may be done in isolation, as an engineer, but it is most powerful when the whole design team works together to document the process. This acts in two ways:

1. It helps establish the reason for the design and avoids reviewing and going over the same processes later in the project, bearing in mind that on most projects, multiple people will work on them, and that while ideally a single person will see a project through from beginning to end, in reality it is likely, especially on large projects that take years, that no one person will be involved from start to finish.
2. It is a way of demonstrating to the client the effort that has been put into producing the elegant design they now see. In my experience, architects are much better versed at going through this process, but that does not mean that we shouldn't also do it and, with practise, we will also be able to articulate the design process. This can often be achieved through a series of simple schematic drawings with supporting text (Figure 12.6).

Figure 12.6: Design development produced for a Stage 2 report

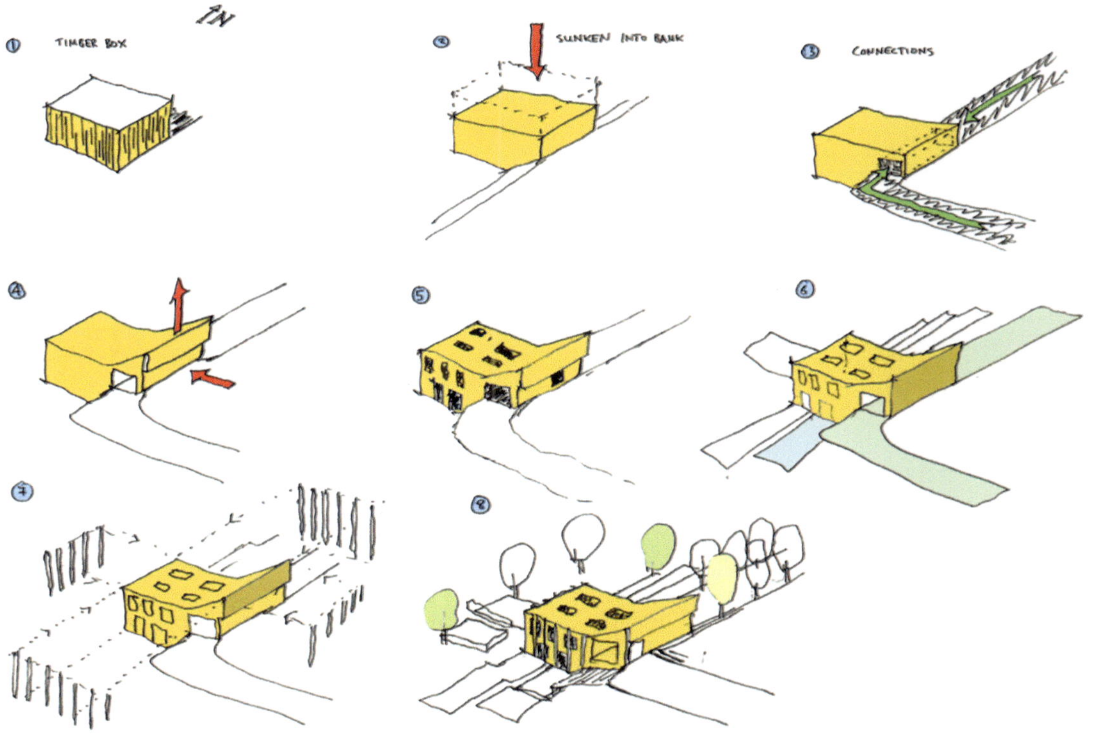

Stage 2 report checklist

Below is a list of Stage 2 content which we would use when writing reports and design philosophies. The list is intentionally quite extensive and we would not cover everything, only what we consider important at each stage. These are typical issues to consider, and are provided not as an exhaustive list but rather to aid you in structuring and considering the content of your report. Each design will be different and will therefore include different elements of this list.

Preliminaries

1. Front cover.
   This is a chance to show off the design. Consider carefully the image you want to place on the cover. It should work with the architects and enhance what they do. It should also be suitable for the client. Figure 12.7 shows two examples — a large university project where a 3D CAD ine drawing was produced, and a primary school, where a simple quick hand-drawing was used. Both are intended to engage with the client, but consider the client's background and expectations.
2. Table of contents.
3. Introduction/description of project.
4. Key project contacts (this is useful information to provide upfront, but could also be placed in an appendix).
5. Responsibility for overall design.
   It is good to make sure everyone is clear on the design responsibility. Depending on the client, it may be good to include early, so that they are clear what is included in the report and, more importantly, what is not. For some clients, this may feel very contractual, and may be in the appendix, or even issued as a separate document.
6. Health and safety, including any site risks e.g. asbestos.
   Health and safety is a key concern and therefore should be at the front of the report. However, the information should justify this prime position. If the project and therefore the health and safety are generic, it may be more useful to cover this later. This section should also include reference to the risk register. The risk register is a live document and should be continuously updated. It should be included as an appendix in its latest form with the Stage 2 report, to ensure the rest of the design team, the client, and anyone else involved in the project are aware of the risks.

Figure 12.7: Front cover designs

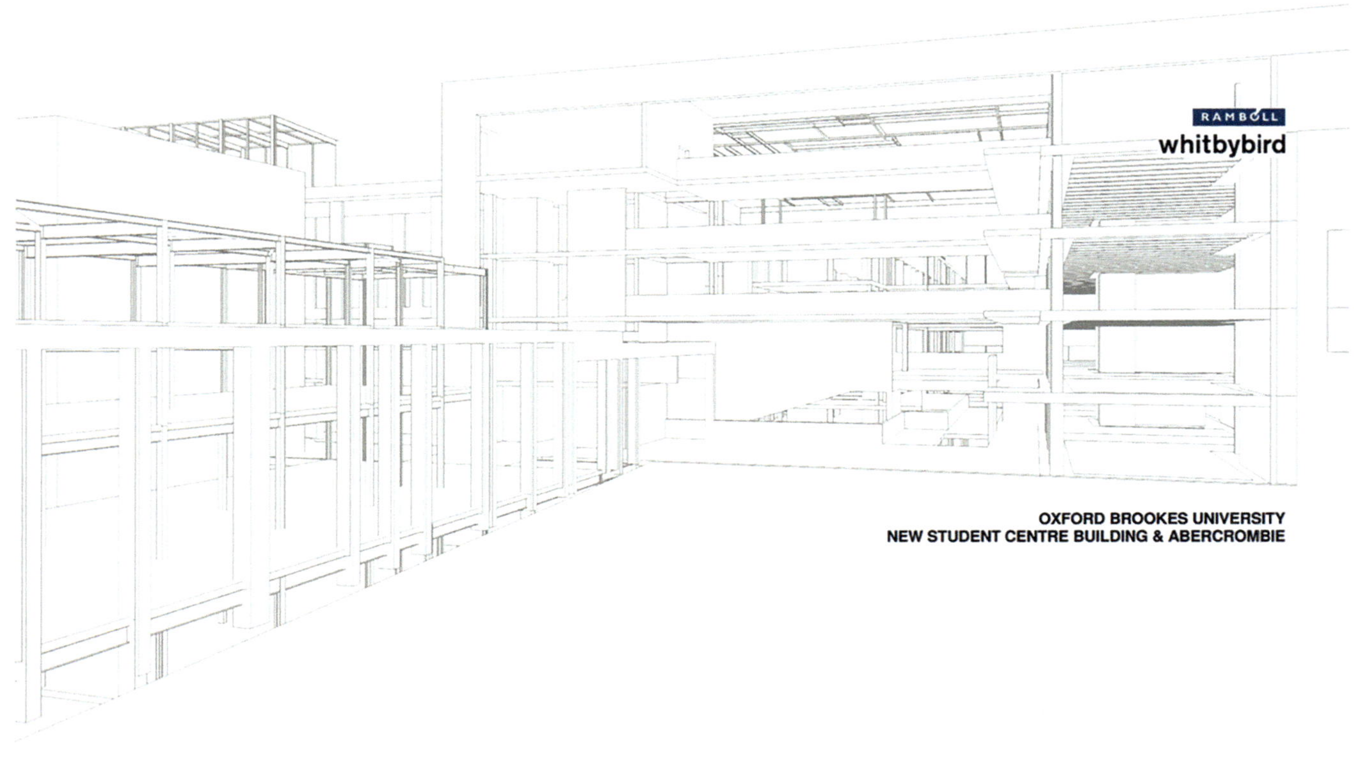

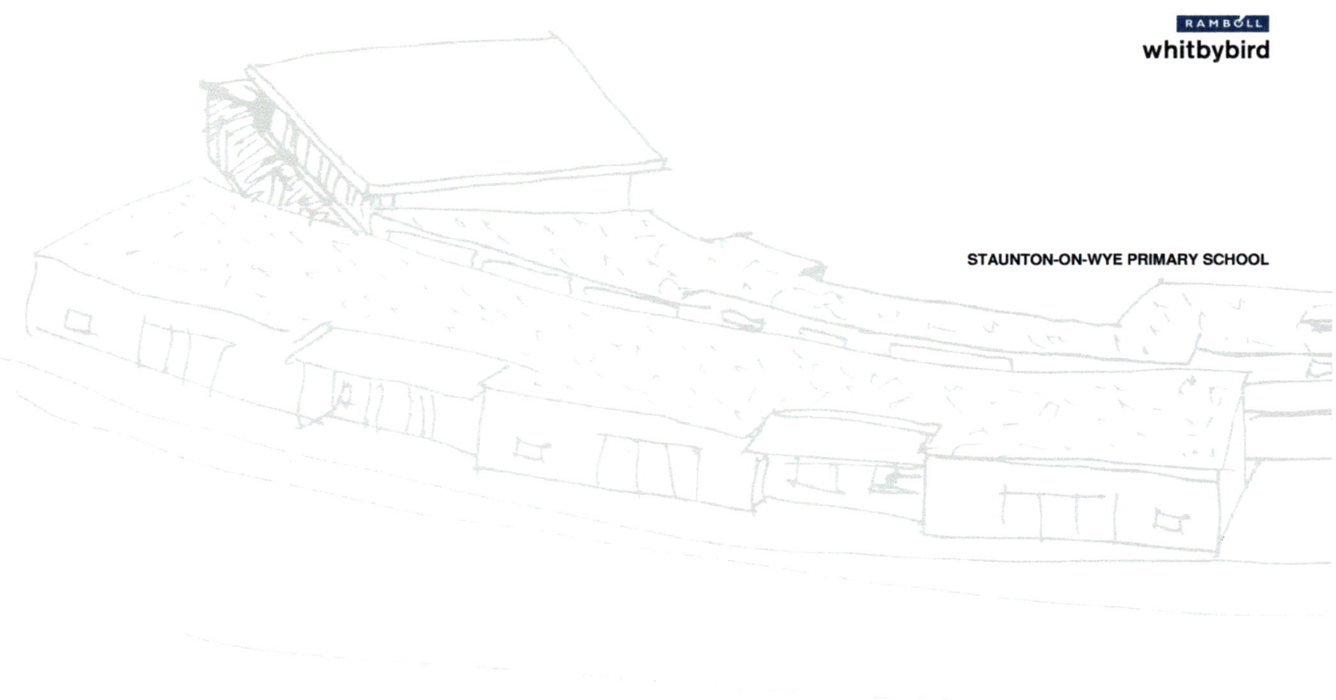

Design development

7. Brief development.
   This section should capture the development of the brief. It may be necessary to write it with the design development, depending on how much interaction there was, and how much this can be rationalised into two separate components.
8. Design development.
   Articulating the design development is helpful both practically and to demonstrate the thought process to the client.
9. Finalised brief.
   If you are producing the brief, this should be captured here.

Final design

The final design should be clearly articulated and rational. While the process of arriving at the design is often anything but linear, a logical framework to show the final design is important. This should start with existing information, then go through the structure, starting at the ground and working up. There are a large number of sections listed here — not all are relevant for every project and caution should be taken not to say too much and lose the client in the detail. If there is a danger of doing this, one approach is to create a design philosophy in tandem, which captures the information that you don't want to include in the report, but is still important to articulate. The list is not exhaustive, but is hopefully a useful starting point:

10. Site information.
    - Location
    - History
    - Description— to include access considerations
    - Geology and hydrogeology
    - Summary of site investigations (including previous, if any)
    - Soil and groundwater profile
    - Sulphates and chlorides
    - Trees
    - Radon
    - Contamination
    - Environmental assessment summary
    - Requirements for further investigation

11. Existing buildings/structures.
    - Condition
    - Age
    - Structural design
    - Key concerns
    - Opening-up works required to resolve concerns
    - Plan for re-use
    - Demolition

12. Substructure design.
    - Foundations
    - Ground floor construction
    - Basement
    - Waterproofing
    - Retaining structures

13. Superstructure design.
    - Movement strategy and movement joints
    - Lateral stability
    - Materials
    - Perimeter wall construction
    - Internal walls

- Upper floor construction
- Stairs
- Roof construction
- Disproportionate collapse/robustness/tying loads
- Construction/demolition considerations — special considerations only
  For complex structures, it is important to demonstrate at least one viable form of construction. This is not to say that this is the way it must be built, but by showing one viable method of construction, we are showing it can be built. If we are unable to show a single way it can be built, it is likely that it cannot be built, and we need to reconsider the design
- Phasing — if affecting design/construction e.g. temporary stability

14. Underground drainage.
    - Existing surface water drainage
    - Existing foul water drainage
    - Proposed surface water drainage (including drainage strategy for planning)
    - Proposed foul water drainage

15. Design criteria.
    - Design life.
      For buildings, this is typically 50 years. It is useful to explain to the client that this does not mean that the building will fall down after 50 years, rather at the end of 50 years, it is important to reassess the loading for the next 50 years, and to carry out a condition survey and that significant maintenance may be required. It is important to make this clear, as there is much misinformation about the design life of buildings. I have heard prominent architects say that the design life is how long a building will stand up. This is much like saying you will need to scrap your car when it has its first MOT after 3 years.
    - Deflection limits and vibration
    - Design loading
    - Fire protection
    - Corrosion protection, material grades
    - Temperature range for external structure or structure exposed to large variations in temperature
    - Design codes and standards and other references

16. Sustainability.
    As noted in Section 5.1.2 the sustainability aspirations of the client can be difficult to articulate. It is therefore important to do so in a way they can understand and that they can dispute if necessary. Remember that the Stage 2 report is not the end of the process, but a marker as we move into the next stage.

    The consideration of sustainability (and health and safety, for that matter) should not be limited to a single section, but should be embedded across the report. This section is a space to summarise the sustainability, cross-referencing the sections where it is mentioned, and also provides a space to cover approaches going forward e.g. the use of BREEAM with input onto suitable credits.

17. Summary and next steps.
    It is important to articulate what you require the client's input on. If further investigation is required, this should be reiterated here so that the client can assess the cost. If decisions are required, these should be articulated here. Often a report is issued and the lack of response from the client is taken as passive agreement of its contents. However, it is much better to ask questions, and these should ideally be closed and in one of two forms e.g.:
    - Option A or option B — would the client like a steel or timber frame?
    - Yes or no, we have suggested that the building is designed for an office loading of 2.5kN/m^2 which will limit future usage to offices and residential — is the client happy with this limitation?

By asking explicit questions, it is quite possible that a larger conversation will follow, but it is useful to be specific, rather than ask vague questions, such as: Is there anything in the report you would like to change? These conversations should then help you to finalise the conceptual design, choose a single solution and progress onto the detailed design.

12.2 Drawings

Alongside the report, a series of drawings should also be produced. At this stage, they can be hand-drawn or CAD. This decision will depend on the type of project, the expectation of the client, and the approach taken by other members of the design team. It is difficult to produce a well-considered set of CAD drawings when the architect is still producing felt tip pen sketches, especially if these are vague.

It is also worth considering mixing-up CAD drawings and hand-drawn sketches. This provides the advantage of the drafts person doing the 'general arrangements' (GAs), while the engineer produces sections and details. This can be advantageous, as the conceptual design process is fast-paced, with changes occurring right up to the moment of issue.

12.2.1 BIM

BIM is discussed extensively in engineering, and there are some advantages, even at the concept design stage. At this early stage, the focus is on information-sharing and the use of the architect's 3D spatial models, to enable quick setting-out of the structure. While this is one facet of BIM, it is unlikely that BIM in its fullest sense will be used with full collaborative working across all the members of the design team. However, the use of 3D models does offer obvious advantages. The risk is that in the perceived precision of this method, due consideration is not given to items such as tolerance, connection sizing and long- and short-term deflection. It is therefore important that these criteria are conveyed alongside the 3D model.

Every practice will have a different view of whether it is too early to use a BIM methodology, and the decision to do so will be a function of the project size, complexity, and whether other members of the design team are also prepared to start the process at this early stage.

One obvious advantage is the use of immersive technology and/or animations. While drawings are adequate for professionals to visualise the proposed buildings to many non-technical clients, a series of lines on a page are difficult for them to conceptualise as their building. Simple 3D models, that they can explore, will help the client better understand what they are getting. However, there is an obvious increase in work to produce this, which should also be considered.

Another advantage of creating 3D models is that we can quite quickly extract material quanitites that can be used to calculate approximate values for the embodied carbon and energy for different options, which can then be reported to the client.

12.2.2 What to include

When pulling together your Stage 2 report, you will want to think carefully about the drawings you create. At this early stage, there will be a number of unknowns, so you don't want to provide a complete set of CAD drawings that look ready for construction. At the same time, you want to include enough information that the scheme can be costed — especially if you have two or more solutions on the table and want to cost them both. It is important to walk the line between being too conservative and ruling an option out based on cost, and being too optimistic and making an unviable option appear viable. At this stage information is key. Both in terms of what we do know, but also what we don't know and any suitable allowance that could be made. The following is a brief list of the types of drawings you might want to include, and what information you may want to present. However, every project is unique and you will want to work out exactly what information you want to include.

Notes drawing

Many companies have a standard notes drawing. It can be tempting to include one at the front of our drawings, but we need to ask what information is really needed. The danger of providing an A1 sheet covered in small text is that no one will read it. If there are a few key messages that don't obviously fit on any other drawings, a notes drawing can be useful, but avoid losing the key messages in all the other text. Generally, I would try not to include a notes drawing at this stage.

GAs (Plans)

GAs are essential — they set out the building design. Every page should include grids and a scale, which despite the note that we should not scale off the drawings, will enable approximate costing to take place. We then need to convey both the substructure and superstructure design. It is worth creating more drawings with less information on, than trying to pack lots of levels of information on one drawing, especially if you are drawing by hand e.g. for the substructure, rather than put everything on one drawing, consider splitting into piles, foundations and ground floor slab. This will ensure everything is clearly visible.

For foundations, provide approximate pile sizes, bearing in mind you probably haven't carried out a site investigation yet. Similarly, we need foundation sizes, ground beams and a ground floor slab, with edge thickening around the perimeter where necessary and possibly an upstand.

For the superstructure, include a limited number of beams and columns to give an approx. volume/weight of structural material. At this stage, the piece count will be more important than the exact specification. Try and show trimmers, cladding support and other steels, not just primary beams — this is particularly relevant at roof level, where you will almost certainly have a parapet and may need a beam at roof level and another beam at the top of the parapet. Mark the slab thickness (whether *in situ*, precast, timber etc.), and clearly label the stability system. Where there is complexity e.g. a mezzanine or half-landing, you may want to do a small GA of this area beside the main GA, or you may just want to note that in this location there will be beams at half-landing level, as well as at floor level. It can also be useful to have a TOS (top of steel) or TOC (top of concrete) on the drawings. Typically, we can provide one value with any variation noted by the beam. We may not have the level relative to the OS datum at this point, so we can measure it from our own datum, possibly the ground floor level, at this stage.

It is also essential that the stability system is marked onto the GA. This has hopefully been agreed across the design team, but may raise some coordination issues. It should be clear what the system is, any design considerations, and any areas where floor penetrations may be impractical, as they prevent transfer of load from the slabs to the stability system. This can be particularly important if you are using walls or bracing around risers to provide the stability.

Any risks that have been highlighted during your risk assessment should ideally also be marked on the GAs and clearly labelled to ensure everyone is aware of them. This doesn't replace the risk assessment (some risks won't be relevant to the GAs), but increases the chances of everyone seeing them. Ensure you mark on the location of any relevant sections and elevations so that they can be easily referenced from one drawing to the next.

Sections
You don't need to draw every section, but a few key details will be useful, especially if they are unusual e.g. exposed connections, moment connections, complex arrangements, steps in level and movement joints. It may also be worth choosing one or two beams and drawing sections through them, showing how the slab connects, and what the tolerance and deflection will be. Providing this information early on will prevent difficult discussions later. A few typical ground sections showing upstands, edge thickenings and wall/column connection details will be helpful. This is also a good place to show what goes on below the slab, especially if you need insulation, or some form or void-former. We would suggest that every project has a minimum of two section drawings — one for substructure and one for superstructure. Depending on the scale of the project, you may want a section drawing for each level, but often at this stage you will be simply duplicating information.

Elevations
Elevations can be very helpful, but I would suggest are a 'nice to have', rather than a necessity at this stage. Most GAs with levels (at this stage measured from the ground floor slab rather than OS datum) will be adequate in confirming the elevations. However, where there is a high level of complexity e.g. a split level or a complex folding roof, elevations can be very useful. We may also find a 3D sketch helpful to articulate highly complex geometry.

Drainage
At this stage, it is useful to provide a drainage drawing (or two). This should summarise what we know about the belowground information e.g. existing manhole locations, buried services etc. We can also highlight areas of uncertainty and how we propose to resolve them, where we want CAT scans and where we want CCTV surveys of drains. We can also start to show how we plan to link our new building into the existing drainage, what strategies are available for SUDS, where we may be able to use permeable paving, where we can add belowground water storage etc. It is particularly helpful if we can then draw foundations, and especially pile layouts, over the top of this information to avoid potential clashes at this early stage.

Geotechnical information

It is unlikely at this stage that you will have carried out a site investigation, but hopefully as part of this report you have started to think about what information you need, and where best to place trial pits, boreholes and soakaways. You may want to include a drawing to show the client the proposed locations for these works, so you can discuss the implications on their site.

In addition, it may be useful, as a record, to include a drawing showing the ground model as you currently understand it, based on the desktop study. It is important to state that this information is preliminary, but it will act as a good record as you progress the design.

Constraints diagram

A constraints diagram can be a very helpful and powerful source of information. This should collect all the different constraints information into one place. This may include, but is not limited to:

- Site boundaries
- Site access
- Any limitations on access, either during construction or in use. This could include neighbouring land that cannot be over-sailed by cranes
- Any in-the-ground information — including all belowground services, known obstructions, other major belowground infrastructure e.g. tunnels and other environmental belowground constraints e.g. in Bath, where there is a natural aquifer, there are limits on belowground works
- All trees with protection orders and any root protection zones
- All existing buildings – including zones where foundations cannot go, or need to be very carefully designed to avoid altering existing load paths
- All existing buildings that are being demolished and the phasing of this works — often demolition requires other sections of a job to be complete and, until this is the case, the buildings need to remain operational
- Any construction operational constraints e.g. noise, traffic movements, vibration (air-borne or ground-borne), or general limitation on site work hours
- Any other constraints you may be aware of

Construction sequencing

On some projects, the construction sequencing is critical to the success of the project. Whether phasing of the construction and demolition works, or for the construction of an element of the building which requires works to be carried out in a particular order, in order for the design to work. This may be covered by your report, or may not be required, but where this information is critical, it is worth including a drawing to outline the proposed sequencing, and any implications it will have on cost, especially in the form of delays on site and temporary works.

12.3 Cost proposal

While a fee was hopefully agreed at the start of a project, the project is often so ill-defined at the conceptual stage that the fee is, at best, provisional going forward, and is often not agreed beyond the end of Stage 2. It is therefore important to agree a fee with the client for the future stages of the project. A fee can take on one of three forms:

- **Hourly rate** — the client pays for every hour the engineer and other staff works on the project. While at first glance this appears fair, clients are often reluctant to accept this approach except on very small projects, as they have little control over the fee and are unable to budget for t
- **Lump-sum** — the client agrees a fee in advanced for the project. While, in theory, this approach is fair, if the project cost rises (and therefore the amount of work increases), the consultant loses out. Clients are often keen to agree a lump sum, but consultants tend to avoid it. In theory, if the project cost goes down, the consultant benefits but this is much less often the case
- **% cost** — this is similar to a lump-sum, except that if the project costs rise/fall, so does the fee. Most UK projects are carried out on this basis. There are still issues, especially around the final costing of projects and the associated fee, but generally both the client and consultant are happy to work on this basis. As a percentage of construction costs, the civil and structural design fee will typically vary between 0.8–2%

The calculation of fees should be carried out by a director, considering the breadth of issues raised through the conceptual design. A crude percentage table is provided in Section 5.2.1. These fees should not be part of the Stage 2 report, but should be discussed in parallel with the submission.

The breakdown of work and how much we would like to be paid at each stage is also important to agree. Ideally, as consultants, we would like to be paid before we do the work (or at least invoice before, bearing in mind quick payers will pay within 30 days, with many large clients stipulating payment terms of up to three months). The client, however, would much rather pay the full fee at the end of the project. In reality, the RIBA provides guidance on fee levels which are summarised in Table 12.1. Along with the fee, this schedule should also be included with a breakdown of fees on a monthly basis. Some clients will ask for a small retainer (3–10%), which they will only pay after a set period after the completion of the project. This is to ensure that any problems that arise following practical completion are taken seriously, and fixed quickly by the consultant.

Table 12.1: Typical breakdown of fees across the duration of a project

RIBA stage	Typical fee	Typical work/tasks
1 Preparation and briefing	5%	Civil and structural input will be fairly limited at this stage. Typical input would be to identify and confirm the client's requirements, and possible constraints on the development. Prepare studies to help inform the client and support decision to proceed and possible procurement routes. Identify others to be engaged in the project
2 Concept design (this is complete at the end of Stage 2)	15%	Working alongside the architect to develop and prepare outline proposals. These might include preliminary materials selection, initial grids and locations of vertical structure
3 Spatial coordination	20–30%	Prepare detailed proposals including grids, detailed material selection, element sizes. Information to inform the cost plan, including identifying key aspects which the design has not yet considered in any detail
4 Technical design	25–30%	Complete detailed design, including all final element sizes, details sufficient for coordination of all components of the project
5 Manufacturing and construction	15%	Produce all detailed design drawings and specifications. Coordination with specialists, submit calculations and drawings to Building Control (in the UK). Review shop drawings and submittals
6 Handover	10%	Site visits and progress reports. Responding to requests for information (RFIs). Updating drawings to 'as-built' (if applicable). Review shop drawings and submittals. Issue sketches where needed to clarify design
7 Use	3–10%	Issue of 'as-built' information for health and safety folder

The final fee should, of course, add up to 100%. Ideally you want to try and make the fee greater at the start of the project and less later, although this will be subject to negotiation.

As with the percentage fee, we would not recommend you create a fee schedule on your own, but that you do it with your director. However, producing a starting template is helpful, as they can advise on where the percentages should be increased and decreased.

12.4 Programme and 'information required schedule'

Some projects will have a clear programme, even at the very earliest stages of a project e.g. a school which needs to build four new classrooms before the start of the next academic year. Most won't have a clear end date at this stage, and will expect to pause the project at different points e.g. while planning permission is granted and finances confirmed. It is unlikely that you will need to produce a full programme at this early stage. That doesn't, however, preclude you from both considering the programme and discussing it with clients. Projects with long drawn-out programmes often cost more to design than those which happen quickly, and this should be considered in the fee proposal.

Along with a programme, an 'information required schedule' (IRS) is a very helpful document. At their worst, they can be seen as a highly contractual document but, at their best, they can generate an honest conversation about delivery dates and what is needed when. This is particularly important when prefabricating structure, as the long lead-ins may require information on the aboveground structure to be produced by the full design team, long before substructure details are required. When used well, an IRS generates conversations about both 'what is needed when' but also the level of detail required. While it is early to produce a full-blown IRS with every package of work listed, a simple IRS that outlines the lead-in for the substructure, superstructure and external works can be very helpful. The IRS can be non-date specific, instead providing the information in terms of weeks prior to construction. This will still help flag up potential information problems, and will highlight to the client when they need to make final decisions.

12.5 Specification

At the end of Stage 2, it is far too early to consider creating a specification for each material used on the project. However, it may be useful to create a single-page outline specification. The main focus of these documents is to capture anything unusual which will have a major impact on program or cost. A simple example is the use of ground granulated blast-furnace slag (GGBS) in concrete, which slows down the rate that concrete sets, so inclusion, especially of high percentages for aesthetic/environmental reasons may extend the strike time of your concrete. You do not need to provide a full concrete specification outlining all tolerances etc. at this stage, but highlighting the intention to use this material, especially to the QS, who will be costing the project, and therefore accounting for site costs due to programme, is helpful. At this stage, I would suggest either creating an outline specification which captures this, or highlighting this information on the drawings. The second option is preferable, as it is more likely to be considered as you enter the next phase of the project.

12.6 Scope of works

As you start to think about the more detailed stages of the project, it is important to check your scope of work. You need to think about two things:

1. Does the scope match what you believe you are doing? Are you being asked to carry out design beyond your area of expertise? Do you have the resource and ability to deliver the work, or do you need to consider employing subcontractors to deliver specialist portions?
2. Are there any gaps in the scope across the project? This can be particularly important on smaller jobs, where the design team is much smaller and maybe doesn't include all areas of expertise. While conversations at this stage can be tricky around scope gaps, especially if you are not offering to fill them or want to be paid more for the work, they are much trickier when you have an irate contractor and client standing in a field, trying to understand why no one designed the drainage under the foundations (which are now cast). It is therefore important to ensure this has been considered. If you are not sure, you should seek advice from your directors who should be able to provide the information required. Most large consultancies have scope of work documents which detail what they will always do, what they can do (for a fee) and what they categorically cannot do. These are a useful starting point.

References

1   Frederick, M. *101 things I learned in architecture school*. Cambridge, MA: MIT Press, 2007
2   RIBA Plan of Work website. Available at: https://www.architecture.com/knowledge-and-resources/resources-landing-page/riba-plan-of-work [Accessed: March 2020]
3   Cobb, F. *Structural engineer's pocket book: Eurocodes (3rd edition)*. Boca Raton, FL: CRC Press, 2015
4   Norman, J. *Structural timber elements: a pre-scheme design guide (2nd edition)*. High Wycombe: Exova BM TRADA, 2018
5   Steel Construction Institute. *Interactive 'Blue book'*. Available at: https://www.steelforlifebluebook.co.uk [Accessed: March 2020] [Interactive version of: Steel Construction Institute. *Steel building design: design data (SCI Publication P363)*. Ascot: SCI, 2015]
6   Goodchild, C.H., Webster, R.M. and Elliott, K.S. *Economic concrete frame elements to Eurocode 2: a pre-scheme handbook for the rapid sizing and selection of reinforced concrete frame elements in multi-storey buildings designed to Eurocode 2*. Camberley: The Concrete Centre, 2009
7   Popovic Larsen, O. *Conceptual structural design: bridging the gap between architects and engineers (2nd edition)*. ICE Publishing, 2016
8   Young, J.W. *A technique for producing ideas*. New York: McGraw-Hill, 2003 [originally published in 1965]
9   Csikszentmihalyi, M. 'Implications of a systems perspective for the study of creativity'. In Sternberg, R.J. ed. *Handbook of creativity*. Cambridge: Cambridge University Press, 1999, pp. 313–338
10 McCann, E. In Think up (2018). *Conceptual design for structural engineers (online) – notes and resources*. Available at https://thinkup.org/conceptual-design-for-structural-engineers-online-notes-and-resources [Accessed: March 2020]
11 Blake, R., & Mouton, J. *The diagnosis and development matrix*. Houston, TX: Scientific Methods, 1972
12 Crawford, M. *The world beyond your head: how to flourish in an age of distraction*. London: Penguin, 2016
13 Collins, M.A., & Amabile, T.M. 'Chapter 15: Motivation and creativity'. In Sternberg, R.J. ed. *Handbook of creativity*. Cambridge: Cambridge University Press, 1999, pp. 297–312
14 Fletcher, A. *The art of looking sideways*. London: Phaidon, 2001
15 Fredrickson, B.L. 'The broaden-and-build theory of positive emotions'. *Philosophical Transactions of the Royal Society B*. 359(10), 2004, pp. 1367–1377. Available at: https://royalsocietypublishing.org/doi/pdf/10.1098/rstb.2004.1512 [Accessed: March 2020]
16 Kline, N. *Time to think: listening to ignite the human mind*. London: Ward Lock, 1999
17 Mlodinow, L. *Elastic: flexible thinking in a constantly changing world*. London: Penguin Books, 2018
18 Lawrence, M. *et al*. 'Design of the HIVE — an innovative research building for the University of Bath'. *The Structural Engineer*, 93(9), September 2015, pp. 24–29
19 BREEAM website. Available at: https://www.breeam.com [Accessed: March 2020]
20 LEED rating system website. Available at: https://www.usgbc.org/leed [Accessed: March 2020]
21 Passivhaus Trust website. Available at: https://passivhaustrust.org.uk [Accessed: March 2020]
22 The Institution of Structural Engineers. *A short guide to embodied carbon in building structures*. London: IStructE Ltd, 2011. Available at: https://www.istructe.org/resources/guidance/embodied-carbon-in-building-structures [Accessed: March 2020]
23 Minimising Energy in Construction [MEICON] website. Available at: https://www.meicon.net [Accessed: March 2020]
24 Ingels, B. *Yes is more: an archicomic on architectural evolution*. Cologne: Taschen, 2009
25 Wright, A. 'Critical method: a pedagogy for design education'. *Design Principles and Practices*, 5(6), 2011, pp. 109–122. Available at: https://purehost.bath.ac.uk/ws/portalfiles/portal/272079/Wright_Alex_DPP_2012_5_6.pdf [Accessed: March 2020]
26 Belardis, P. *Why architects still draw*. Cambridge, MA: MIT Press, 2014
27 Jeffers, O. *Stuck*. London: HarperCollins. 2012
28 Rogers, R. *Architecture: a modern view*. London: Thames & Hudson, 1990
29 Andrew, R.M. 'Global CO_2 emissions from cement production'. *Earth System Science Data*, 10, 2018, pp. 195–217. Available at: https://www.earth-syst-sci-data.net/10/195/2018/essd-10-195-2018.pdf [Accessed: March 2020]

30   Watts, J. 'Concrete: the most destructive material on earth'. *The Guardian*, 25 February 2019. Available at:
https://www.theguardian.com/cities/2019/feb/25/concrete-the-most-destructive-material-on-earth
[Accessed: March 2020]

31   The World Bank (2019). *Forest area (% of land area)*. Available at: https://data.worldbank.org/indicator/ag.lnd.frst.zs
[Accessed: March 2020]

32   Ramage, M. *et al.* 'The wood from the trees: the use of timber in construction'. *Renewable and Sustainable Energy Reviews*, 68(1), 2017, pp. 333–359

33   Ella & Nicki. *Building a Martian house*. Available at: https://www.ellaandnicki.com/index.html
[Accessed: March 2020]

34   British Geological Survey. *The Geology of Britain*. Available at: http://mapapps.bgs.ac.uk/geologyofbritain3d
[Accessed: March 2020]

35   Chandler, R.J. and David, A.G. *Further work on the engineering properties of Keuper Marl (CIRIA Report 47)*. London: CIRIA, 1973

36   British Geological Survey. *Borehole scans*. Available at: https://www.bgs.ac.uk/data/boreholescans
[Accessed: March 2020]

37   *BS 5930:1981: Code of practice for site investigations*. London: BSI, 1981

38   *CP 2001:1957: Site investigations*. London: BSI, 1957

39   *BS 5930:2015: Code of practice for ground investigations*. London: BSI, 2015

40   Brown, E.T. ed. *Rock characterization testing and monitoring: ISRM suggested methods*. Oxford, Pergamon, 1981

41   *BS EN ISO 14689:2018: Geotechnical investigation and testing. Identification, description and classification of rock. Identification and description*. London: BSI, 2018

42   *BS 8004:2015+A1:2020: Code of practice for foundations*. London: BSI, 2015

43   Bieniawski, Z.T. *Engineering rock mass classifications: a complete manual for engineers and geologists in mining, civil and petroleum engineering*. New York: John Wiley, 1989

44   Stroud, M.A. 'The standard penetration test: its application and interpretation'. *Institution of Civil Engineers conference on penetration testing, Birmingham, 1989*. London: Thomas Telford, 1989, pp. 29–49

45   Parry, D.N. and Chiverrell, C.P. *Abandoned mine workings manual (CIRIA C758)*. London: CIRIA, 2019

46   *BS EN 1997-1:2004+A1:2013: Eurocode 7. Geotechnical design. General rules*. London: BSI, 2004

47   *BS 8004:1986: Code of practice for foundations*. London: BSI, 1986

48   Tomlinson, M.J. *Foundation design and construction (7th edition)*. Harlow: Prentice-Hall, 2001

49   Burland, J.B., Broms, B.B. and De Mello, V.F.B. *Behaviour of foundations and structures (BRE Current Paper CP 51/78)*. Garston: BRE, 1978

50   Broms, B. 'Methods of calculating the ultimate bearing capacity of piles: a summary'. *Sols-Soils*, 5(18–19), 1966, pp. 21–31

51   Gaba, M. *et al. Guidance on embedded retaining wall design (CIRIA C760)*. London: CIRIA, 2017

52   *BS 8002:2015: Code of practice for earth retaining structures*. London: BSI, 2015

53   *BS 8081:2015+A2:2018: Code of practice for grouted anchors*. London: BSI, 2015

54   Preene, M., Roberts, T.O.L. and Powrie, W. *Groundwater control: design and practice (CIRIA C750) (2nd edition)*. London: CIRIA, 2016

55   Garvin, S.L. *Soakaway design (BRE Digest 365) (2nd edition)*. Bracknell, IHS BRE Press, 2016

56   Woods Ballard, B. *et al. The SuDS manual (CIRIA C753)*. London: CIRIA, 2015

57   Highways England. *Design manual for roads and bridges*. Available at:
http://www.standardsforhighways.co.uk/ha/standards/dmrb/index.htm [Accessed: March 2020]

58   *BS 8485:2015+A1:2019: Code of practice for the design of protective measures for methane and carbon dioxide ground gases for new buildings*. London: BSI, 2015

59   UKRadon website. Available at: https://www.ukradon.org [Accessed: March 2020]

60   NHBC (2019). *NHBC Standards. 4.2: Building near trees*. Available at:
https://nhbc-standards.co.uk/4-foundations/4-2-building-near-trees [Accessed: March 2020]

61   *BS 5837:2012: Trees in relation to design, demolition and construction. Recommendations*. London: BSI, 2012

62   *BS 10175:2011+A2:2017: Investigation of potentially contaminated sites. Code of practice. Code of practice*. London: BSI, 2011

63   Environment Agency (2015). *Contaminated land exposure assessment (CLEA) tool*. Available at:
https://www.gov.uk/government/publications/contaminated-land-exposure-assessment-clea-tool
[Accessed: March 2020]

64 *BS EN 1992-1-1:2004+A1:2014: Eurocode 2: Design of concrete structures. General rules and rules for buildings*. London: BSI, 2004

65 *BS EN 1993-1-1:2005+A1:2014: Eurocode 3. Design of steel structures. General rules and rules for buildings*. London: BSI, 2005

66 *BS EN 1995-1-1:2004+A2:2014: Eurocode 5: Design of timber structures. General. Common rules and rules for buildings*. London: BSI, 2004

67 *BS EN 1996-1-1:2005+A1:2012: Eurocode 6. Design of masonry structures. General rules for reinforced and unreinforced masonry structures*. London: BSI, 2005

68 *NA+A2:14 to BS EN 1992-1-1:2004+A1:2014: UK National Annex to Eurocode 2. Design of concrete structures. General rules and rules for buildings*. London: BSI, 2005

69 *NA+A1:2014 to BS EN 1993-1-1:2005+A1:14: UK National Annex to Eurocode 3. Design of steel structures. General rules and rules for buildings*. London: BSI, 2008

70 *NA to BS EN 1995-1-1:2004+A2:2014: UK National Annex to Eurocode 5: Design of timber structures. General. Common rules and rules for buildings*. London: BSI, 2019

71 *NA to BS EN 1996-1-1:2005+A1:2012: UK National Annex to Eurocode 6. Design of masonry structures. General rules for reinforced and unreinforced masonry structures*. London: BSI, 2007

72 *BS 5268-2:2002: Structural use of timber. Code of practice for permissible stress design, materials and workmanship*. London: BSI, 2002

73 *BS 5628-1:2005: Code of practice for the use of masonry. Structural use of unreinforced masonry*. London: BSI, 2005

74 The Institution of Structural Engineers. *Manual for the design of concrete building structures to Eurocode 2*. London: IStructE Ltd, 2006

75 The Institution of Structural Engineers. *Manual for the design of plain masonry in building structures to Eurocode 6 (2nd edition)*. London: IStructE Ltd, 2018

76 The Institution of Structural Engineers. *Manual for the design of steelwork building structures to Eurocode 3*. London: IStructE Ltd, 2010

77 The Institution of Structural Engineers. *Manual for the design of timber building structures to Eurocode 5 (2nd edition)*. London: IStructE Ltd, 2020

78 Merrick, J. 'High density, low carbon'. *Architects Journal*, 244(19), 5 October 2017, pp. 30–38

79 Fast, P. and Jackson, R. 'The TallWood House at Brock Commons, Vancouver'. *The Structural Engineer*, 96(10), October 2018, pp. 18–25

80 HM Government (2013). *The Building Regulations 2010. Approved Document A: Structure*. Available at: https://www.gov.uk/government/publications/structure-approved-document-a [Accessed: March 2020]

81 British Council for Offices. *Guide to specification: best practice for offices*. London: BCO, 2019

82 Heywood, H. *101 Rules of thumb for low energy architecture*. London: RIBA Publishing, 2012

83 Heywood, H. *101 Rules of thumb for sustainable buildings and cities*. London: RIBA Publishing, 2015

84 Ranasinghe, K. *Eurocode 5 span tables: for solid timber members in floors, ceilings and roofs for dwellings (4th edition)*. High Wycombe: BM TRADA, 2014

85 Brooker, O. *Concrete buildings scheme design manual: a handbook for the IStructE chartered membership examination, based on Eurocode 2*. Camberley: MPA – The Concrete Centre, 2009

86 Forterra. *Beam and block floors: load-span tables*. Available at: https://www.forterra.co.uk/bison-precast-concrete/beam-and-block-floors/load-span-tables-1 [Accessed: March 2020]

87 Forterra. *Hollowcare range*. Available at: https://www.forterra.co.uk/bison-precast-concrete/hollowcore-floors [Accessed: March 2020]

88 Tata Steel. *ComFlor® 80*. Available at: https://www.tatasteelconstruction.com/en_GB/Products/structural-buildings-and-bridges/Composite-floor-deck/Comflor%C2%AE-80 [Accessed: March 2020]

89 Tata Steel. *ComFlor® 225*. Available at: https://www.tatasteelconstruction.com/en_GB/Products/structural-buildings-and-bridges/Composite-floor-deck/Comflor%C2%AE-225 [Accessed: March 2020]

90 Buildoffsite (2013). *Offsite construction: sustainability characteristics*. Available at: https://www.buildoffsite.com/content/uploads/2015/03/BoS_offsiteconstruction_1307091.pdf [Accessed: March 2020]

91 Kier Group (2019). *Offsite and modern methods of construction*. Available at: https://www.kier.co.uk/media/3011/offsite-mmc-brochure-2019-digital.pdf [Accessed: March 2020]

92 The Institution of Structural Engineers. *Practical guide to structural robustness and disproportionate collapse in buildings*. London: IStructE Ltd, 2010

93 Elghazouli A. *Seismic design of buildings to Eurocode 8 (2nd edition)*. Boca Raton, FL: CRC Press, 2017

94 The Institution of Structural Engineers. *Manual for the seismic design of steel and concrete buildings to Eurocode 8*. London: IStructE Ltd, 2010

95 *NA to BS EN 1991-1-4:2005+A1:2010: UK National Annex to Eurocode 1. Actions on structures. General actions. Wind action*. London: BSI, 2005

96 *BS EN 1991-1-4:2005+A1:2010: Eurocode 1. Actions on structures. General actions. Wind actions*. London: BSI, 2005

97 Cook, N.J. *The designer's guide to wind loading of building structures. Part 2: Static structures*. London: Butterworth, 1990

98 Brady, S. 'Citicorp Center Tower: failure averted'. *The Structural Engineer*, 92(2), February 2014, pp. 14–15

99 HM Government (2019). *The Building Regulations 2010. Approved Document B: Fire safety. Volume 1: Dwellings; Volume 2: Buildings other than dwellings*. Available at: https://www.gov.uk/government/publications/fire-safety-approved-document-b [Accessed: March 2020]

100 *BS EN 1994-1-2:2005+A1:2014: Eurocode 4. Design of composite steel and concrete structures. General rules. Structural fire design*. London: BSI, 2005

101 HM Government (2015). *The Construction (Design and Management) Regulations*. Available at: http://www.legislation.gov.uk/uksi/2015/51/contents [Accessed: March 2020]

102 Carpenter, J. 'Temporary Works Toolkit. Part 2: CDM 2015 and the responsibilities of permanent works designers with regard to temporary works'. *The Structural Engineer*, 94(11), November 2016, pp. 34–36

103 Baker, W. 'Taller'. In *Taller, longer, lighter: IABSE-IASS Symposium, London, September 20–23, 2011*. Zurich: IABSE, 2011

104 *BS EN 1991-1-1:2002: Eurocode 1. Actions on structures. General actions. Densities, self-weight, imposed loads for buildings*. London: BSI, 2002

105 *NA to BS EN 1991-1-1:2002: UK National Annex to Eurocode 1. Actions on structures. General actions. Densities, self-weight, imposed loads for buildings*. London: BSI, 2002

106 Morrison, H. *Structural engineering art and approximation (3rd edition)*. [s.l.]: Paragon Publishing, 2016

107 Chilton, J. *Heinz Isler*. London: Thomas Telford, 2000

108 Byfield, M. *Structural design from first principles*. Boca Raton, FL: CRC Press, 2018

109 Millais, M. *Building structures: understanding the basics (3rd edition)*. Abingdon: Routledge, 2017

110 MacLeod, I.A. *Principles for computer analysis of structure (Essential Knowledge Series 14)*. London: IStructE Ltd, 2016

111 MacLeod, I.A. *Modern structural analysis: modelling process and guidance*. London: Thomas Telford, 2005

112 Debney, P. *Computational engineering*. London: IStructE Ltd [in production]

113 Brohn, D. *Understanding structural analysis (3rd edition)*. London: New Paradigm Solutions, 2005

114 Brettle, M.E. *et al. Steel building design: concise Eurocodes. In accordance with Eurocodes and the UK National Annexes (Revised edition) (Publication P362)*. Ascot: SCI, 2017

115 Davison, B. and Owens, G.W. eds. *Steel designers' manual (7th edition)*. Chichester: Wiley-Blackwell, 2012

116 Narayanan, R.S., Goodchild, C.H. *Concise Eurocode 2: for the design of in-situ concrete framed buildings to BS EN 1992-1-1:2004 and its UK National Annex: 2005*. Camberley: The Concrete Centre, 2006

117 Sandaker, B.N., Eggen, A.P. and Cruvellier, M.R. *The structural basis of architecture (3rd edition)*. Abingdon: Routledge, 2019

118 The Concrete Centre. *RC spreadsheets version 4C*. Available at: https://www.concretecentre.com/Publications-Software/RC-Spreadsheets-v4C.aspx [Accessed: March 2020]

119 Steelconstruction.info website. Available at: https://www.steelconstruction.info [Accessed: March 2020]

120 Steel Construction Institute. *NCCI: Initial design of composite beams (SN022a-EN-GB)*. Ascot: SCI, 2006

121 *PD 6697:2019: Recommendations for the design of masonry structures to BS EN 1996-1-1 and BS EN 1996-2*. London: BSI, 2019

The front cover image

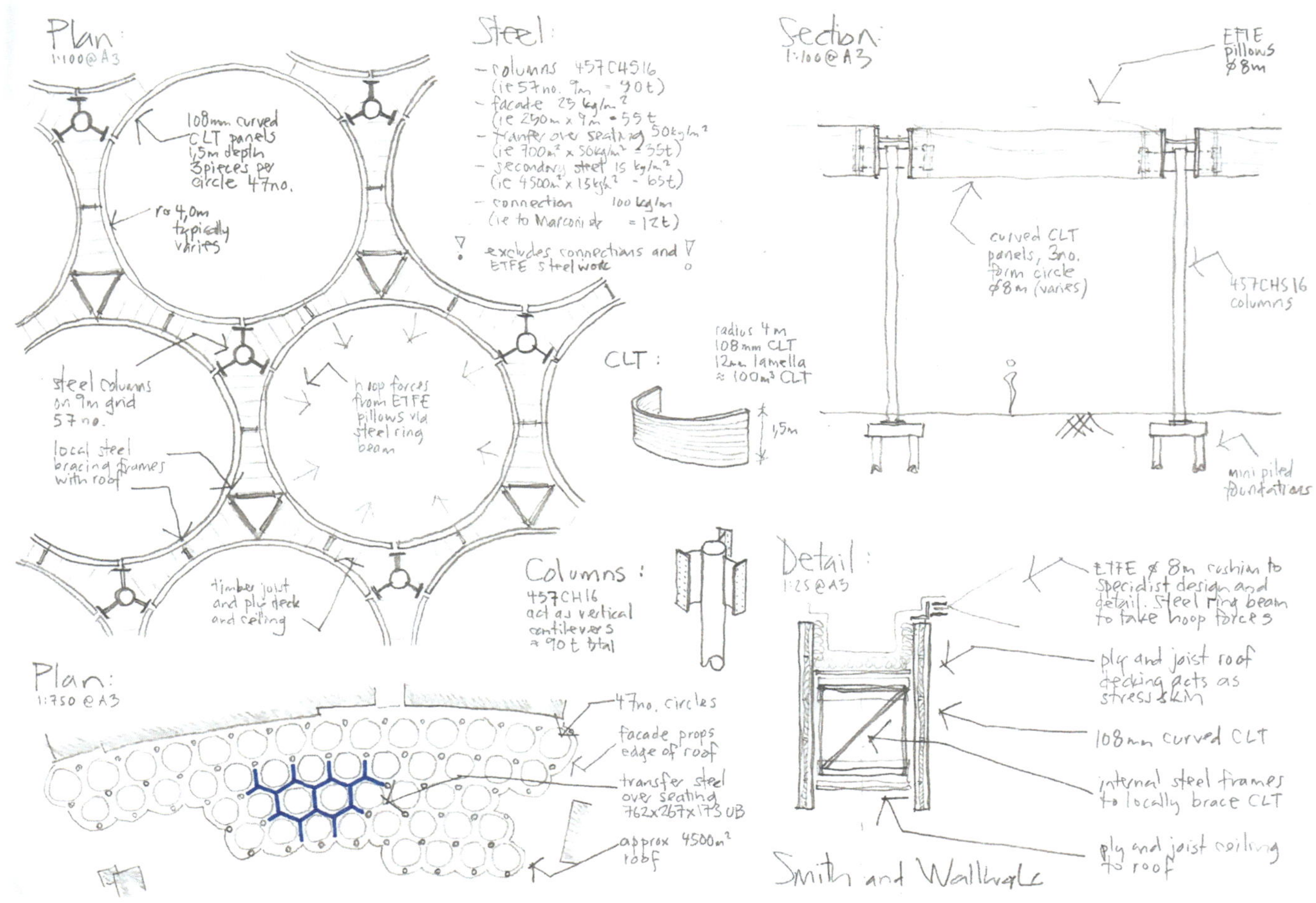

The sketch was one of a number of roof structure concepts generated in collaboration with Grimshaw architects to respond to Anglia Ruskin University's wish to create a new social and events space at their Chelmsford Campus. At the heart of the concept was a repeating roof form suspended 10m above ground level; in this instance 47 circular roof lights — comprising curved CLT beams that support 8m diameter ETFE roof cushions. The 1:750 roof plan highlights the requirement for a column-free area for seating an audience of 750 for student graduation events.

Lateral stability of the structure was proposed via vertical cantilever action of the columns, albeit this would require careful detailing at the facade and existing building interface.

Important with all concepts were some engineering numbers to guide cost checks and challenge ourselves in terms of lean design — in this instance a breakdown of CLT volume and steel tonnage is given. At a steel use of 57kg/m^2 the next stage would be to refine the design to get below 50kg/m^2 — probably through exploring opportunities to reduce column weights by introducing an alternative lateral stability system.

Simon Smith
Smith & Wallwork